ADVANCES IN PHOTOCHEMISTRY

Volume 20

ADVANCES IN PHOTOCHEMISTRY

Volume 20

Editors

DOUGLAS C. NECKERS
Center for Photochemical Sciences, Bowling Green State University, Bowling Green, Ohio

DAVID H. VOLMAN
Department of Chemistry, University of California, Davis, California

GÜNTHER VON BÜNAU
Physikalische Chemie, Universität Siegen, Siegen, Germany

A WILEY-INTERSCIENCE PUBLICATION

JOHN WILEY & SONS, INC.

New York • Chichester • Brisbane • Toronto • Singapore

This text is printed on acid-free paper.

Library of Congress Cataloging in Publication Data:

Library of Congress Catalog Card Number: 63-13592
ISBN 0-471-11469-3

Printed in the United States of America

10 9 8 7 6 5 4 3 2 1

CONTRIBUTORS

Tatsuo Arai
Department of Chemistry
University of Tsukuba
Tsukuba, Ibaraki 305, Japan

Karl Heinz Drexhage
Fachbereich Chemie
Universität Siegen
57068 Siegen, Germany

Alexander Eychmüller
Institut für Physikaliche Chemie
Universität Hamburg
20146 Hamburg, Germany

Joseph S. Francisco
Department of Chemistry
Wayne State University
Detroit, Michigan 48202

M. Matti Maricq
Research Laboratory
Ford Motor Company
Dearborn, Michigan 48120

Katsumi Tokumaru
Department of Chemistry
University of Tsukuba
Tsukuba, Ibaraki 305, Japan

Horst Weller
Institut für Physikaliche Chemie
Universität Hamburg
20146 Hamburg, Germany

Bilha Willner
Institute of Chemistry
The Hebrew University of Jerusalem
Jerusalem 91904, Israel

Itamar Willner
Institute of Chemistry
The Hebrew University of Jerusalem
Jerusalem 91904, Israel

Christoph Zander
Fachbereich Chemie
Universität Siegen
57068 Siegen, Germany

PREFACE

Volume 1 of *Advances in Photochemistry* appeared in 1963. The stated purpose of the series was to explore the frontiers of photochemistry through the medium of chapters written by pioneers who are experts. As editors we have solicited articles from scientists who have strong personal points of view, while encouraging critical discussions and evaluations of existing data. In no sense have the articles been simply literature surveys, although in some cases they may have also fulfilled that purpose.

In the introduction to Volume 1 of the series, the editors noted developments in a brief span of prior years which were important for progress in photochemistry: flash photolysis, nuclear magnetic resonance, and electron spin resonance. Since then two developments have been of prime significance: the emergence of the laser from an esoteric possibility to an important light source; the evolution of computers to microcomputers in common laboratory use for data acquisition. These developments have strongly influenced research on the dynamic behavior of excited state and other transients.

With an increased sophistication in experiment and interpretation, photochemists have made substantial progress in achieving the fundamental objective of photochemistry: Elucidation of the detailed history of a molecule which absorbs radiation. The scope of this objective is so broad and the systems to be studied are so many that there is little danger of exhausting the subject. We hope that the series will reflect the frontiers of photochemistry as they develop in the future.

DOUGLAS C. NECKERS

Bowling Green, Ohio
February 1995

CONTENTS

ADVANCES IN PHOTOCHEMISTRY

Volume 20

PRESENT STATUS OF THE PHOTOISOMERIZATION ABOUT ETHYLENIC BONDS

Tatsuo Arai and Katsumi Tokumaru
Department of Chemistry, University of Tsukuba,
Tsukuba, Ibaraki 305, Japan

CONTENTS

Advances in Photochemistry, Volume 20, Edited by Douglas C. Neckers, David H. Volman, and Günther von Bünau
ISBN 0-471-11469-3

I. INTRODUCTION

Photochemical cis → trans isomerization has extensively been investigated [1–17]. Among ethylenic compounds the mechanism and potential energy surfaces of the isomerization of stilbene have been studied not only by steady-state irradiation, but also by transient kinetic spectroscopy [1–13]. Thus, on direct irradiation, stilbene undergoes mutual isomerization between the cis (^{1}c) and trans (^{1}t) isomers by way of the deactivation of a perpendicular excited singlet state (^{1}p*) accompanied by cyclization to dihydrophenanthrane (DHP) from excitation of the cis isomer [1–3, 7–13]. In this case the excited singlet state of cis-stilbene undergoes either isomerization to ^{1}p* or cyclization to DHP in a 70:30 ratio [11]; the deactivation from ^{1}p* gives ^{1}c and ^{1}t in an almost equal ratio. Similarly, on triplet sensitization, the quantum yields of cis → trans ($\Phi_{c\to t}$) and trans → cis isomerization ($\Phi_{t\to c}$) are both about 0.5, and deactivation of the triplet state takes place from the perpendicular excited triplet state (^{3}p*) to give ^{1}c and ^{1}t with nearly equal probability [1–6].

During this decade it has been revealed that replacement of a phenyl group of stilbene with a 2-anthryl group brings about a novel one-way cis → trans isomerization unaccompanied by the reverse trans → cis isomerization [4, 18–22]. This one-way isomerization takes place as an adiabatic process in the excited triplet state through a quantum chain process; $\phi_{c \to t}$ increases to the order of 10 with increasing cis-isomer concentration

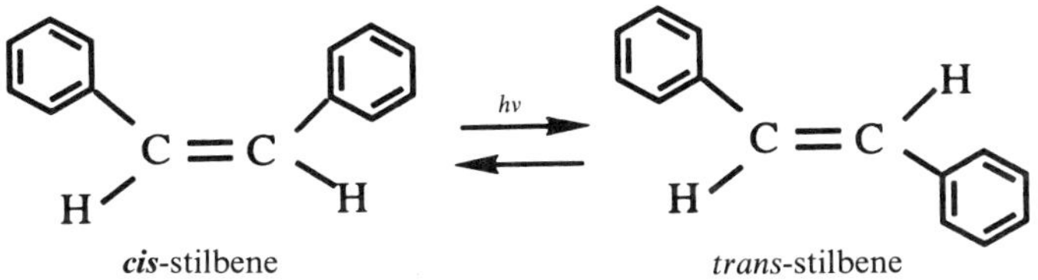

cis-stilbene *trans*-stilbene

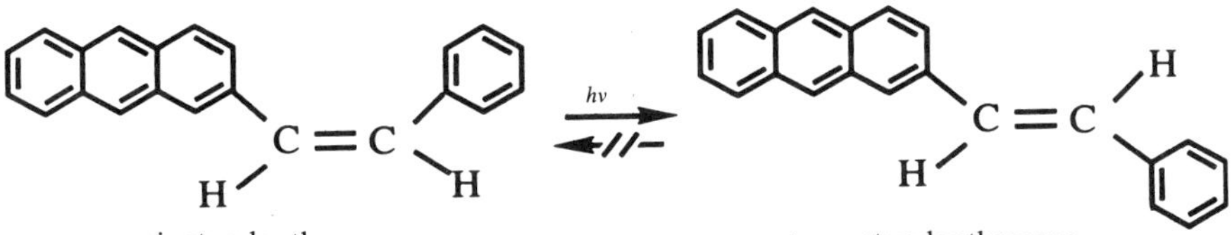

cis-styrylanthracene *trans*-styrylanthracene

Table 1 compares the typical features of one- and two-way photoisomerization in the triplet excited state [1–4, 18–65]. Briefly, stilbene undergoes typical two-way isomerization between cis and trans isomers in the triplet state; the triplet state deactivates in a relatively short lifetime (60 ns) from $^3p^*$ to the ground state to give ^1c and ^1t in almost the same ratio [15]. However, 2-styrylanthracene undergoes one-way cis → trans isomerization in the triplet excited state by an adiabatic conversion from the cis ($^3c^*$) to the trans conformer ($^3t^*$); the deactivation takes place only from $^3t^*$ with a long lifetime on the order of 100 μs to give ^1t either by a unimolecular process or by a bimolecular energy transfer to ^1c, regenerating $^3c^*$ and accomplishing the quantum chain process [18–23].

TABLE 1 Features of Typical One- and Two-Way Isomerization of Arylethenes in the Triplet State

Mode of the Isomerization	One-Way	Two-Way
Example	2-Styrylanthracene	Stilbene
At the photostationary state	100% trans	Cis + trans
Quantum yield	$\phi_{c \to t} \gg 1$	$\phi_{c \to t}$ = ca. 0.5
of isomerization	$\phi_{t \to c} = 0$	$\phi_{t \to c}$ = ca. 0.5
Triplet lifetime, τ_T	ca. 100 μs	ca. 100 ns
Intermediate for deactivation	Trans triplet ($^3t^*$)	Twisted triplet ($^3p^*$)

Studies for various ethylenes substituted with aromatic groups of varying triplet energies revealed that a lowering in the triplet energy (E_T) of the aromatic nucleus altered the mode of the isomerization from the conventional two-way isomerization to a novel one-way isomerization

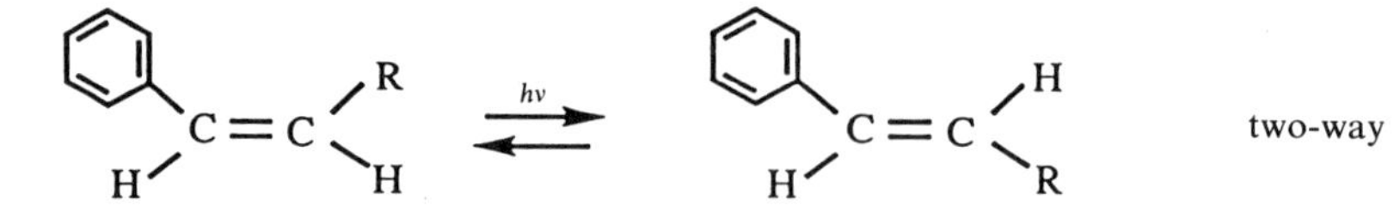

cis-**1**, a: R = ^{t}Bu, b: R = Ph *trans*-**1**

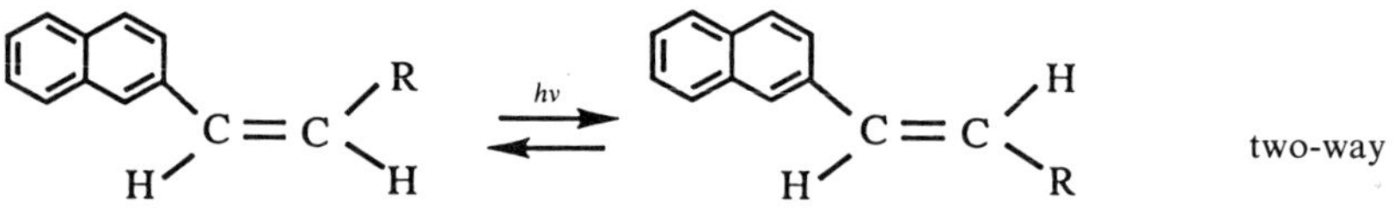

cis-**2**, a: R = ^{t}Bu, b: R = Ph *trans*-**2**

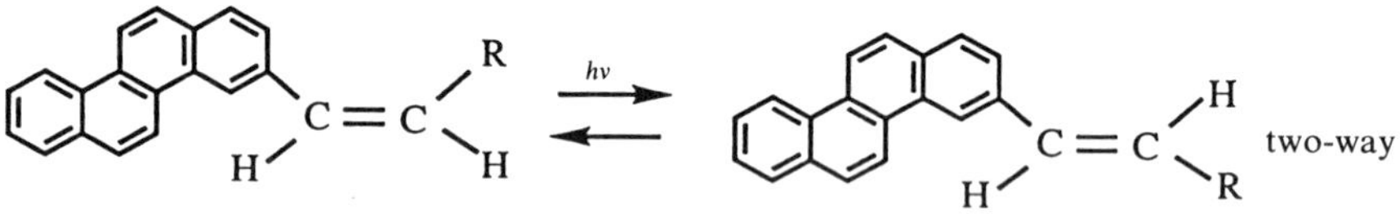

cis-**3**, a: R = ^{t}Bu, b: R = Ph *trans*-**3**

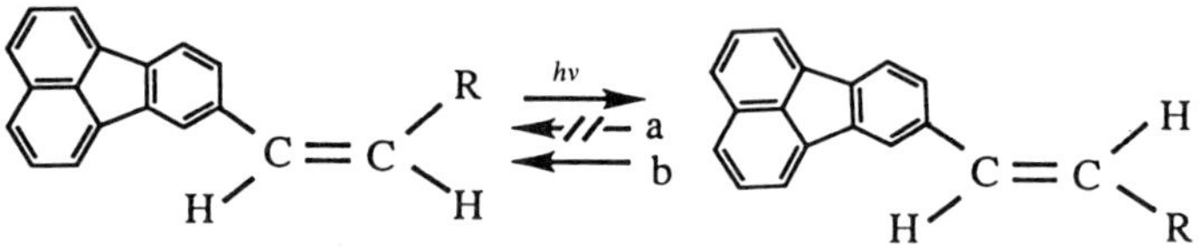

cis-**4**, a: R = ^{t}Bu, b: R = Ph *trans*-**4**

R = ^{t}Bu: one-way R = Ph: two-way (dual)

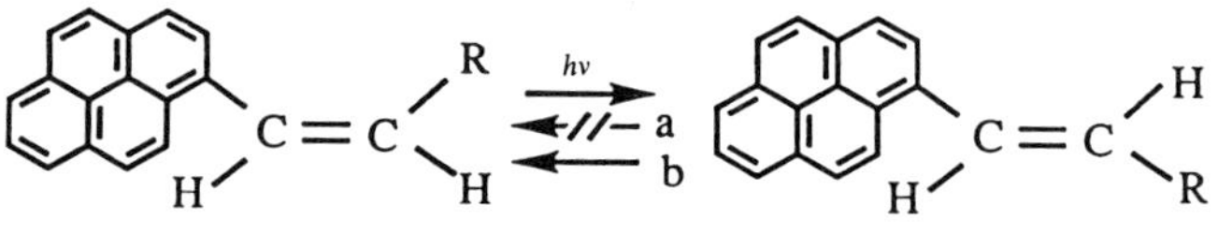

cis-**5**, a: R = ^{t}Bu, b: R = Ph *trans*-**5**

R = ^{t}Bu: one-way R = Ph: two-way (dual)

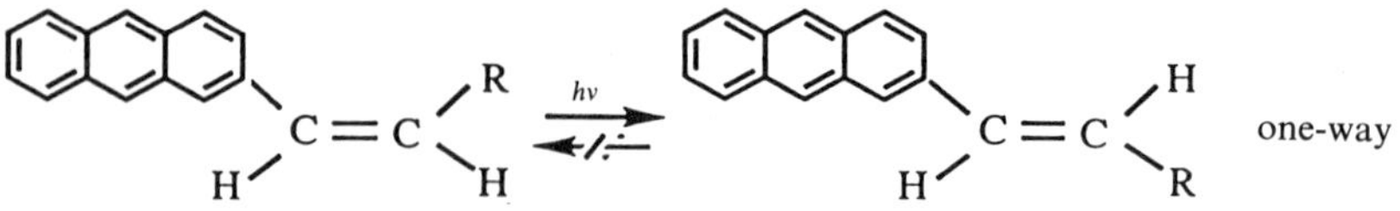

cis-**6**, a: R = ^{t}Bu, b: R = Me, *trans*-**6**
c: R = Ph, d: R = 2-naphthyl

Scheme 1

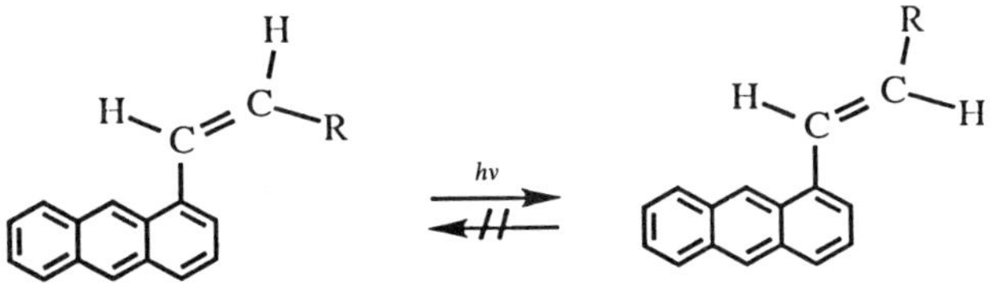

cis-**7**, a: R = ^{t}Bu, b: R = Ph *trans*-**7**

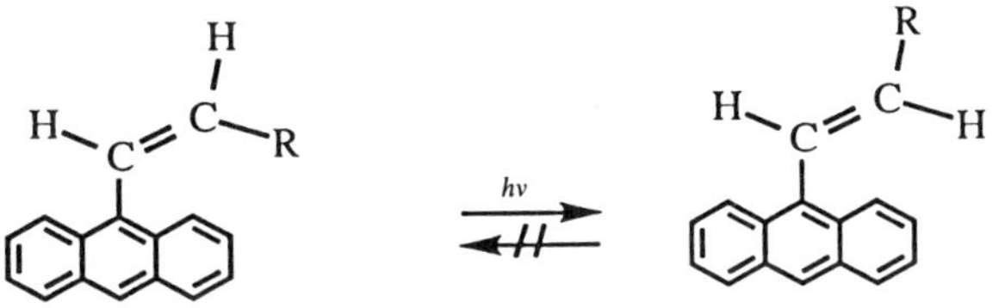

cis-**8**, a: R = ^{t}Bu, b: R = Ph *trans*-**8**

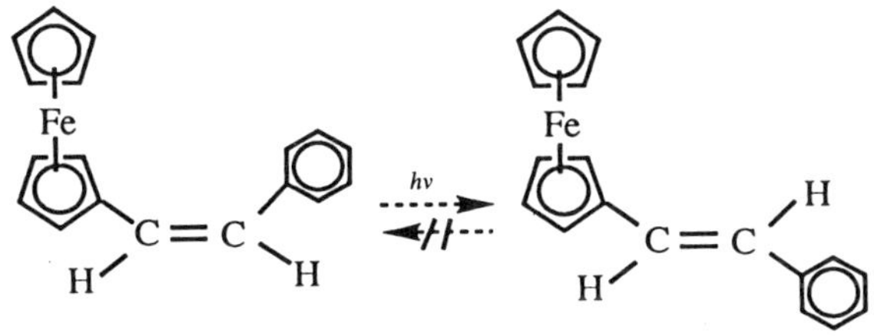

cis-**9** *trans*-**9**

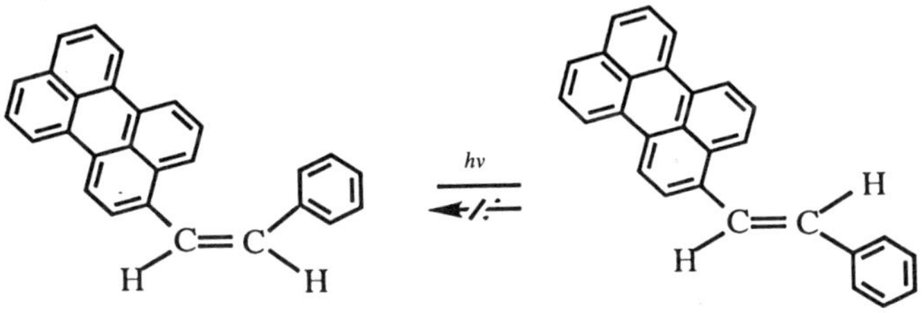

cis-**10** *trans*-**10**

Scheme 1 (*continued*)

when E_T = 54–56 kcal mol^{-1} in the series ArCH═CH^{t}Bu and when E_T = 42–48 kcal mol^{-1} in the series ArCH═CHPh [4]. Scheme 1 and Table 2 summarize the present status of the effect of aromatic substituents on the mode of the isomerization.

In this chapter we describe the features of one- and two-way isomerization of arylethenes and the factors controlling the modes of the isomerization. Furthermore, the effect of aromatic substituents on the isomerization

TABLE 2 Modes of the Isomerization and the Triplet Lifetimes of ArCH=CHR

Ar	Triplet Energy of ArH ($kcal\ mol^{-1}$)	$\tau_T(\mu s)$ [λ_{max}(T–T) (nm)] R = ^{t}Bu	R = Ph	Refs.
Phenyl	84.3	Two-way	Two-way 0.063 (<360)	15, 50
9-Phenanthryl	61.9		Two-way 0.43 (460)	59–61
2-Naphthyl	60.9	Two-way 0.13 (420, 570)	Two-way 0.14 (400, 500)	49, 56, 57
3-Chrysenyl	56.6	Two-way 0.36 (600)	Two-way 0.14 (<400)	30
8-Fluoranthenyl	54.2	One-way 25 (440, 580)	Two-way (dual) 0.50 (480, 600)	27, 28
1-Pyrenyl	48.2	One-way 54 (445)	Two-way (dual) 27 (470, 520)	25
1-Anthryl	42	One-way ~ 100 (440, 500)	One-way ~ 100 (570)	43
2-Anthryl	42	One-way 280 (450, 540)	One-way 190 (460, 620)	19, 20a
9-Anthryl	42	One-way ~ 100 (410, 435)	One-way ~ 100 (325, ≈450)	43, 71–73
Ferrocenyl	40		Inefficient one-way	62, 63
3-Perylenyl	35		One-way ~ 100 (540, 580)	64, 65

of polyenes, the role of intramolecular hydrogen bonding in the isomerization, the photochemical behavior of cyclic ethenes, the internal rotation between the rotamers of arylethenes, and isomerization in the radical cations will be argued.

II. ONE- AND TWO-WAY ISOMERIZATION OF ARYLETHENES IN THE TRIPLET STATE

Until 10 years ago it had been well accepted that the unsaturated bonds generally photoisomerize between the cis and trans isomers in a mutual way. A finding of one-way cis → trans isomerization of anthrylethenes occurring in an adiabatic pathway provided a basis for more thorough understanding of the nature of the photoisomerization in terms of the potential energy surface. Now the features of one-way and conventional two-way isomerization and the factors controlling the modes of one- and two-way isomerization are discussed.

In Figure 1 we compare the potential energy surfaces of isomerization of ethenes with varying substituted aromatic nuclei. In the series of styrylarenes, ArCH=CHPh, stilbene (**1b**) [1–3, 5, 6, 15], 2-styrylnaphthalene (**2b**) [56, 57], and 3-styrylchrysene (**3b**) [30] undergo two-way isomerization (Figure 1*a*). However, 2- (**6c**), 1- (**7b**), and 9-styrylanthracene (**8b**) [4, 18–23] and 3-styrylperylene (**10**) [64] undergo one-way isomerization (Figure 1*b*), while styrylpyrene (**5b**) [25, 26] undergoes two-way isomerization showing features of one-way isomerization (Figure 1*c*). Furthermore, isomerization scarcely proceeds when either deactivation of the initially resulting excited state to the ground state is accelerated in preference to the isomerization to the perpendicular excited state (Fig. 1d) or the barrier for rotation of the double bond is too high to be overcome within the lifetime of the excited state (Fig. 1e). Styrylferrocene [62–63] undergoing an inefficient one-way cis → trans isomerization with a considerably low $\phi_{c\to t}$ value ($\approx 10^{-3}$) belongs to the former case (Fig. 1d), and deuterated vinylanthracene (**6e**) [42] belongs to the latter (Fig. 1e), where practically no isomerization takes place at 7°C.

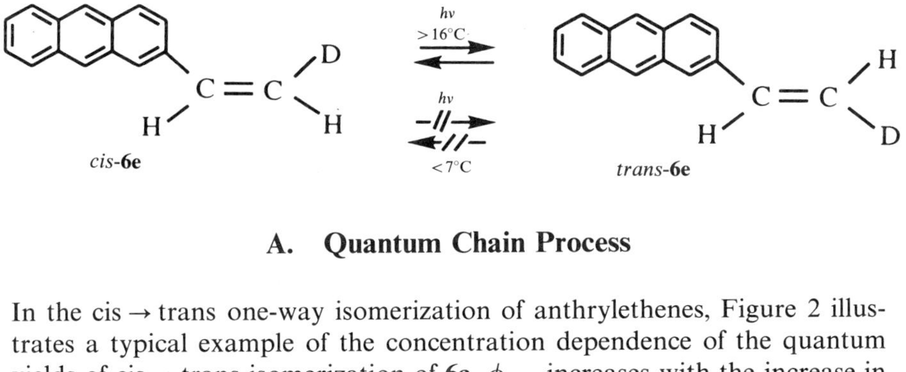

cis-**6e** *trans*-**6e**

A. Quantum Chain Process

In the cis → trans one-way isomerization of anthrylethenes, Figure 2 illustrates a typical example of the concentration dependence of the quantum yields of cis → trans isomerization of **6a**. $\phi_{c\to t}$ increases with the increase in cis-isomer concentration [4, 18–20]. The isomerization proceeds in the quantum chain mechanism described in Scheme 2, where $^3S^*$ is the sensitizer triplet and the k's are the rate constants for the corresponding processes.

According to Scheme 2, $\phi_{c\to t}$ is expressed by Eq. (1) and increases linearly with cis-isomer concentration as observed (Figure 2) [4, 19, 20]. In Eq. (1) ϕ_T^S is the quantum yield of the intersystem crossing of the sensitizer; $k_{td}(=1/\tau_T)$ and k_{tc} are the rate constant of unimolecular deactivation from

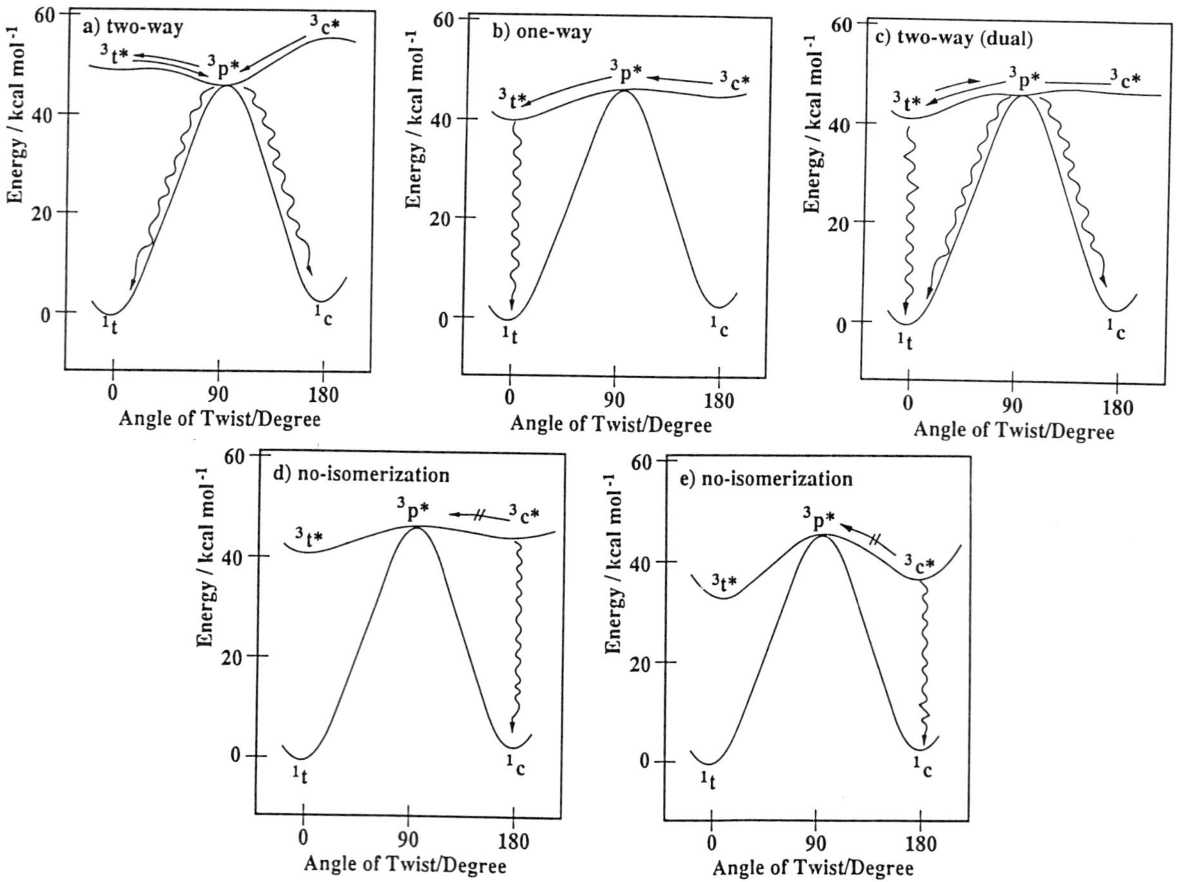

Figure 1. Potential energy surfaces of photoisomerization in the triplet state.

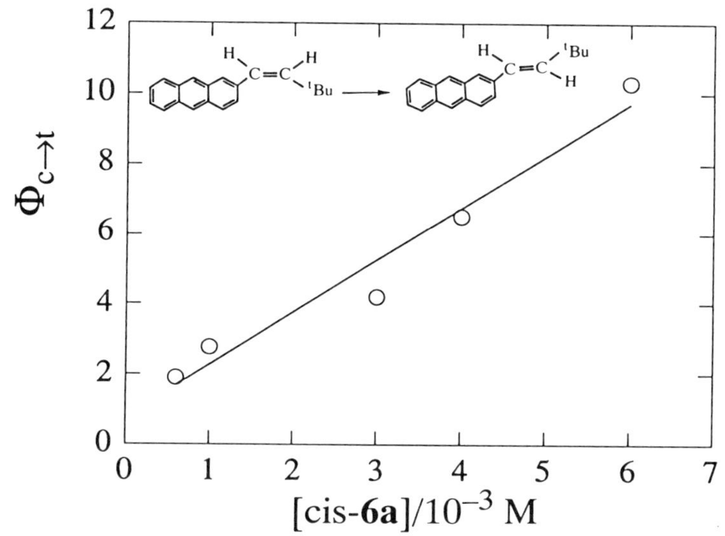

Figure 2. Effect of concentration of *cis*-**6a** on the quantum yields of cis → trans isomerization of **6a** on Michler's ketone sensitization.

^{3}t* and the rate constant of energy transfer from ^{3}t* to ^{1}c, respectively.

$$\phi_{c\to t} = \phi_T^S\left(1 + \frac{k_{tc}[\text{cis}]}{k_{td}}\right)$$
$$= \phi_T^S(1 + k_{tc}\tau_T[\text{cis}]) \qquad (1)$$

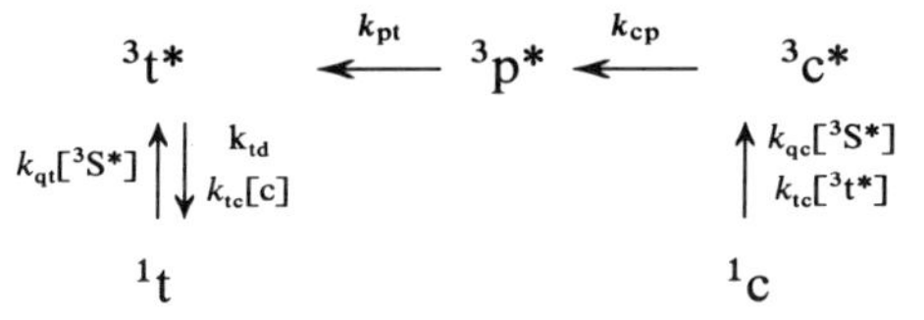

Scheme 2

It is to be noted that the chain carrier in the quantum chain process is the excited triplet state of the product of the reaction, ^{3}t*.

B. Observation of the Excited Triplet State

Laser excitation of either the cis or trans isomer of anthrylethylenes (**6**) [4, 19–21] in degassed benzene at room temperature affords the same T–T absorption spectra nearly 10 μs after laser excitation, as illustrated in

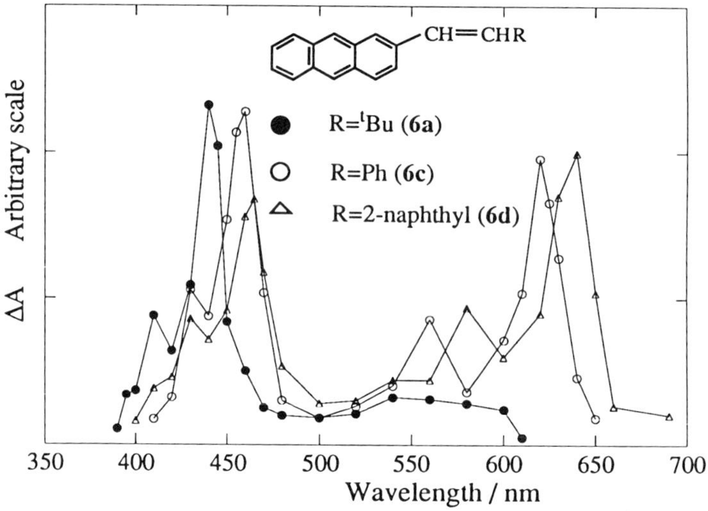

Figure 3. T–T absorption spectra of anthrylethenes at room temperature [4, 19, 20].

Figure 3. The spectra observed are assigned not to $^3p^*$ but to $^3t^*$, on the following grounds. First, the spectra observed are shifted to a longer wavelength in the order R = alkyl (^{t}Bu and Me), phenyl, then 2-naphthyl. The intensity at a longer-wavelength region increases in the same order. If the triplets observed retained a perpendicular conformation, **6** would exhibit essentially the same T–T absorption at the longest wavelength due to the largest chromophore, the anthrylmethyl moiety, regardless of the β-substituent R. Second, the triplet states observed have lifetimes of hundreds of microseconds, suggesting a large energy gap between the triplet and the ground state. The perpendicular triplet state ($^3p^*$) is very close in energy to ^1p and therefore deactivates to the ground state with a shorter lifetime, typically 10 to 100 ns (cf. stilbene at 60 ns) [15]. Third, the results of semiempirical calculation of the wavelengths and the oscillator strengths of the T–T absorptions support assignment of the triplet state to $^3t^*$, not to $^3p^*$ [21b]. Furthermore, for the quantum chain process to take place, the $^3t^*$ must lie at an energy minimum in the triplet potential energy surface.

C. $^3p^*$ as an Energy Barrier or Energy Minimum

The presence of an activation barrier for $^3c^* \rightarrow {}^3t^*$ isomerization is definitely demonstrated by the temperature effect on the cis → trans isomerization of **6a**, **7a**, and **8a** from the transient spectroscopy immediately after the laser excitation [21,43]. With **6a**, the absorption ($\lambda_{max} = 443$ nm) observed immediately after the laser excitation of *cis*-**6a** in benzene at room temperature evolved into a longer-wavelength absorption ($\lambda_{max} = 445$ nm) at 500 ns after the excitation. The initial absorption was assigned mostly to $^3c^*$ and the later one to $^3t^*$. In the time profile of the transient absorption, the initial decay observed at 440 nm was accompanied by an absorption rise, with the same time constant of nearly 500 ns observed at 450 nm. Measurement at various temperatures gave an activation energy of 6 kcal mol^{-1} and a frequency factor of 5×10^{10} s^{-1}.

As to **6a**, the activation energy combined with the energies of $^3c^*$ (E_T = 42.5 kcal mol^{-1}) and $^3t^*$ (42.5 kcal mol^{-1}) determined from their phosphorescence spectra and the energy difference between cis and trans at the ground state estimated as 5 kcal mol^{-1} enables one to draw the potential energy surface for **6a** as shown in Figure 4. Similarly, the barriers for $^3c^* \rightarrow {}^3t^*$ conversion [$E_a(^3c^* \rightarrow {}^3t^*)$] were determined as 4.6 and 3.1 kcal mol^{-1}, for **7a** and **8a**, respectively. Thus, in the series of ArCH=CH^{t}Bu, the value of $E_a(^3c^* \rightarrow {}^3t^*)$ decreases in the order 2-, 1-, and 9-anthryl substitution, probably due to the increase in conjugation between the anthracene

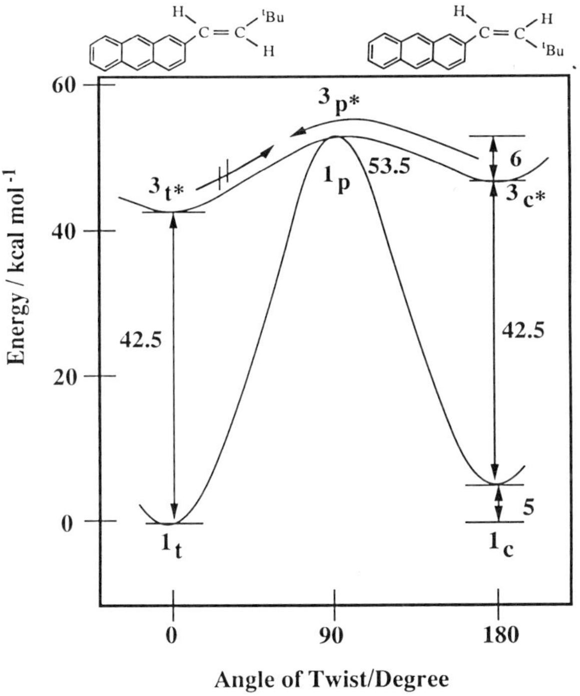

Figure 4. Potential energy surfaces of one-way isomerization of **6a** [21].

nucleus and the double bond, which reduces the energy required for breaking the double bond.

With **6c**, $^3c^* \rightarrow {}^3t^*$ isomerization takes place too rapidly to be followed even on a nanosecond time scale. However, lowering the triplet energy of the aromatic group in the styrylarene series made it possible to observe $^3c^* \rightarrow {}^3t^*$ conversion in 3-styrylperylene (**10**) [64]. The conversion of $^3c^* \rightarrow {}^3t^*$ on excitation of *cis*-**10** measured at 205.4 K by the growing of the absorbance at 580 nm and the concomitant decrease in absorbance at 510 to 520 and 550 to 560 nm in the microsecond time scale is shown in Figure 5. The conversion rate constant $[k(^3c^* \rightarrow {}^3t^*)]$ increases with increased temperature, to give an activation energy E_a and a frequency factor A for $^3c^* \rightarrow {}^3t^*$ isomerization of 6.6 kcal mol^{-1} and $2.1 \times 10^{12}\ s^{-1}$, respectively.

D. Effect of Aromatic Nuclei on the Isomerization Mode

If we assume that energy minima exist only at $^3p^*$ and $^3t^*$, the isomerization mode will be determined by whether the deactivation takes place from $^3p^*$ (two-way) or $^3t^*$ (one-way). The triplet lifetime generally depends on the equilibrium constant K_{tp} between $^3p^*$ and $^3t^*$ ($K_{tp} = [^3p^*]/[^3t^*]$) and increases with decreased K_{tp} even though the mode of the isomerization is two-way.

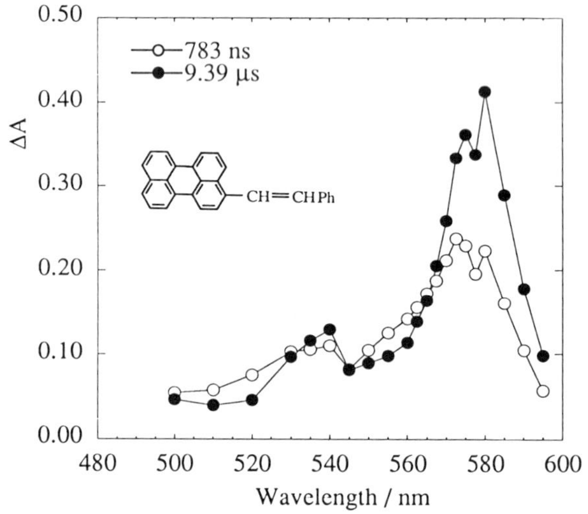

Figure 5. T–T absorption spectra determined on excitation of *cis*-**10** on 308 nm laser at 205.4 K showing the $^3c^* \rightarrow {}^3t^*$ conversion in methylcyclohexane [64].

The lifetime τ_T of the triplet states comprising the equilibrating mixture of $^3t^*$ and $^3p^*$ is expressed as

$$\tau_T = \frac{1 + K_{tp}}{K_{tp}k_{pd} + k_{td}} \tag{2}$$

For example, if the $^3t^*$ and $^3p^*$ are nearly the same in energy, $^3t^*$ and $^3p^*$ exist with nearly the same probability ($K_{tp} = 1$); therefore, the lifetime of these triplet states should be two times that of those where only $^3p^*$ is populated.

On benzil sensitization, 1-pyrenylethene (**5**) undergoes one-way isomerization when R = ^{t}Bu (**5a**); however, when R = Ph (**5b**), a photostationary mixture of cis and trans isomers results (e.g., 97.7% trans at 1.5×10^{-3} *M*). The remarkably high trans content suggests that the isomerization occurs with the dual character of both two- and one-way isomerization [25].

Compound **5a** shows the typical behavior of a one-way isomerization, and its triplet state deactivates solely from $^3t^*$ [25]. Thus $\phi_{c \rightarrow t}$ of *cis*-**5a** on benzil sensitization increases nearly linearly with concentration in degassed benzene, attaining more than 20 at [*cis*-**5a**] = 3.3×10^{-3} *M*. The stable triplet state is $^3t^*$ (λ_{max} = 445 nm) with a lifetime of 54 μs.

On the other hand, in the triplet state of **5b**, $^3t^*$ is the most stable and is in equilibrium with the slightly less stable $^3p^*$ [25]; therefore, the triplet state deactivates from both $^3p^*$ and $^3t^*$, several thousand times faster from $^3p^*$ ($k_{dp} = 10^7$ s^{-1}) than from $^3t^*$ ($k_{dt} = 10^4$ s^{-1}). The deactivation from $^3p^*$ gives trans and cis isomers as the typical two-way mode, and that from $^3t^*$ gives solely the trans isomer by either unimolecular deactivation or transferring energy to the cis isomer to regenerate $^3c^*$, leading to quantum chain isomerization of cis to trans isomers. The isomerization proceeds with the dual mechanism depicted in Scheme 3.

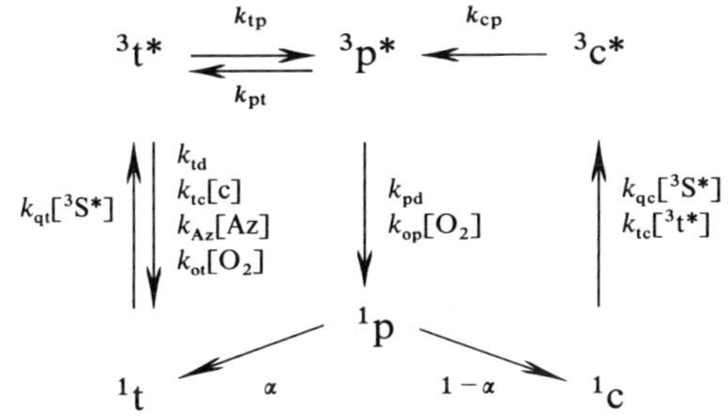

Scheme 3

In Scheme 3, when the cis and trans isomers quench the triplet state of a sensitizer with equal rate constants, that is $k_{qc} = k_{qt}$ and equilibrium is established between ^{3}t* and ^{3}p* ($K_{tp} = k_{tp}/k_{pt}$), the photostationary isomer ratio, $([t]/[c])_s$, and $\phi_{c\to t}$ are expressed by Eqs. (3) and (4), respectively. Here, $[\mathbf{5b}] = [t]_s + [c]_s$ is the total concentration of **5b**, and ϕ_T is the quantum yield for *cis*-**5b** triplet formation. These equations explain the concentration dependence of $([t]/[c])_s$ and $\phi_{c\to t}$.

$$([t]/[c])_s = \frac{\alpha K_{tp}k_{pd} + k_{td} + k_{tc}[\mathbf{5b}]}{(1-\alpha)K_{tp}k_{pd}} \tag{3}$$

$$\phi_{c\to t} = \phi_T \times \frac{\alpha K_{tp}k_{pd} + k_{td} + k_{tc}[cis\text{-}\mathbf{5b}]}{K_{tp}k_{pd} + k_{td}} \tag{4}$$

K_{tp} can generally be determined by measuring τ_T in the presence of varying concentrations of azulene and oxygen according to Eqs. (5) to (7), since azulene works as a quencher of solely ^{3}t* and oxygen quenches ^{3}p* three times more efficiently than ^{3}t*. In these equations, k_q^{Az} and k_q^{Ox} are the apparent quenching rate constants of the triplet state as a mixture of ^{3}t* and ^{3}p* by azulene and oxygen, respectively, and k_{az} [ca. (0.8–1) × $10^{10}\ M^{-1}\ s^{-1}$] and k_{to} [ca. $3 \times 10^9\ M^{-1}\ s^{-1}$] are quenching rate constants of ^{3}t* by azulene and oxygen, respectively, and k_{po} (= ca. $9 \times 10^9\ M^{-1}\ s^{-1}$) is the quenching rate constant of ^{3}p* by oxygen [15,24].

$$\frac{1}{\tau_T} = \frac{k_{td} + K_{tp}k_{pd}}{1 + K_{tp}} + k_q^{Az}[\mathrm{Az}] + k_q^{Ox}[\mathrm{O_2}] \tag{5}$$

$$k_q^{Az} = \frac{k_{az}}{1 + K_{tp}} \tag{6}$$

$$k_q^{Ox} = \frac{k_{to} + K_{tp}k_{po}}{1 + K_{tp}} \tag{7}$$

Whether arylethenes undergo one or two-way isomerization is governed by the triplet excitation energies of the aromatic nuclei substituted on the unsaturated bonds [4,28,31,63]. Table 2 summarizes the effects of the triplet energy of aromatic nuclei on the mode of the isomerization and on the triplet lifetime for two series of olefins, ArCH=CH^{t}Bu and ArCH=CHPh. In the series of ArCH=CH^{t}Bu, aromatic groups with triplet energies higher than chrysene (56 kcal mol^{-1}) lead to the two-way mode; however, those with a lower triplet energy than that give the one-way mode. In the series of ArCH=CHPh, aromatic groups with triplet energies lower than anthracene (42 kcal mol^{-1}) result in one-way isomerization, and Ar =

1-pyrenyl with the triplet energy of 48 kcal mol^{-1} behaves with dual one- and two-way character, as described above. The triplet states of ethylenes undergoing two-way isomerization have lifetimes in the range of tens of nanoseconds to tens of microseconds, which increase with the increase of the population of $^3t^*$ relative to $^3p^*$.

As described above, the isomerization behavior of ArCH=CHR is determined primarily by the triplet energy of the main aromatic group Ar and, in addition, by the conjugative property of the second substituent R. Thus the triplet energy surfaces of arylethenes can be interpreted as a compromise of the two effects proposed. The key conformations, $^3p^*$ and $^3t^*$, respond to substituents in different manners: (1) the energy of $^3t^*$ decreases with decreasing triplet energy of the main aromatic group Ar, and (2) the energy of $^3p^*$ is not so dependent on the triplet energy of Ar but is reduced when the second substituent R is phenyl.

E. Conformational Change Along the Single Bond in One-Way Isomerization

Because of steric factors, cis isomers adopt conformations where the aromatic nucleus and the double bond are not coplanar but the aromatic nucleus is rotated more than 30° around the single bond from the plane of the double bond. According to the x-ray analysis of the crystalline *cis*-**5a**, the angle between the pyrenyl ring and the double bond is 90.0°; that is, it adopts a totally perpendicular arrangement of the pyrenyl group to the double bond [58]. Therefore, the one-way isomerization of **5a** should take place through some kind of intramolecular excitation transfer from the initially produced excited state of the pyrenyl group to the olefinic part during the twisting of the double bond to the perpendicular triplet state ($^3p^*$), finally twisting to the most stable trans triplet ($^3t^*$). For this process to occur, conjugation between the pyrenyl part and the double bond is necessary, and thus torsional vibration around the single bond connecting the pyrene nucleus and the double bond should take place concurrently to double-bond twisting, since this reduces steric repulsion between the pyrene nucleus and the *tert*-butyl group.

F. Spectroscopic and Relaxed Triplet Energies

In the triplet states of ethenes, the energies of $^3t^*$ and sometimes $^3c^*$ can be determined by S_0–T_1 absorption and phosphorescence spectroscopy. However, the triplet energy for $^3p^*$, the relaxed triplet states, are generally difficult to be determined by spectroscopy.

Caldwell et al. determined the relaxed triplet energies of a series of arylethenes by means of photoacoustic calorimetry (PAC) and compared these with the spectroscopic energies determined by S_0–T_1 absorption or phosphorescence spectra [37]. In compounds undergoing two-way isomerization with flexible C=C double bonds such as styrenes and stilbene they observed a considerable difference between the spectroscopic triplet energy [E_T(spectroscopic)] and the relaxed triplet energy [E_T(relaxed)]. For example, E_T(spectroscopic) is 64.9 and 60.5 kcal mol^{-1}, respectively, for styrene and β-methylstyrene, while E_T(relaxed) is 51.2 and 53.2 kcal mol^{-1}, respectively. Thus, in the triplet potential energy surface of these ethylenes, not only $^3c^*$, but also $^3t^*$ is stabilized by relaxation to $^3p^*$.

On the contrary, in the one-way isomerizing compound, **5a**, E_T(spectroscopic) determined by phosphorescence spectroscopy (44 kcal mol^{-1}) is similar to the E_T(relaxed) determined by the PAC method (43 kcal mol^{-1}), which shows that in **5a** the spectroscopic and relaxed triplet states have the same conformation as the transoid triplet state [58].

The foregoing methodology enables one to draw the potential energy surfaces of the photoisomerization of ethenes more precisely.

G. Isomerization Around a C=N Double Bond

Like arylethenes, the triplet potential energy surface for the isomerization around the C=N double bond in ArCR=$NOCH_3$ is governed by the triplet excitation energy of the aromatic groups (Ar) [66, 67]. While Ar = phenyl or 2-naphthyl brought about two-way isomerization as in styrene and the naphthylethenes, an aromatic nucleus with a low triplet energy, Ar = anthryl, changes the isomerization mode from two-way to one-way.

The E and Z isomers of *N*-methoxy-1-(2-anthryl)ethanimine (**11**) are stable at room temperature in the dark. (*Z*)-**11** isomerizes to the corresponding E isomer via the triplet state on direct (366 nm) as well as on benzil-sensitized irradiation (436 nm) in degassed benzene. In contrast, the E isomer gave no Z isomer, resulting in one-way Z → E isomerization [68]. The quantum yield for Z → E isomerization of **11** on benzil sensitization increases with the concentration of *cis*-**11** and reaches 24 at 1.3×10^{-3} *M*. On laser flash photolysis in benzene in the presence of benzil at room temperature, both (*Z*)- and (*E*)-**11** afford a transient absorption assigned to $^3E^*$ with a lifetime of 90 μs.

Furthermore, like styrylpyrene (**5b**), pyrene-substituted imine (**12**) undergoes mutual isomerization between Z and E isomers via the triplet excited state, and the quantum yield for the Z → E isomerization increases with Z-isomer concentration by way of the dual mechanism.

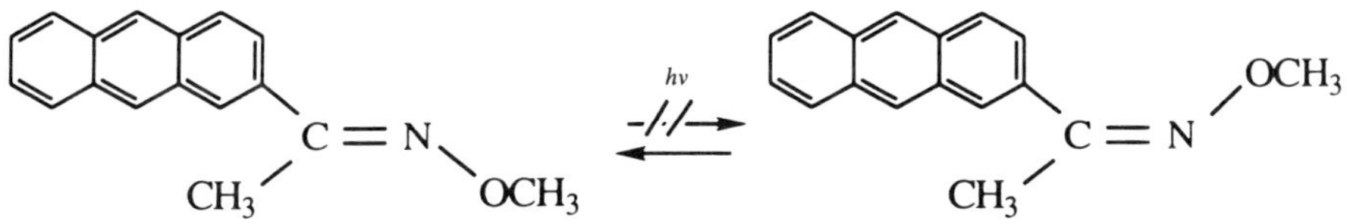

(*E*)-**11** (*Z*)-**11**

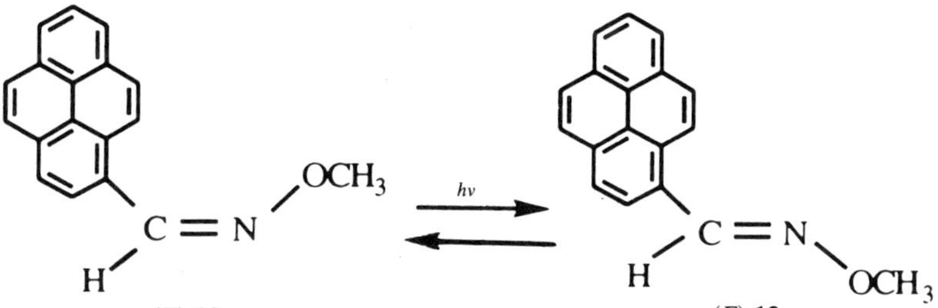

(*Z*)-**12** (*E*)-**12**

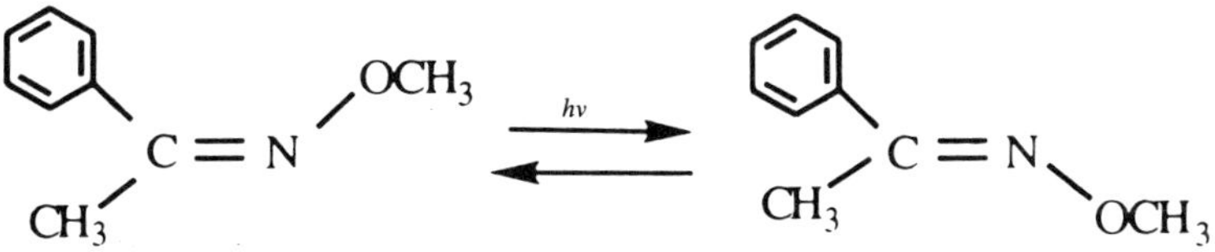

(*Z*)-**13** (*E*)-**13**

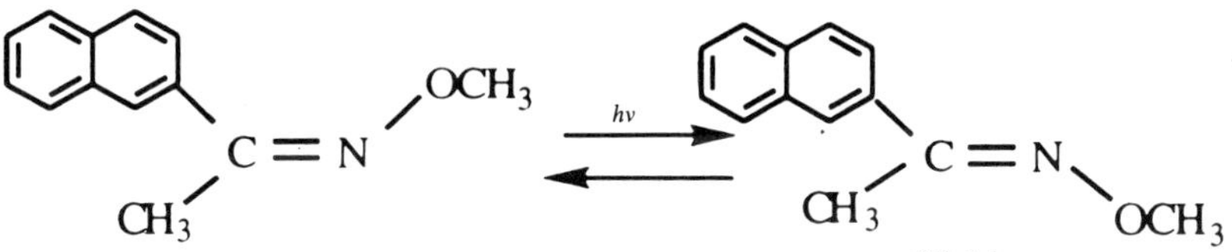

(*Z*)-**14** (*E*)-**14**

Table 3 summarizes the effect of aromatic nuclei of varying triplet energies on the mode of the isomerization of ArCR=NOCH_3.

Substitution of a 9-anthryl group on the carbon of the C=N bond **(15a)** induces two-way and one-way cis → trans isomerization in the S_1 and T_1 states, respectively [69a].

When aryl groups are substituted on the nitrogen of the C=N bond, the E isomers are stable at room temperature and undergo photoisomerization to Z isomers, which thermally revert to the E isomers [70]. For example, (*E*)-**15b** isomerizes to (*Z*)-**15b** on direct as well as triplet sensitization, giving photostationary state mixtures of Z and E isomers, and (*Z*)-**15b** reverts to

TABLE 3 Effect of Aromatic Substituents on the Mode of the Isomerization of ArCR=NOCH$_3$ in the Triplet State

Ar	R	Triplet Energy of ArH (kcal mol^{-1})	Mode of Isomerization, $\tau_T(\mu s)$ [λ_{max}(T–T)(nm)]	$\Phi_{Z \to E}$ ([Z])	Refs.
Phenyl	CH_3	84.3	Two-way	0.35 (3×10^{-2} *M*)	66, 67a
2-Naphthyl	CH_3	60.9	Two-way	0.51 (5×10^{-2} *M*)	66, 67b
1-Pyrenyl	H	48.2	Two-way 10 (440)	3.2 (1.3×10^{-3} *M*)	69a
2-Anthryl	CH_3	42	One-way 90 (450, 540)	24 (1.3×10^{-3} *M*)	68
9-Anthryl	H	42	One-way 58 (< 420)	2.8 (1.47×10^{-3} M)	69a

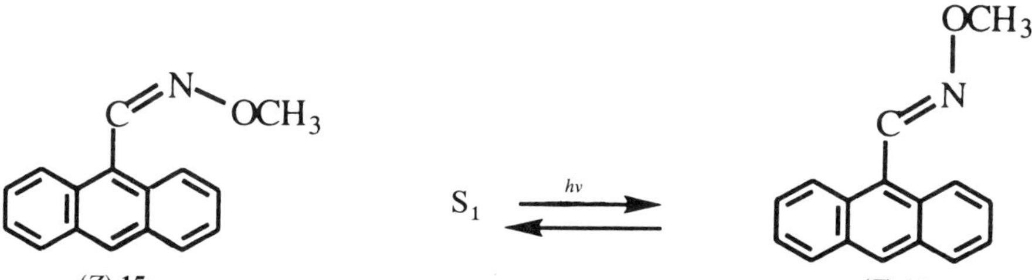

(*Z*)-**15a** (*E*)-**15a**

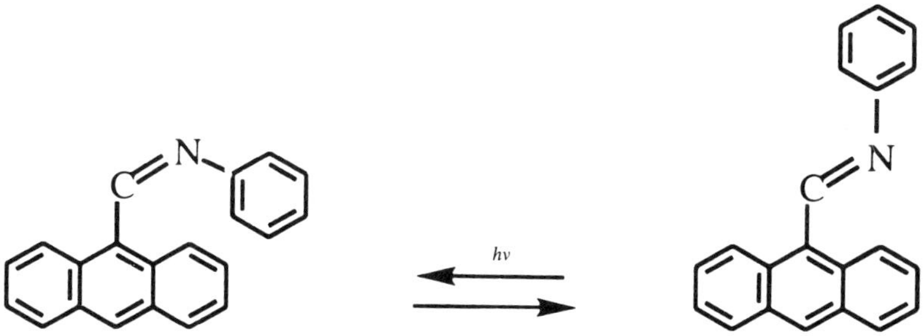

(*Z*)-**15b** (*E*)-**15b**

(*E*)-**15b** in the ground state, overcoming an activation barrier of 15 to 20 kcal mol^{-1} [70].

H. Adiabatic or Diabatic Photoisomerization

The potential energy surface of the isomerization is discussed in terms of adiabatic and diabatic processes [1–4, 12, 18, 71]. Two-way isomerization in the triplet manifold without a quencher takes place as a diabatic process by deactivation at ^{3}p*. However, as mentioned above, photochemical cis → trans one-way isomerization in the triplet state proceeds by an adiabatic process where the excited state of a starting material, ^{3}c*, undergoes adiabatic conversion to the excited state of the product, ^{3}t*, followed by either unimolecular deactivation to the ground state of the product, ^{1}t, or energy transfer to ^{1}c to give ^{1}t and ^{3}c*. The isomerization of **5b** also proceeds partly by way of an adiabatic process. Deactivation from ^{3}t* occurs as an adiabatic process, but that from ^{3}p* proceeds as a diabatic process [25]. Therefore, two-way photoisomerization usually takes place as a diabatic process, whereas one-way photoisomerization and isomerization via a dual mechanism, take place totally and partly as an adiabatic process, respectively. One-way isomerization need not be an adiabatic process, as seen in compounds where isomerization takes place in one direction but is prohibited in the opposite direction by structural effects such as intramolecular hydrogen bonding, as described in Section V. Furthermore, in two-way isomerization of olefins such as stilbene, isomerization from the cis to the trans isomer can be regarded partly as an adiabatic process in the presence of azulene as a quencher of $^3t^*$. Azulene quenches $^3t^*$ to give solely ^1t in an adiabatic process; however, unimolecular deactivation from $^3p^*$ to give ^1c takes place diabatically.

As to isomerization in the excited singlet state, the potential energy surface of the isomerization may be governed by the singlet excitation energy of the aromatic group, similar to that in the triplet state. Thus decreased singlet energy in the aromatic nucleus lowers the energy of planar $^1c^*$ and $^1t^*$ more than that of $^1p^*$. Therefore, changing the aromatic group from phenyl to anthryl is expected to change the profile of the isomerization in the excited singlet state on direct irradiation from two-way to one-way: $^1c^*$ of **6** does not isomerize in the excited singlet state, probably due to the increased energy of the isomerization barrier, but undergoes either fluorescence emission or intersystem crossing to the triplet state, from which one-way isomerization occurs. The quantum yields of the intersystem crossing from $^1c^*$ to $^3c^*$ are 0.59, 0.17, and 0.22 for **6a**, **6c**, and **6d**, respectively [23]. Pyrenylethylene (**5b**) is intermediate between these two cases. The pyrene group considerably lowers the energy of $^1c^*$ and $^1t^*$ to decrease compared to that of $^1p^*$; as a result, $^1p^*$ is located at an isomerization barrier and does not deactivate. However, the barrier is not high enough to inhibit $^1c^* \rightarrow {}^1t^*$ isomerization. These results, together with those for 9-styrylanthracene, are described in Section III.

III. ADIABATIC PHOTOISOMERIZATION OF ARYLETHENES IN THE SINGLET STATE

With the recent finding that adiabatic one-way isomerization of some of the arylethenes proceeds in the triplet state as described in Section II, attention has turned to examining the adiabatic nature of the isomerization that takes place in the singlet excited state. Now, adiabatic isomerization of ethenes other than those argued in Section II, particularly in the excited singlet state, is described.

A. 9-Styrylanthracenes

Becker et al. reported that **8b** undergoes one-way cis → trans isomerization on direct irradiation as well as with biacetyl sensitization and proposed that the isomerization took place as an adiabatic process in both the excited singlet and triplet states [71]. The decay of the fluorescence emission observed on excitation of the cis isomer fits a dual rather than a single exponential decay kinetics. One of the time constants observed was the same as that obtained by excitation of the trans isomer. Thus $^1c^*$ underwent isomerization to $^1t^*$ in the excited singlet manifold, thus showing fluorescence emission of both $^1c^*$ and the resulting $^1t^*$, although the quantum yield of $^1c^* \rightarrow {}^1t^*$ isomerization in the excited singlet state remained at 0.03, as estimated from the fluorescence data.

The quantum yields of the cis → trans isomerization of **8b** in cyclohexane increased from 0.22 to 0.87 with increased *cis*-**8b** concentration from 1.7×10^{-4} to 4.9×10^{-3} *M*. Accordingly, the one-way cis → trans isomerization also proceeded in the triplet manifold after intersystem crossing. At infinite dilution, where the quantum chain process did not work in the excited triplet state, $\phi_{c \rightarrow t}$ was estimated as 0.20. This value gave the ratio of the one-way isomerization via excited singlet and triplet states in dilute solution as 0.03-0.17. Figure 6 depicts the potential energy surfaces of the one-way isomerization of **8b** in the excited singlet and triplet states.

Goerner reported a slightly different mechanism based not only on fluorescence spectroscopy and quantum yield measurements, but also on laser flash photolysis experiments [72]. Thus it is probable that $^1c^*$ may undergo either isomerization in the excited singlet state giving $^1t^*$ directly, or intersystem crossing to $^3c^*$ followed by isomerization to $^3t^*$, and that the isomerization in both states takes place adiabatically in cyclohexane [71–74].

Similarly, 9-anthrylethenes with either a 1-naphthyl (**16a**) or an 9-anthryl group (**16b**) at another ethylenic carbon exhibit similar one-way cis → trans

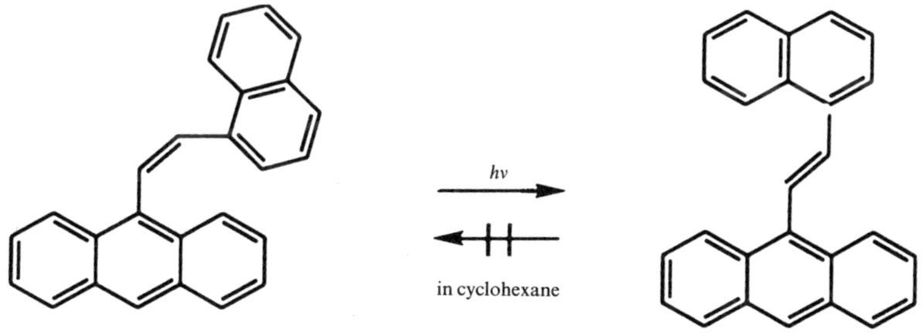

cis-**16a** *trans*-**16a**

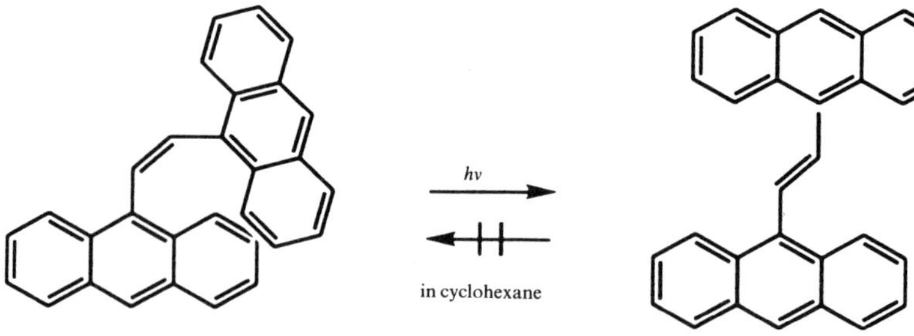

cis-**16b** *trans*-**16b**

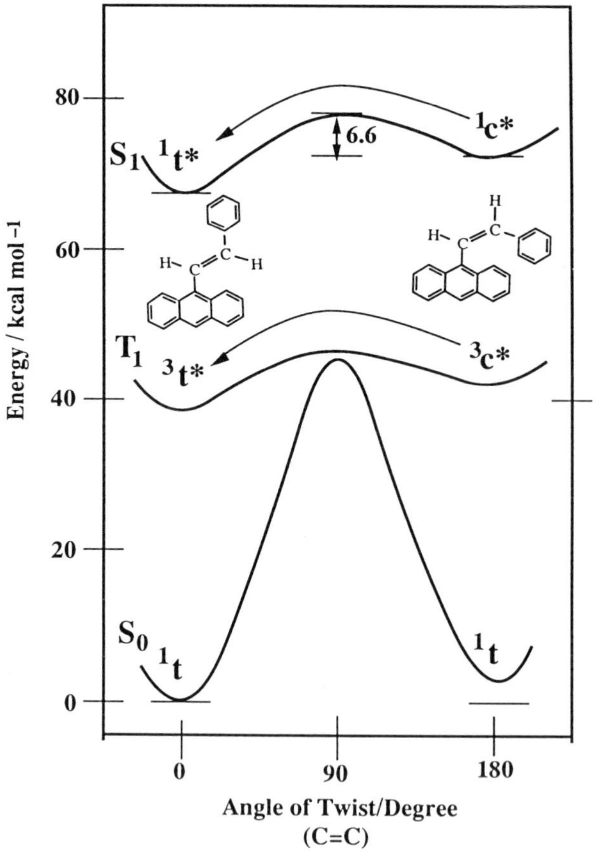

Figure 6. Potential energy surfaces of one-way photoisomerization of **8b** in the excited singlet and triplet states [71].

isomerizations in cyclohexane in the excited singlet state, although in very low quantum yields: 7×10^{-4} and $\ll 1 \times 10^{-4}$ for $\phi_{c\to t}$ and $\phi_{t\to c}$ of **16b**, respectively [75].

Solvents and polar substituents affect the isomerization mode and quantum yields [76]. For example, **17a** undergoes almost purely one-way cis → trans isomerization in cyclohexane but two-way isomerization in acetonitrile, with $\phi_{c\to t}$ and $\phi_{t\to c}$ values of 0.35 and 0.38, respectively. Goerner et al. recently reported similar solvent effects for the isomerization of 9-styrylanthracene with either a 4-methyl (**17b**) or a 4-methoxy substituent (**17c**) on the benzene nucleus. For example, $\phi_{t\to c}$ for *trans*-**17c** increased from 0.02 to 0.33 with increasing solvent polarity from cyclohexane to acetonitrile [76].

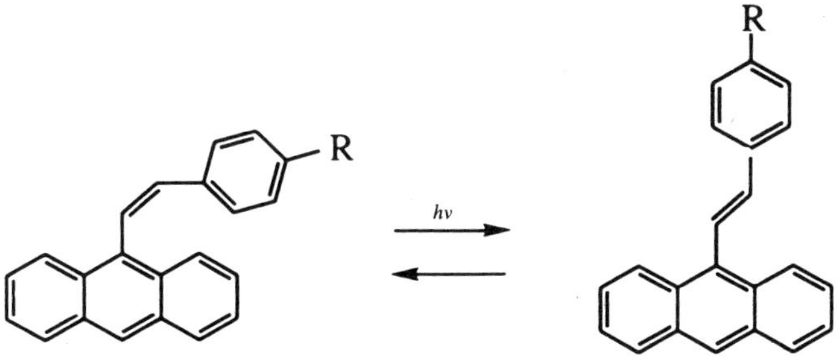

cis-**17** *trans*-**17**

a: R = CHO, b: R = CH_3, c: R = OCH_3

B. 1-Styrylpyrene

As described earlier, on benzil sensitization, 1-styrylpyrene (**5b**) undergoes two-way photoisomerization in the excited triplet state with a quantum chain process in the cis → trans direction. However, on direct excitation, one-way cis → trans isomerization takes place on the excited singlet surface as an adiabatic conversion from $^1c^*$ to $^1t^*$ as reported by Mazzucato et al. [77a].

cis-**5b** and *trans*-**5b** exhibit similar fluorescence spectra at room temperature. However, at lower temperature, *cis*-**5b** shows a slightly broader fluorescence spectrum than that of *trans*-**5b**. On excitation of *cis*-**5b** in methylcyclohexane-3-methylpentane(9-1) at room temperature, the resulting fluorescence decays with rising and decaying time constant components of 1.3 and 4.4 ns; the latter is the same for the lifetime observed on excitation of *trans*-**5b**. The shorter time constant results from adiabatic $^1c^* \to {}^1t^*$ isomerization. Increased temperature decreased the lifetime of the shorter component to give 0.27 ns at 354 K. Thus the activation barrier of the $^1c^* \rightarrow {}^1t^*$ isomerization of *cis*-**5b** was obtained as 7.7 kcal mol^{-1}. The $\phi_{c \rightarrow t}$ isomerization was 0.50 and that of $^1c^* \rightarrow {}^1t^*$ was 0.61. Moreover, increase of the polarity of the solvents reduces the activation barrier by stabilizing the $^1p^*$ conformation [77b].

Similar to 9-styrylanthracenes (**17**), introduction of polar substituents on the benzene nucleus of 1-styrylpyrene as well as use of polar solvents alters the isomerization mode. Thus *trans*-1-[2(4-nitrophenyl)ethenyl]pyrene (**18a**) is difficult to isomerize from the trans to the cis isomer in acetronitrile on direct irradiation ($\phi_{t \rightarrow c} < 10^{-5}$) [78], although **18a**, **18b**, and **18c** undergo mutual isomerization with a quantum chain process in the cis → trans direction on triplet sensitization [79].

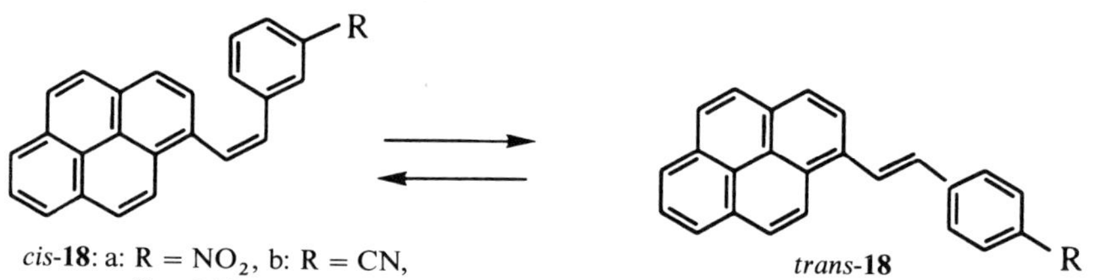

cis-**18**: a: R = NO_2, b: R = CN,
c: R = OMe

trans-**18**

C. Stilbene

Very careful work by Saltiel et al. on stilbene demonstrates clearly that direct excitation of the cis isomer results in fluorescence emission from both $^1t^*$ and $^1c^*$, suggesting that cis → trans isomerization proceeds partly adiabatically [12].

cis-Stilbene exhibits fluorescence spectra composed of the emission from $^1c^*$ and $^1t^*$ with quantum yields of about 8×10^{-5} and 9×10^{-5}, respectively; production of $^1t^*$ is attributed not to excitation of *trans*-stilbene resulting from the photoisomerization, but to direct production from $^1c^*$ with a quantum yield of 2×10^{-3} for $\phi_{^1c^* \rightarrow {}^1t^*}$.

Thus cis → trans isomerization in the excited singlet state proceeds by deactivation from $^1t^*$ as well as $^1p^*$, even though the contribution from the former deactivation ($\phi = 10^{-4}$) is very small compared to that of the latter ($\phi = 0.33$). The potential energy surfaces of the isomerization of stilbene are depicted briefly in Figure 7.

IV. ONE-WAY ISOMERIZATION AND QUANTUM CHAINING OF POLYENES

Substitution of an ethylenic group on an unsaturated bond is as effective as that of a phenyl group to lower the triplet excitation energy of the molecule

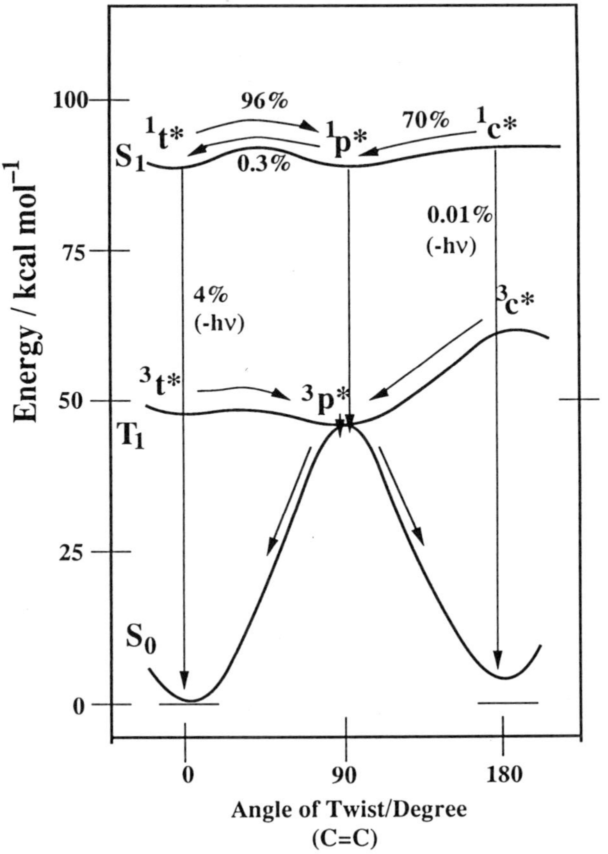

Figure 7. Potential energy surfaces of isomerization of stilbene in the excited singlet and triplet states [12].

as seen in those of 1,3-butadiene and styrene. Therefore, conjugated polyenes are expected to behave more or less similarly to unsaturated bonds that have been replaced by aromatic nuclei with low excitation energies undergoing adiabatic one-way isomerization. The isomerization of polyenes is described next in terms of the aspects discussed above.

A. Dienes, Trienes, and Polyenes Without Aromatic Substituents

The quantum chain process in the photoisomerization of the C═C double bond was first reported by Hammond et al. [80a], followed by Hurley and Testa [80b] and Caldwell [81]. Subsequently, the former work [80a] was shown by Saltiel [82] to be erroneous. Thus 1,3-pentadiene and

2,4-hexadiene were found to undergo mutual isomerization among the isomers, and the quantum yields of their cis → trans isomerization increased with increased diene concentration. Although the quantum yields were different in the various reports, mechanisms involving energy transfer from the diene triplet to the diene ground state were proposed [81, 82].

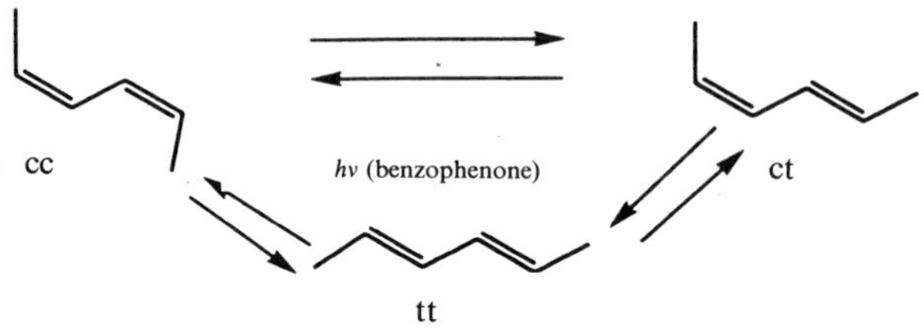

2,4-Hexadiene isomerized among cis-cis (cc), cis-trans (ct), and trans-trans (tt) isomers with quantum chain processes for cc→ct, cc→tt, tt→ct, and tt→cc [82]. Thus, on benzophenone sensitization, $\phi_{cc\rightarrow ct}$, $\phi_{cc\rightarrow tt}$, $\phi_{tt\rightarrow ct}$, and $\phi_{tt\rightarrow cc}$ increased from 0.47, 0.17, 0.51, and 0.32, respectively, at 0.088 *M* concentration to 0.94, 0.34, 1.88, and 1.04, respectively, at 7.83 *M*. Similarly, 1,3,5-hexatriene, 2,6-dimethyl-1,3,5-hexatriene, and 1,5-dimethyl-1,3,5-heptatriene photoisomerized with quantum chain processes [83].

As the origin of the occurrence of the quantum chain process in the isomerization of the dienes and trienes above, energy transfer from their excited triplet states to their ground states of the s-cis form was proposed. Compared to s-trans rotamers, s-cis rotamers have lower triplet energies and are more readily excited, although less populated in the ground state. For example, the following reaction may proceed [82]:

$$^3tt^* + \text{s-cis-ct} \rightarrow tt + {}^3ct^*$$

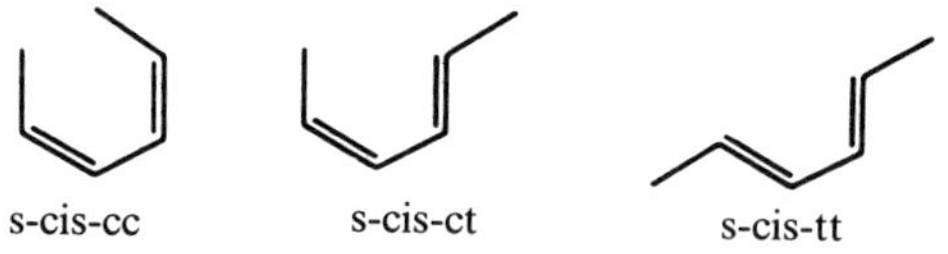

Upon anthracene sensitization of 7-*cis*, 9-*cis*, 13-*cis*, 15-*cis*-, or all-*trans*-β-carotene, the relative quantum yields for the disappearance increased in the order all-*trans* < 7-*cis* < 9-*cis* ≪ 13-*cis* < 15-*cis* [84]. The cis → trans isomerizations from 13-cis and 15-cis to the all-trans isomer proceeded as adiabatic processes to give the all-trans triplet state as revealed by transient absorption and Raman spectroscopy. A photostationary state is composed of mostly all-trans isomer: all-trans (85%), 7-cis (0%), 9-cis (10%), 13-cis (4%), 15-cis (1%).

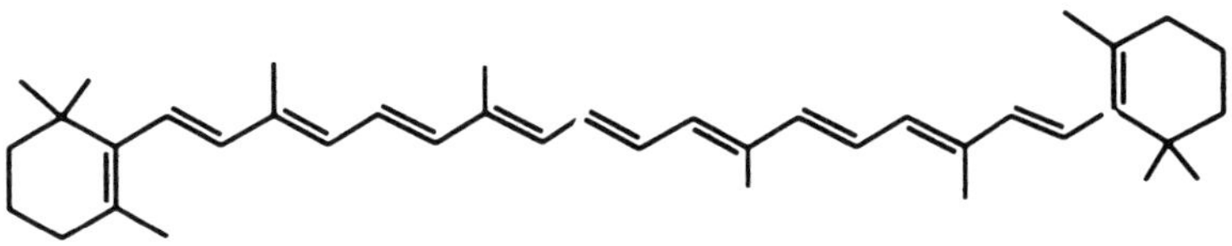

all-*trans*-β-carotene

B. Isomerization of Retinals

Ramamurthy and Liu [85] reported on the effect of chain length on the potential energy surfaces of triplet-sensitized isomerization of polyenes of retinal homologs. Thus dienes and trienes undergo unimolecular deactivation from the twisted triplet because the planar trans triplet and the twisted triplet are in equilibrium and are very close in energy. On the other hand, in tetraenes and higher polyenes, deactivation proceeds from the planar triplet states, which are much more stable than the twisted triplet states.

β-Ionol undergoes only trans → cis isomerization on sensitization with sensitizers of low triplet energies such as 1-acetylnaphthalene, while with sensitizers of higher triplet energies such as acetophenone, isomerization occurs in both directions [85]. This is because acetophenone can sensitize both trans and cis isomers; however, 2-acetylnaphthalene sensitizes mostly trans isomers.

Among the isomers of retinal, the 11-cis isomer isomerizes in hexane from the triplet state to give the all-trans isomer accompanied by more than 5% of 13-cis and 9-cis isomers at the photostationary state; the reaction is largely a one-way isomerization. The isomerization proceeds through a common excited state, the all-trans triplet state, as observed by transient Raman spectroscopy [86–90]. The quantum yield of the 11-cis → all-trans isomerization on triplet sensitization increases linearly with 11-cis isomer concentration, exceeding unity as in a typical one-way isomerization [91–93] (Scheme 4). Moreover, even all-*trans*-retinal undergoes adiabatic isomerization to the cis isomers through quantum chain processes in the excited triplet state [92, 93]. On triplet sensitization the cis isomers give not only all-trans but also other cis or di-cis isomers, attributable to one-photon/two (or multiple)-bond isomerization at the initial stage of the photoisomerization. For example, the 9-cis isomer affords 13-cis, 9,13-di-cis, and all-trans isomers in a ratio 12, 3, and 84, respectively, as the initial products [93].

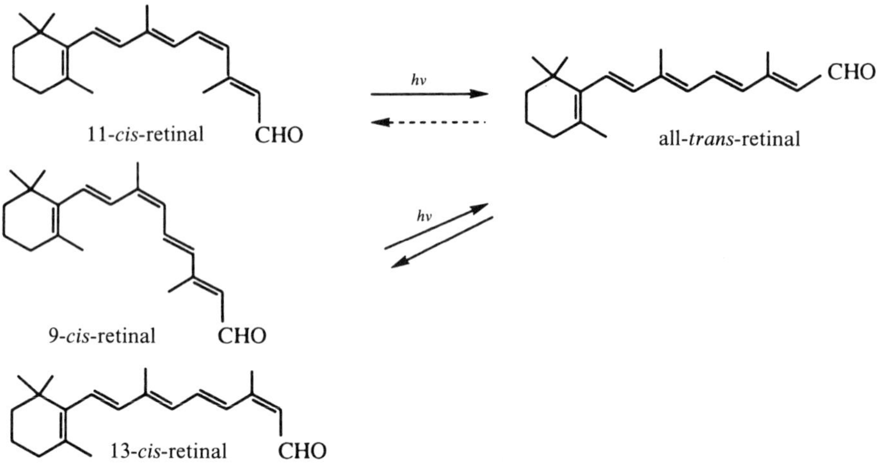

Scheme 4

C. Dienes with Aromatic Substituents

1,4-Diphenylbutadiene (**19**) undergoes ZZ → ZE and ZZ → EE one-way photoisomerization in the triplet state in a quantum chain process [94]. On the other hand, ZE and EE isomers undergo mutual isomerization. However, ZE → EE isomerization proceeds in a quantum chain process, since its quantum yield increases linearly with the total concentration of **19** on 9-fluorenone sensitization. On 9-fluorenone sensitization, all the isomers exhibited the same T–T absorption ($\lambda_{max} = 390$ nm, $\tau_T \approx 1.6\ \mu s$) attributable to ^{3}EE* at 1 μs after the laser pulse [94, 95]. Furthermore, the photostationary mixture becomes richer in EE with increased initial ZE concentration. Therefore, the isomerization between ZE and EE takes place similarly to that of **5b** with a dual mechanism. Thus the triplet state of **19** is composed of ^{3}EP* and ^{3}EE* in equilibrium, where EP means the perpendicular geometry at one double bond and E geometry at the other double bond of the diene (Scheme 5); unimolecular deactivation from ^{3}EP* gives EE and ZE, and bimolecular deactivation from ^{3}EE* with ZE isomer gives solely EE.

Similar to the effect of substitution of a 2-anthryl group on a double bond, substitution of a 2-anthryl group on the terminal carbon of conjugated dienes leads to one-way cis → trans isomerization (Scheme 6). Thus *ct*-**20** and *cis*-**21** isomerize solely to *tt*-**20** and *trans*-**21**, respectively, in benzene with benzil sensitization, while no reverse isomerization takes place from trans to any of the cis isomers [96]. The quantum yields of benzil-sensitized isomerization from *ct*- to *tt*-**20** ($\phi_{ct \to tt}$) and *cis*- to *trans*-**21** ($\phi_{c \to t}$)

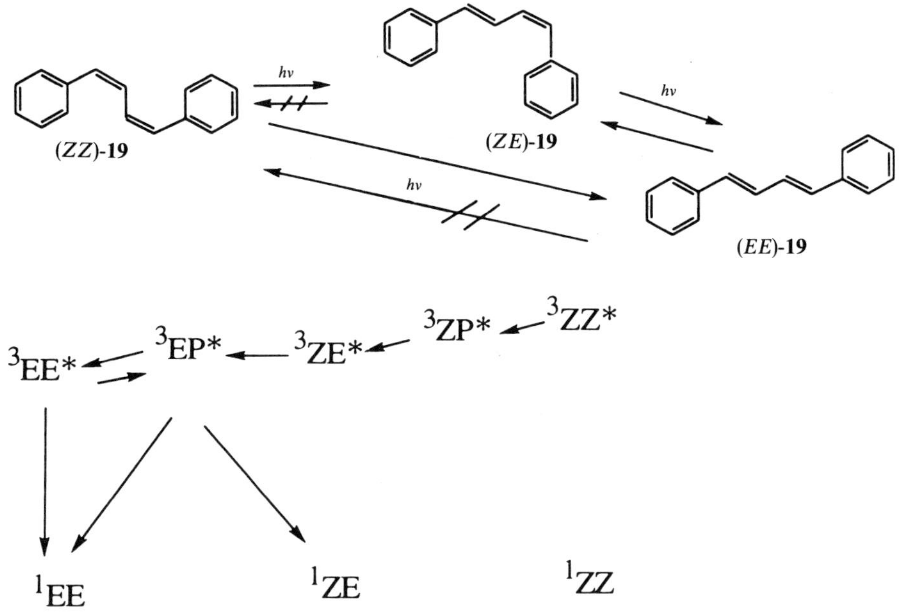

Scheme 5

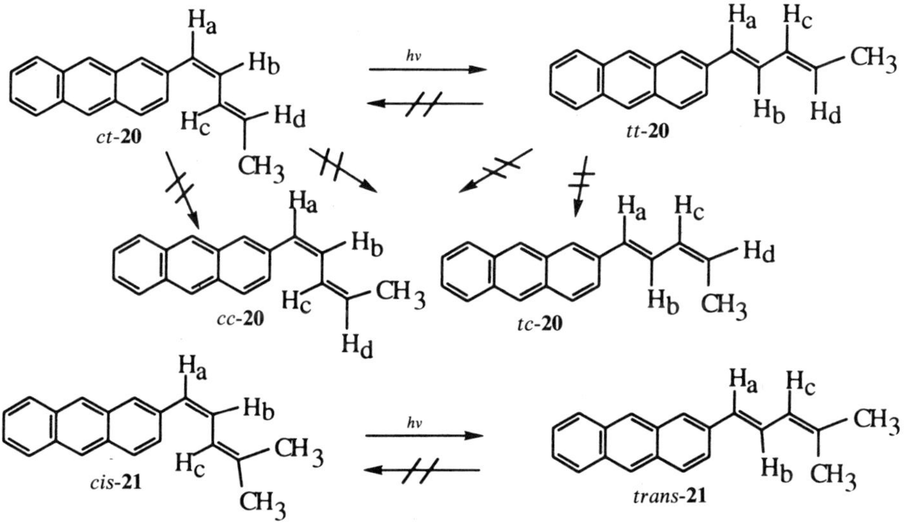

Scheme 6

increase with cis-isomer concentration to attain values as high as 24 and 11 at 5×10^{-3} *M* of *ct*-**20** and 2×10^{-3} *M* of *cis*-**21**, respectively. Excitation of both *ct*- and *tt*-**20** affords essentially the same transient absorptions as those assigned to the *tt* triplet state. Thus **20** and **21** undergo one-way cis → trans isomerization in the triplet state by an adiabatic process followed by energy transfer from the resulting ^{3}*tt*-**20*** or ^{3}*trans*-**21*** to *ct*-**20** or *cis*-**21** to regenerate ^{3}*ct*-**20*** or ^{3}*cis*-**21***, respectively. Scheme 7 depicts the triplet-sensitized reaction mechanism for **20**, where S denotes the triplet sensitizer, benzil.

$$^3S^* + ct\text{-}\mathbf{20} \xrightarrow{\phi(^3ct\text{-}\mathbf{20}^*)} S + {^3ct}\text{-}\mathbf{20}^* \tag{8}$$

$$^3ct\text{-}\mathbf{20}^* \longrightarrow {^3tt}\text{-}\mathbf{20}^* \tag{9}$$

$$^3tt\text{-}\mathbf{20}^* \xrightarrow{1/\tau_T} tt\text{-}\mathbf{20} \tag{10}$$

$$^3tt\text{-}\mathbf{20}^* + ct\text{-}\mathbf{20} \xrightarrow{kq} tt\text{-}\mathbf{20} + {^3ct}\text{-}\mathbf{20}^* \tag{11}$$

Scheme 7

The behavior of **20** and **21** contrasts sharply with that of 1,3-pentadiene and 2,4-hexadiene. Therefore, substitution of an anthracene nucleus on the diene can control the isomerization mode leading to a highly specific isomerization of very large quantum efficiency by completely suppressing the quantum chain isomerization from a trans to a cis double bond as observed in 2,4-hexadiene [82].

D. Arylethenes of Bis-stilbene Structures

Arylethenes carrying bis-stilbene units undergo not only one- but also twofold adiabatic cis → trans isomerization. Thus 4-styrylstilbene (**22**) undergoes adiabatic Z → E isomerization from the ZZ isomer to the ZE and EE isomers in the excited singlet state as revealed by fluorescence spectroscopy [97]. The fluorescence from the ZZ isomer decays with components of 0.19 ps, 0.65 ps, and 1.25 ns at 298 K, assigned to the fluorescence of ZZ, ZE, and EE isomers. The quantum yields of the adiabatic isomerization from ^{1}ZZ* to ^{1}ZE* and to ^{1}EE* are 0.038 and 0.019, respectively.

On biacetyl sensitization **22** undergoes adiabatic one-way ZZ → ZE and ZZ → EE isomerizations in the triplet manifold (Scheme 8) [98]. The ZE isomer also isomerizes to EE. The quantum yield of ZE → EE isomerization ($\phi_{ZE \rightarrow EE}$) and the photostationary state isomer ratios, $([EE]/[ZE])_{pss}$, increase with increasing diene concentration. The $\phi_{ZE \rightarrow EE}$ value is about 14 at

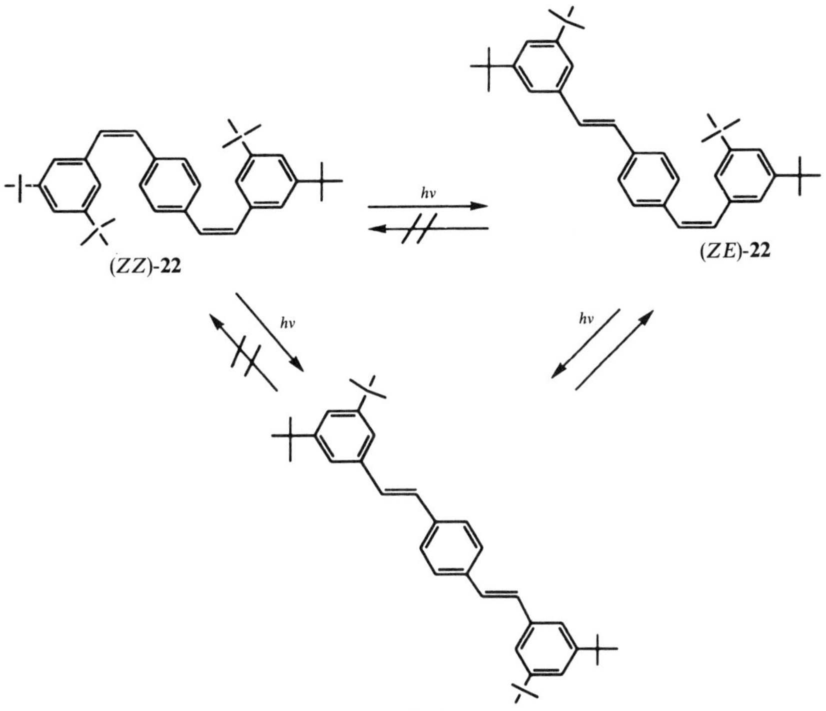

(EE)-22

Scheme 8

[ZE] = 0.033 *M*. Figure 8 illustrates the potential energy surfaces of **22** in the excited singlet as well as the triplet state.

Bis-stilbenes connected with an ethane bridge (**23**) behave similarly to **22** (Scheme 9). The ZZ isomer undergoes almost exclusive one-way isomerization, while the ZE and EE isomers conduct two-way isomerization in the singlet excited state on direct irradiation [99]. The one-way ZZ → ZE isomerization is thought to occur by way of intramolecular excitation transfer from the ZE*, where the E chromophore is excited, to Z*E, where the Z chromophore is excited.

Sundahl et al. extended the scope of the isomerization of this type of molecule to the photoisomerization of oligoenes with three stilbene-like chromophores [100]. Although there are six isomers, the photostationary state consists of >99% **24f**, indicating that the photoisomerization of **24** always gives trans isomers in a one-way manner (Scheme 10). Furthermore, the conversions **24a** → **24d**, **24b** → **24e**, and **24c** → **24f** take place adiabatically, as revealed by fluorescence spectroscopy. For example, the quantum yield of **24c** → **24f** isomerization by the adiabatic route is 0.066 at 25°C in toluene.

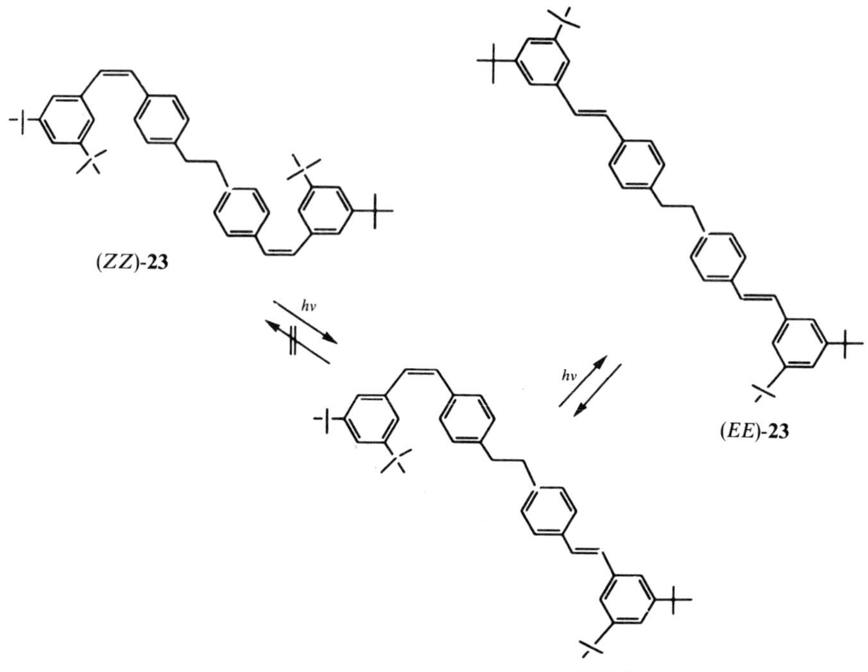

Scheme 9

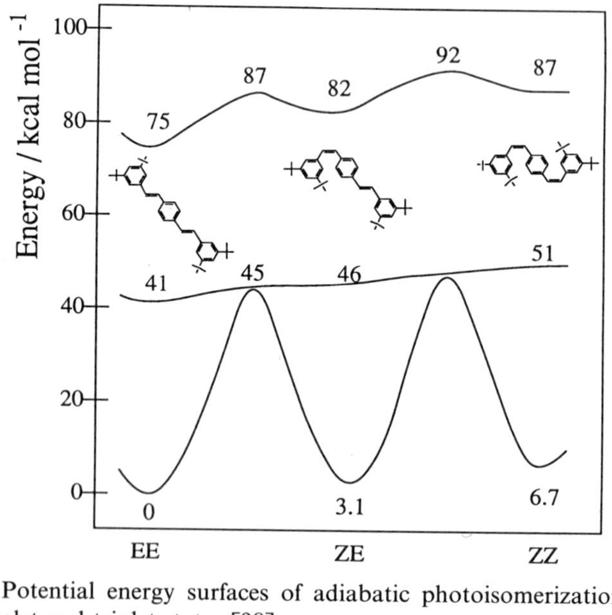

Figure 8. Potential energy surfaces of adiabatic photoisomerization of **22** in the excited singlet and triplet states [98].

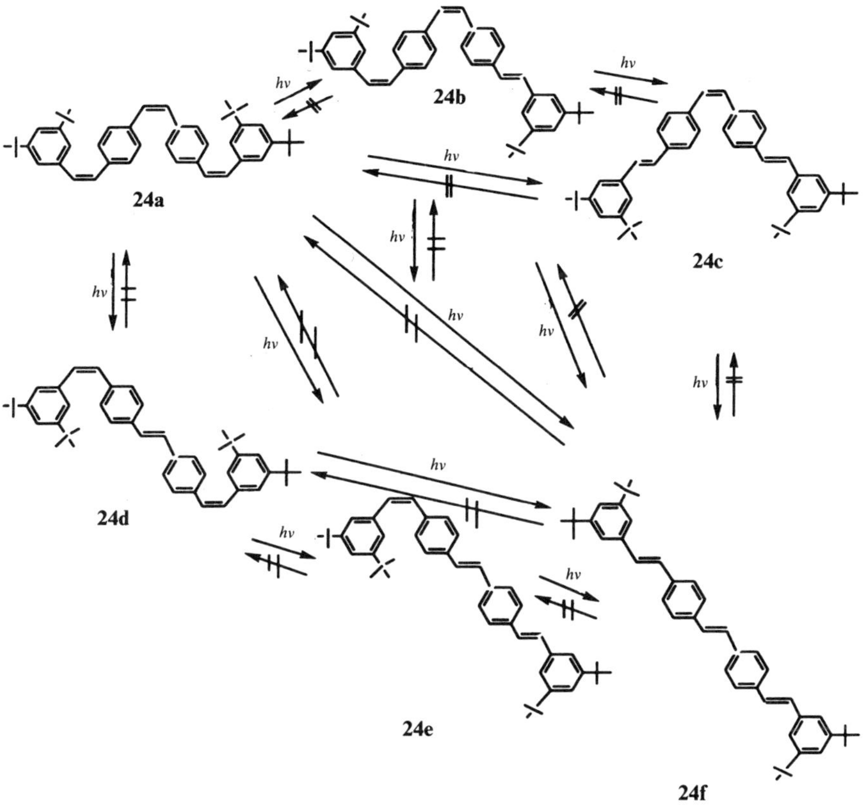

Scheme 10

A cyclic polyene (**25**) undergoes adiabatic all-cis to all-trans sixfold isomerization in the triplet state, while no isomerization from all-trans to all-cis occurs [101]. Excitation of either isomer at room temperature affords the same T–T absorption around 520 nm, decaying with a lifetime of 45 μs, which is assigned to the all-trans triplet.

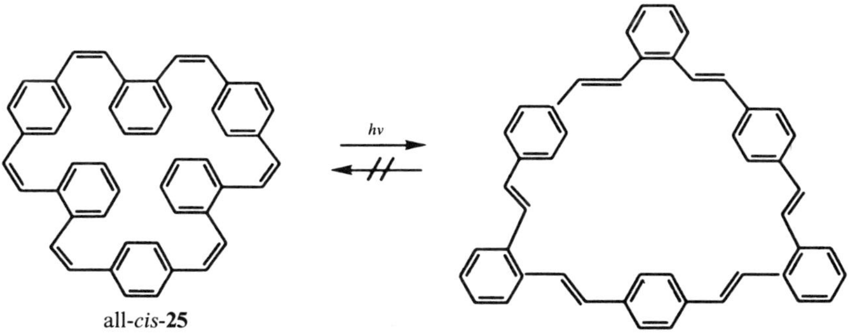

all-*cis*-**25**

all-*trans*-**25**

V. EFFECT OF INTRAMOLECULAR HYDROGEN BONDING

A. One-Way Trans → Cis Isomerization

As described before, substitution of an aromatic group with a low triplet excitation energy on an unsaturated bond leads to one-way cis → trans isomerization of the unsaturated bond in an adiabatic pathway. Here we deal with intramolecular hydrogen bonding between two heteroaromatic groups substituted on the cis side of an unsaturated bond which can result in one-way trans → cis isomerization.

In nonpolar solvents such as benzene and hexane, *cis*-**26** and *cis*-**27** form intramolecular hydrogen bonds [102–104]. **26** undergoes one-way trans → cis isomerization almost exclusively in CH_2Cl_2, but undergoes two-way isomerization in solvents of high-hydrogen-bond-accepting capability, such as $(CH_3)_2SO$ [102]. Therefore, the intramolecular hydrogen bonding in *cis*-**26** increases a barrier for C═C rotation or accelerates nonradiative decay in the excited singlet state.

On direct irradiation with 366-nm light in benzene, *trans*-**27** undergoes isomerization to *cis*-**27**, while *cis*-**27** does not isomerize at all to *trans*-**27** [103, 104]. These facts show that the intramolecular hydrogen bond in *cis*-**27** is still effective in the excited state and apparently inhibits isomerization around the C═C double bond.

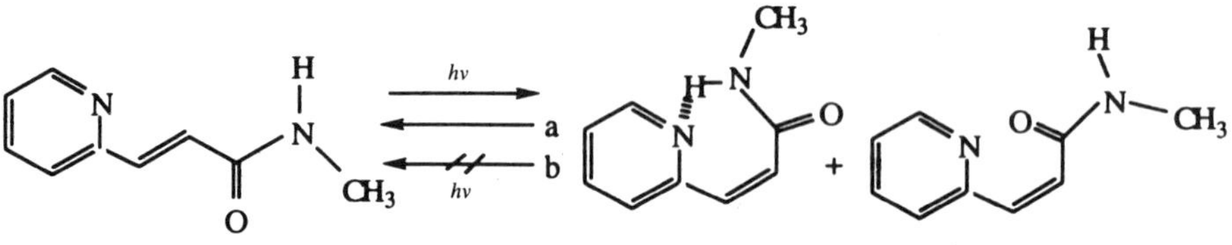

trans-**26**   a: in CH_3CH_2OH or $(CH_3)_2SO$ *cis*-**26**
b: in CH_2Cl_2

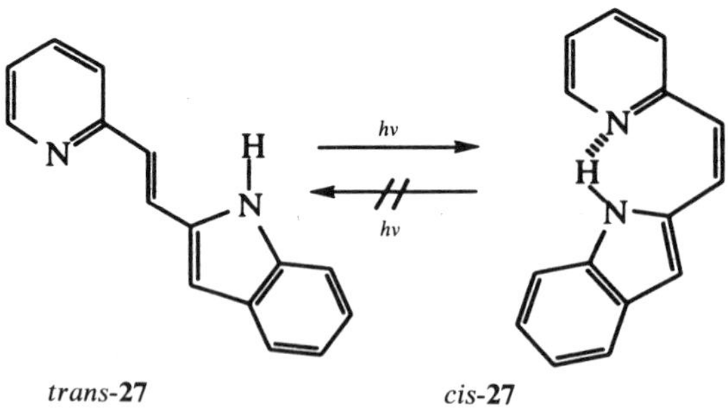

trans-**27**   *cis*-**27**

This one-way trans → cis isomerization seems to take place through a diabatic process via deactivation at the perpendicular excited singlet state ($^1p^*$) [process (1) in Figure 9] or through adiabatic $^1t^* \rightarrow {}^1c^*$ conversion followed by deactivation to the ground-state cis isomer [processes (2) and (3) in Figure 9] [103].

cis-1-(2-Pyrrolyl)-2-(2-quinolyl)ethene (*cis*-**28**) also forms an intramolecular hydrogen bond [105]. Contrary to **26** and **27**, **28** underwent mutual isomerization between the cis and trans isomers on direct irradiation in benzene, giving a very high ratio of cis to trans isomers (94:6) at the photostationary state on 400-nm light irradiation. *cis*-**28** exhibits fluorescence at 600 nm with a very much larger Stokes shift than that of *trans*-**28**. Thus *cis*-**28** undergoes hydrogen atom transfer in the singlet excited state to give its tautomer (*cis*-**28′**) in the excited singlet state, which deactivates to the ground state of *cis*-**28′** emitting fluorescence around 600 nm. Alternatively, this may induce isomerization of the double bond to *trans*-**28**. The resulting *cis*-**28′** reverts to *cis*-**28** with a 300-ns lifetime at ambient temperature, overcoming an activation energy of 9.4 kcal mol^{-1}, as revealed by laser photolysis.

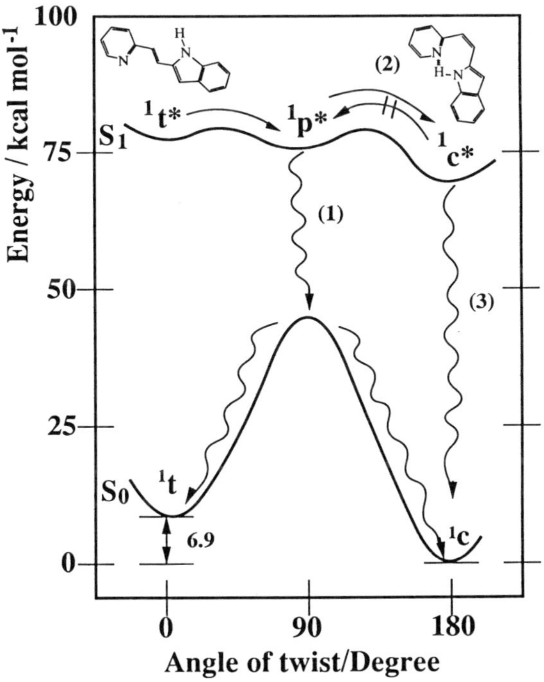

Figure 9. Potential energy surfaces of one-way trans → cis isomerization of **27** [103].

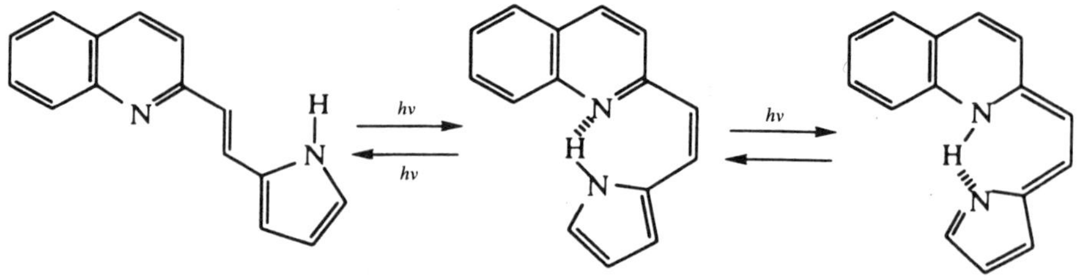

trans-**28** *cis*-**28** *cis*-**28′**

B. Photoisomerization of Bilirubins

Intramolecular hydrogen bonding plays a significant role in the Z-E isomerization of (4*Z*, 15*Z*)-bilirubin IXα (ZZ-BR), which is the final metabolic product of hemoglobin in living bodies [106]. This compound is composed of two pyrromethenone moieties carrying a propionic acid chain. Each propionic acid group forms strong intramolecular hydrogen bonds with lactam CO/NH and pyrrole NH groups of the adjacent pyrromethenone moiety. These intramolecular hydrogen bonds lead ZZ-BR to a folded conformation without any free polar groups oriented toward the outside. Therefore, ZZ-BR is lipophilic and stable in the dark. However, on irradiation with visible light, ZZ-BR isomerizes into E isomers (ZE-BR, EZ-BR, EE-BR) and a cyclized product (lumirubin, LR) (Scheme 11). The quantum yields of the isomerization depend very strongly on the properties of solvents and media, as shown in Table 4 [107].

TABLE 4 Quantum Yields for Isomerization and Photostationary-State Isomer Compositions of BR in Various Solvents

	Quantum Yield			Composition (%)		
Solvent	ϕ_{ZZ-ZE}	ϕ_{ZZ-EZ}	ϕ_{ZZ-LR}	ZZ	ZE	EZ
$CHCl_3$	3×10^{-3}	6×10^{-4}	$< 1 \times 10^{-5}$	77.1	17.3	5.6
THF	7×10^{-3}	9×10^{-4}	1×10^{-4}	80.1	15.5	4.4
DMSO	5×10^{-3}	2×10^{-3}	2×10^{-4}	81.1	13.7	5.2
$CHCl_3$–Et_3N (1:1)	3×10^{-2}	6×10^{-3}	1×10^{-4}	73.3	20.7	6.0
$CHCl_3$–$(C_8H_{17})_3N$ (1:1)	5×10^{-2}	1×10^{-2}	3×10^{-4}	66.6	25.8	7.5
H_2O (pH 7.4) with HSA	0.11	< 0.02	2×10^{-3}	63.0	35.5	1.7

Source: Ref. 107.

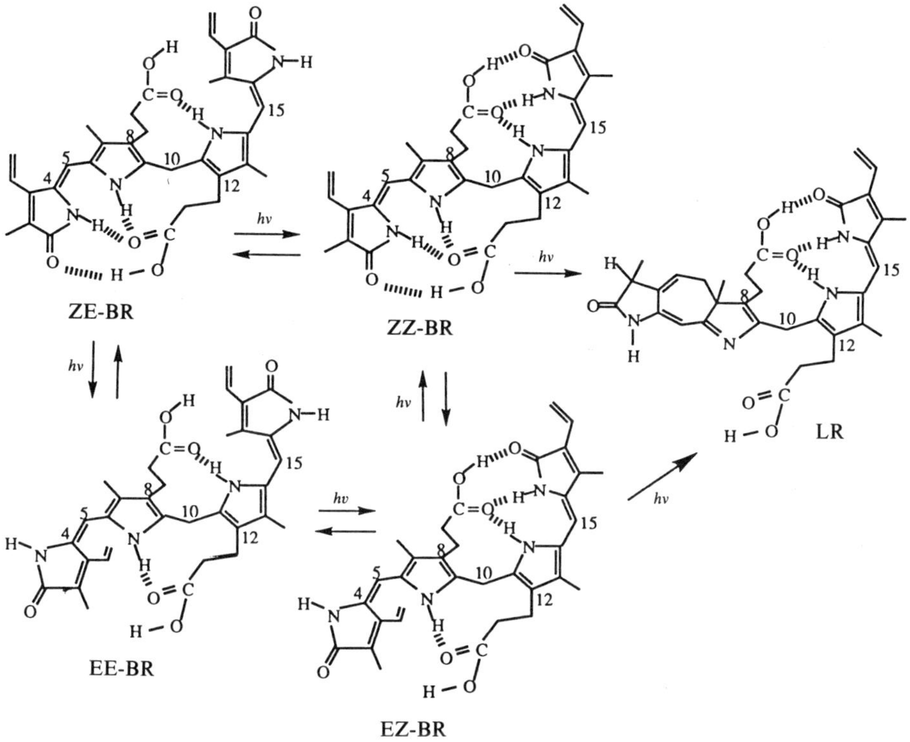

Scheme 11

Table 4 indicates that the strength of the intramolecular hydrogen bonds in ZZ-BR governs the efficiency of its isomerization. The more the hydrogen bonds are weakened, the more efficiently the isomerization takes place. Thus solvents that break the intramolecular hydrogen bonds of ZZ-BR tend to induce the isomerization.

In human serum albumin (HSA) solution ZZ-BR undergoes a specific photoisomerization to ZE. The incorporation of ZZ-BR into HSA through formation of hydrogen bonds or salt bridges with its amino acid residue will lead to a BR conformation different from that in organic solvents, such as an extended conformation, which will easily undergo isomerization to ZE.

VI. PHOTOCHEMICAL BEHAVIOR OF CYCLIC ETHENES

Cyclic ethenes are resistant to isomerization of the double bond unless the members are sufficiently large. Instead, the photochemical behavior of molecules with these fixed double bonds provides us with valuable information concerning the excited states of unsaturated molecules, which is not available from the acyclic unsaturated molecules as described below.

A. Importance of Double-Bond Torsion in Photochemical and Photophysical Processes of Styrene and Stilbene Derivatives

Nonvertical triplet excitation transfer was proposed by Hammond et al. based on their observation that several sensitizers ($^3S^*$) with triplet energies lower than that of *cis*-stilbene (57 kcal mol^{-1}) underwent energy transfer to *cis*-stilbene with much higher rate constants than expected from the spectroscopic triplet energies [5, 6]. Thus the endothermic excitation transfer from $^3S^*$ to ^1c was supposed to produce not $^3c^*$ but $^3p^*$ accompanying torsion of the C=C double bond.

Gorman et al. unambiguously revealed that torsion around the single bond connecting the phenyl ring and the ethylenic double bond in arylethenes played an important role in nonvertical triplet excitation transfer [108, 109]. The relaxed triplet energy of a rigid model compound of *cis*-stilbene, 2,3-diphenylnorbornene, where double-bond twisting was inhibited and only phenyl-vinyl torsion was possible, was determined as 47 kcal mol^{-1}. This was about 10 kcal mol^{-1} lower than the most reliable spectroscopic energy of the triplet state of *cis*-stilbene (57 kcal mol^{-1}) [108]. Moreover, quenching-rate constants of the triplet states of sensitizers with varying triplet excitation energies by 2,3-diphenylnorbornene fitted not the classical but the nonvertical profile. Therefore, the nonvertical excitation transfer was attributed to the consequence of relaxation around the single bond between the phenyl and ethylenic double bond.

As to deactivation from the perpendicular triplet state to the ground state, Caldwell et al. stressed that the precisely 90° twisted conformation in the triplet state cannot undergo intersystem crossing because of the small spin-orbit coupling [110]. Both the spin-orbit coupling and T_1–S_0 energy gap increase with decreased twist angle around the double bond from 90°. Therefore, the deactivation from nearly perpendicular forms—for example, from the 88° or 89° twisted conformation to the ground state—can occur

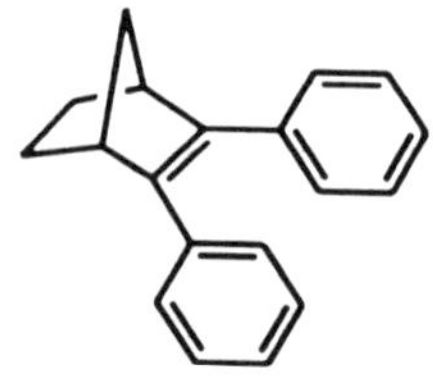

by the compromise of spin-orbit coupling and the T_1–S_0 energy gap. An important role of pyramidalization was proposed for deactivation from T_1 to S_0 in cyclic ethenes such as norbornene ($\tau_T = 250$ ns) and 1-phenylcyclohexene ($\tau_T = 67$ ns) [110, 111], where the triplet lifetime is shorter than expected from the intrinsic restriction of double-bond twisting in these olefins. Furthermore, substitution of a chlorine atom as a heavy atom on an unsaturated bond enhances intersystem crossing of aralkene triplet state to the ground state when the triplet can take a nonplanar conformation [112]. As Table 5 indicates, substitution of a chlorine atom on indene is not effective, but substitution on more flexible dihydronaphthalene and very flexible diphenylethenes effectively reduces the triplet lifetime by enhancing intersystem crossing from the triplet state to the ground state, due to the increase in T_1–S_0 spin-orbit coupling.

On direct excitation of *cis*-stilbene, both cyclization to produce dihydrophenanthrene and double-bond isomerization to $^1p^*$ occur competitively in the ratio 30:70. The lifetime of the excited singlet *cis*-stilbene is estimated as about 1 ps at room temperature in solution as measured via femtosecond spectroscopy by Hochstrasser et al. [9] and Fleming et al. [10]. This

TABLE 5 Effect of Chlorine Atom on the Lifetime of Aralkene triplets

	τ_T/(ns)	
	X = H	X = Cl
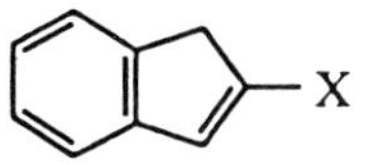	1400	1300
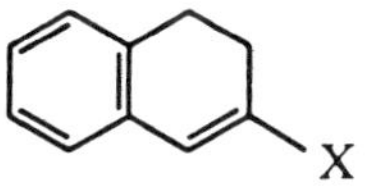	6800	14
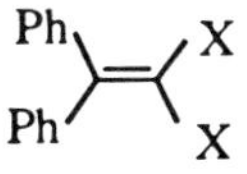	38	<1.5
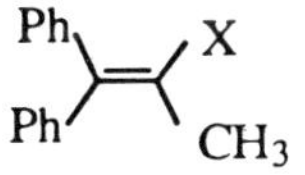	51	<1.5
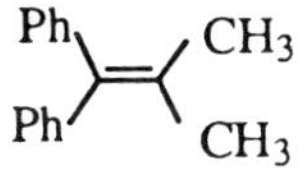	113	

Source: Ref. 112.

indicates the absence of an energy minimum at the excited singlet state of cis conformation. On the contrary, as described before (IIIc), Saltiel et al. observed fluorescence from the cis isomer [12]. Moreover, Yoshihara et al. found that in argon clusters *cis*-stilbene emits fluorescence with a 17.2-ns lifetime, indicating the presence of a shallow energy minimum at $^1c^*$ [11]. From studies with model compounds with four- to seven-membered cycloalkenes inhibiting the isomerization as shown below, Yoshihara et al. proposed the potential surfaces for cis → trans isomerization as well as cyclization to produce dihydrophenanthrene in the excited singlet state [11]. The conceptual picture of the S_1 potential surface of *cis*-stilbene shows that a small part of the wave packet finds its way to vibrationally excited DHP, while most of the wave packet ends up in the perpendicular conformation to induce cis → trans isomerization.

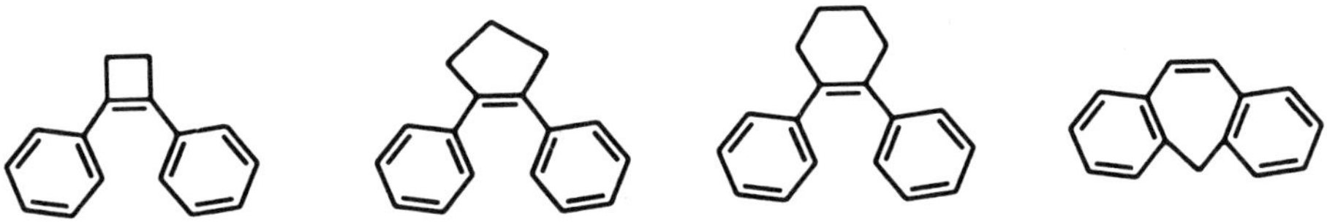

B. Photochromism

Diarylethenes incapable of undergoing double-bond isomerization can give cyclized products when properly substituted. Irie et al. reported that whereas dihydrophenanthrene resulting from photocyclization of diphenylethenes revert easily to the starting materials at room temperature in the dark, cyclized products from irradiation of diarylethenes with heterocyclic rings were stable and reversibly give the starting compounds on irradiation with longer-wavelength light than that for the cyclization [113, 114]. The distribution of conformers around the single bond depending on solvent polarity affects the efficiency of photocyclization of **29**. Since cyclized products show absorption at longer wavelengths, these compounds exhibit photochromism and attract much attention for their practical application for photoimaging and so on. Among the compounds examined, the system **30** ⇄ **30′** is the most stable to prolonged, repeated irradiation. Another type of photochromic material, fulgide, has been investigated by Heller et al. [115] and Yokoyama et al. [115c]. It is to be noted that Irie's procedure for preparing stable cyclized products was applied not only to the preparation of photochromic compounds but also to photosensitive compounds exhibiting photoelectric properties [115d].

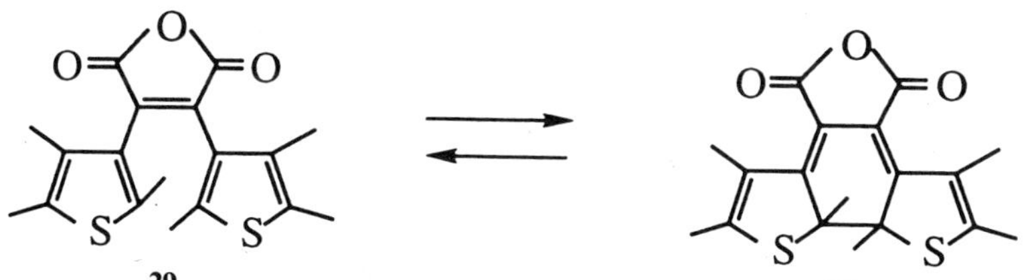

29 **29′**

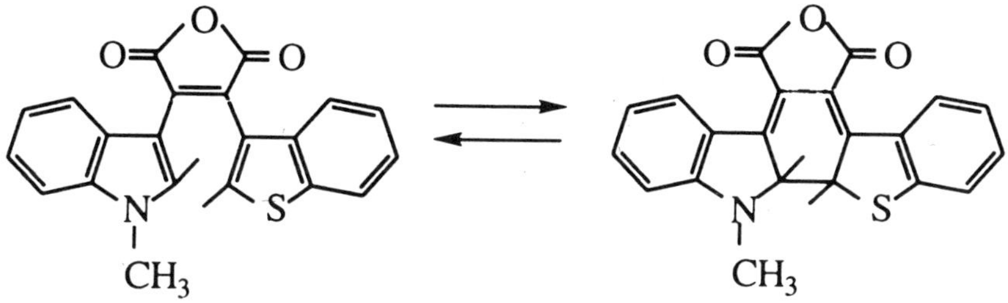

30 **30′**

C. Photosensitized Enantiodifferentiating Isomerization of a Cyclic C═C Double Bond

trans-Cyclooctene (*trans*-**31**) resulting from irradiation of *cis*-**31** is known to have two enantiomeric isomers. In an attempt to produce selectively an optical isomer, Inoue et al. reported that direct irradiation of *cis*-**31**

cis-**31** (−)-(*R*)-*trans*-**31** (+)-(*S*)-*trans*-**31**

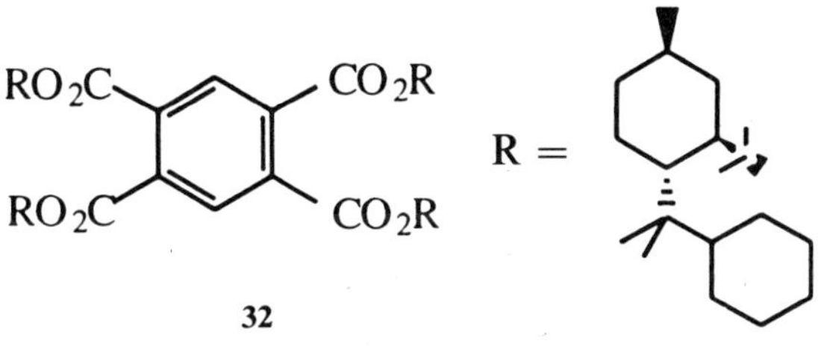

32

as an inclusion complex with β-cyclodextrin led to *trans*-**31** with 0.24% optical purity [116]. However, sensitization with optically active benzenepolycarboxylates **(32)** yielded *trans*-**31** with optical purities as high as 64% at 89°C and 49% at 25°C [117].

VII. INTERNAL ROTATION BETWEEN THE ROTAMERS OF ARYLETHENES

Many arylethenes, such as 2-styrylnaphthalene, have rotational isomers around the single bond connecting a condensed aromatic ring and the unsaturated bond. In the ground state the internal rotation between these rotamers occurs too rapidly to be followed spectroscopically. Each rotamer shows an intrinsic absorption and fluorescence spectrum. Therefore, these mixtures of rotamers show fluorescence spectra changing with the excitation wavelength [118–137]. However, in the excited singlet state, internal rotation between excited rotamers had not been observed until recently; therefore, the principle of nonequilibration of excited rotamers (NEER principle) [119] had long been believed to hold generally. Excitation of arylethenes generally lowers the double-bond character of the double bonds and increases it at the adjacent single bond connecting the aromatic rings, which suppresses internal rotation around the single bond in excited states.

However, it has recently been found that one-way-isomerizing 2-anthryl-substituted unsaturated compounds undergo internal rotation in the excited singlet state as well as in the triplet state [133–138]. No occurrence of

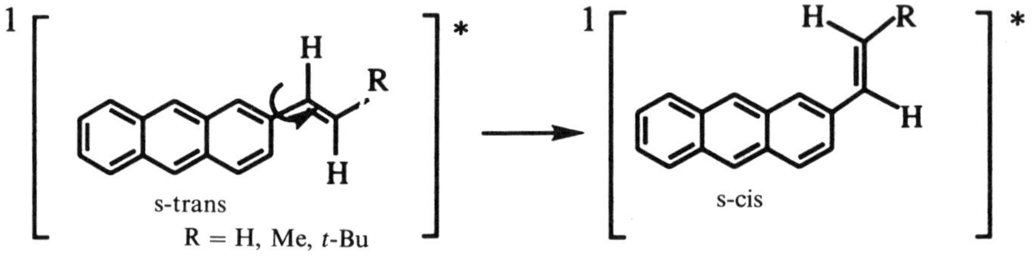

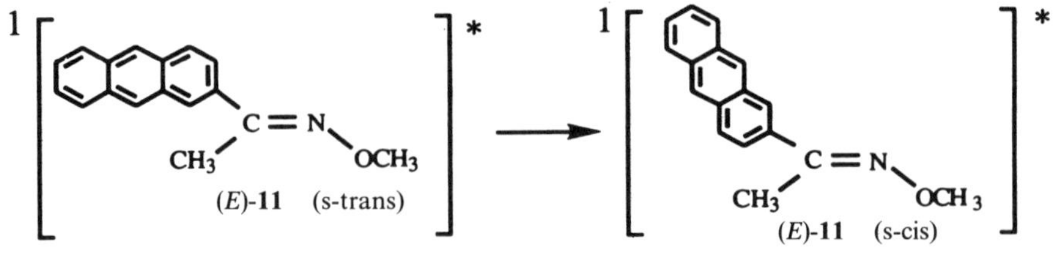

double-bond isomerization of **6** in the singlet excited state suggests that excitation is localized in the anthryl group of **6** in the singlet excited state, and therefore the electronic properties of the single bond linking the anthryl group and the double bond are similar to those in the ground state compared with those of 2-styrylnaphthalene.

Furthermore, excitation of cis isomers of arylethenes generally results in double-bond isomerization, which is accompanied by internal rotation of the aromatic ring around the single bond connected to the unsaturated bond. Therefore, the subject of internal rotation concerns the mechanism of double-bond isomerization as well as the reactivity of excited rotamers toward various reactions.

The internal rotation of trans isomers of some of 2-anthryl-substituted unsaturated compounds takes place in a very specific way: one-way from one rotamer to another in the excited singlet and triplet states.

A. *N*-methoxy-1-(2-anthryl)ethanimine (11)

(*E*)-**11** has two rotamers, s-trans and s-cis, as revealed by the effect of the excitation wavelength on the fluorescence spectra. The s-trans and s-cis rotamers exhibit absorption and fluorescence spectra at shorter and longer wavelength, respectively [137].

The time-resolved fluorescence spectrum from (*E*)-**11** observed 0 to 250 ps after a 316-nm laser excitation is due primarily to s-trans (λ_{max} = 394 nm). The spectrum shifts to that of s-cis (λ_{max} = 407 nm) in a nanosecond time scale at 58°C (Figure 10*a*), indicating that ^{1}(s-trans)* converts to ^{1}(s-cis)* in a one-way manner (Scheme 12). Decrease of temperature retards the conversion, and finally at −90°C in toluene no spectral change occurs at all, due to complete suppression of the conversion (Figure 10*b*). The ^{1}(s-trans)* → ^{1}(s-cis)* conversion proceeds with an activation energy of

$$\text{s-trans} \xrightarrow{316\ \text{nm}} {}^1\text{s-trans}^* \xrightarrow{k_1} \text{s-trans} + h\nu_{\text{s-trans}}$$

$$\xrightarrow{k_2} \text{s-trans or } {}^3\text{s-trans}^*$$

$$\xrightarrow{k_3} {}^1\text{s-cis}^* \xrightarrow{k_4} \text{s-cis} + h\nu_{\text{s-cis}}$$

$$\xrightarrow{k_5} \text{s-cis or } {}^3\text{s-cis}^*$$

Scheme 12 (at room temperature)

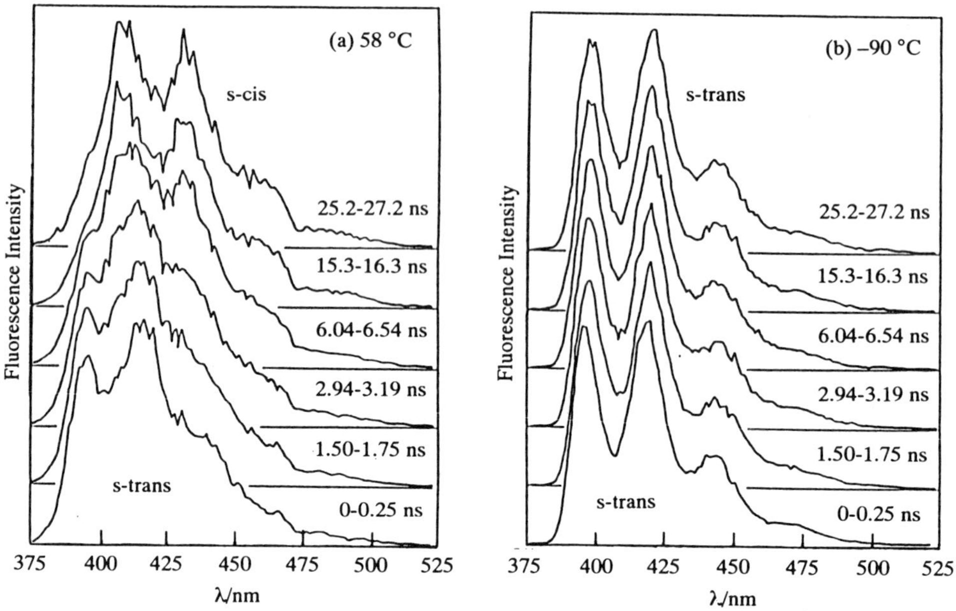

Figure 10. Time-resolved fluorescence spectra of (E)-**11** on 316-nm laser excitation in toluene at 58°C (*a*) and at −90°C (*b*) [137].

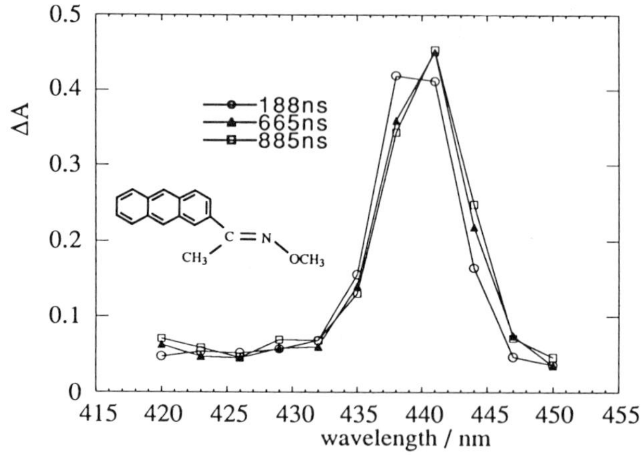

Figure 11. Time-resolved T–T absorption spectra of (E)-**11** observed on 308-nm laser excitation at 173 K [69].

$$\text{s-trans} \xrightarrow{316\ \text{nm}} {}^{1}\text{s-trans}^{*} \xrightarrow{k_{isc}} {}^{3}\text{s-trans}^{*} \xrightarrow{k_{1}} \text{s-trans}$$

$$\xrightarrow{k_{2}} {}^{3}\text{s-cis}^{*} \longrightarrow \text{s-cis}$$

Scheme 13 (at $-100°C$)

4.6 kcal mol^{-1} and a frequency factor of $1.8 \times 10^{11}\ s^{-1}$, and ^{1}(s-cis)* is at least 1.4 kcal mol^{-1} more stable than ^{1}(s-trans)* at room temperature.

In the excited singlet state, internal rotation takes place only above $-90°C$. However, in the excited triplet state, internal rotation from the s-trans triplet [^{3}(s-trans)*] [λ_{max}(T–T) = 438 nm at 173 K] to the s-cis triplet [^{3}(s-cis)*] [λ_{max}(T–T) = 441 nm at 173 K] (Scheme 13) takes place at lower temperature as revealed by transient spectroscopy. A typical example is shown in Figure 11 observed at 173 K. The ^{3}(s-trans)* → ^{3}(s-cis)* conversion takes place with an activation energy of 4.6 kcal mol^{-1} and a frequency factor of $8.9 \times 10^{11}\ s^{-1}$ [69]. Figure 12 depicts the potential energy surfaces of the single-bond rotation of **11** in the ground, excited singlet, and triplet states, and Schemes 12 and 13 indicate the phenomena occurring at room temperature and $-100°C$, respectively. In both the excited singlet and triplet states, the internal rotation proceeds in a one-way manner from s-trans to s-cis. As shown in Figure 12, in the ground state, s-trans is more stable than s-cis; however, in the excited states, s-cis is more stable than s-trans; therefore, the conversion takes place from s-trans to s-cis in the excited states.

B. 2-Vinylanthracene and Its β-Alkyl Derivatives

In benzene the time-resolved fluorescence spectra of 2-vinylanthracene show one-way rotation from a rotamer showing shorter-wavelength fluorescence to another with longer-wavelength fluorescence [136] similar to **11**. Substitution of a *t*-butyl and methyl group on the terminal carbon of 2-vinylanthracene, that is, *trans*-**6a** and **6b**, leads to similar internal rotations in the excited singlet state. The activation energies and the frequency factors for the internal rotation are 3.9, 4.3, and 4.3 kcal mol^{-1} and 1.7, 1.8, and $1.8 \times 10^{11}\ s^{-1}$, for 2-vinylanthracene, **6a**, and **6b**, respectively.

However, substitution of a phenyl or a naphthyl group on the terminal carbon of 2-vinylanthracene inhibits single-bond rotation in the excited singlet state. Irradiation of *trans*-**6c** and **6d** with 316-nm laser light excites both s-trans and s-cis rotamers to give fluorescence of both rotamers. The

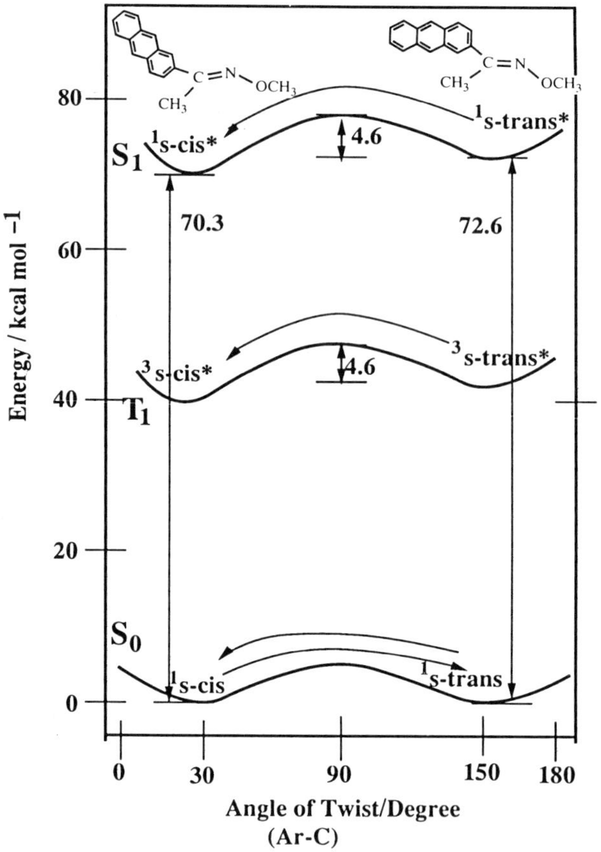

Figure 12. Potential energy surfaces of one-way internal rotation in the excited singlet and triplet states of (E)-**11**.

fluorescence spectrum decays in a nanosecond time scale without exhibiting any rise of fluorescence due to formation of another excited rotamer. Furthermore, the decay time constants do not change at all with temperature. Accordingly, the activation barrier for the internal rotation in their excited singlet states must be higher than 7 kcal mol^{-1}, which is too high to be overcome by the excited molecules competing with their deactivation. The introduction of a conjugated group such as phenyl or 2-naphthyl on the terminal carbon of 2-vinylanthracene must enhance conjugation between the anthryl part and the olefinic part, increasing the double-bond character of the single bond in the excited singlet state. This accordingly elevates the barrier for the single-bond rotation of **6c** and **6d** in the singlet excited state [45, 118].

In the triplet excited state, the T–T absorption of *trans*-**6c** changes in the millisecond time range at −130°C [39]. Calculations support the supposition that this is due to internal rotation in the excited triplet state. Internal rotation of **6c** and 2-(4-bromostyryl)anthracene takes place in the excited triplet state by overcoming an activation barrier of about 7 kcal mol^{-1} [138]. The occurrence of rotational isomerization of **6c** in the triplet state, in contrast to the lack of occurrence in the singlet excited state, is attributed to the fact that the triplet state has a 1000-fold longer lifetime than the singlet state.

VIII. ISOMERIZATION OF RADICAL IONS OF ARYLETHENES

As described above, upon photoexcitation, stilbene undergoes a two-way isomerization. However, upon 9,10-dicyanoanthracene (DCA) sensitization in acetonitrile, *cis*-stilbenes undergo essentially one-way isomerization via a radical cation chain mechanism. Either a unimolecular cis$^{+\cdot}$ → trans$^{+\cdot}$ or a bimolecular process involving addition of cis$^{+\cdot}$ to the neutral stilbene may lead to trans isomers [139–156].

Lewis et al. reported almost one-way cis → trans isomerization of stilbenes on DCA sensitization based on the observed increase of both quantum yield of cis → trans isomerization and the photostationary-state isomer composition, $([t]/[c])_s$, with increasing concentration of *cis*-stilbene [139, 140]. Table 6 summarizes values of $\phi_{c\to t}$, photostationary-state trans-isomer percentage, and the efficiency of free stilbene radical cation formation, ϕ_{sep}, arising from quenching of ^{1}DCA* by *cis*-stilbenes. The $\phi_{c\to t}/\phi_{sep}$ values all exceed unity, although dependent on the substituents. Accordingly, the isomerization proceeds in a chain mechanism to give photostationary mixtures highly enriched in the trans isomer (<96%).

The chain reaction can proceed by two mechanisms. The first is electron transfer between the trans-radical cation produced by the unimolecular cis$^{+\cdot}$ → trans$^{+\cdot}$ isomerization and the neutral cis isomer to give trans isomer and cis$^{+\cdot}$ (Scheme 14) [139]. The second is addition – elimination between radical cation and neutral cis isomer to induce cis → trans isomerization (Scheme 15) [140, 145, 155].

The cis → trans isomerization of stilbenes by way of radical cations or anions was first reported by Shida and Hamill in their radiolytic work [148] and was subsequently followed by other workers [145, 146, 149–151]. To examine whether or not cis$^{+\cdot}$ isomerizes to trans$^{+\cdot}$ in the excited state, the effect of photoexcitation of the resulting cis$^{+\cdot}$ was examined. Ebbesen [152] found that excitation of the *cis*-stilbene radical cation (λ_{max} = ~510 and ~760 nm) [140, 148] generated by pulse radiolysis with 500-nm light at 298 K effected cis → trans isomerization ($\phi_{c\to t} = 0.25$), whereas excitation with 720-nm light did not. Therefore, the energy of the lowest excited state of cis$^{+\cdot}$ produced on 720-nm light irradiation is too low to induce isomer-

TABLE 6 Quantum Yields of Dicyanoanthracene-Sensitized Isomerization of Stilbenes in Acetonitrile[a]

Stilbene	$\phi_{c \to t}$	ϕ_{sep}	$\phi_{c \to t}/\phi_{sep}$	% Trans at pss
4-Methoxy	0.12	0.032	6.0[b]	
4,4-Dimethyl	0.35	0.048	7.3	
4-Methyl	0.14	0.022	2.8	99.9
Stilbene	0.32	0.14	2.5	98.8
4-Chloro	2.5	0.019	17	96.0
4-Cyano	13	0.39	34	98.8
4-Trifluoromethyl				98.8

Source: Ref. 140.

[a] [*cis*-stilbene] = 0.05 *M* unless otherwise indicated.

[b] [*cis*-stilbene] = 0.02 *M*.

$$\mathrm{DCA} + \mathrm{cis} \xrightarrow{h\nu} \mathrm{DCA}^{-\cdot} + \mathrm{cis}^{+\cdot}$$

$$\mathrm{cis}^{+\cdot} \longrightarrow \mathrm{trans}^{+\cdot}$$

$$\mathrm{trans}^{+\cdot} + \mathrm{cis} \rightleftharpoons \mathrm{trans} + \mathrm{cis}^{+\cdot}$$

$$\mathrm{trans}^{+\cdot} + \mathrm{DCA}^{-\cdot} \longrightarrow \mathrm{trans} + \mathrm{DCA}$$

Scheme 14

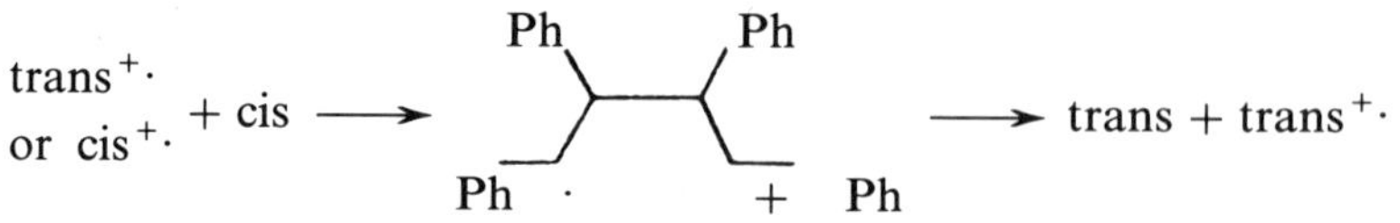

Scheme 15

ization, but the higher excited state produced on 500-nm light irradiation has sufficient energy for cis$^{+\cdot}$ → trans$^{+\cdot}$. In two-step laser excitation of DCA sensitization of *cis*-stilbene, the first laser (425 nm) excites DCA to generate cis$^{+\cdot}$, by electron transfer from cis to ^{1}DCA*, and the second laser (500 nm) excites the resulting cis$^{+\cdot}$, leading to one-way cis → trans isomerization in the excited state of radical cations [156].

Substitution with 4,4′-dimethoxy groups enhances unimolecular cis$^{+\cdot}$ → trans$^{+\cdot}$ isomerization, which proceeds in a time range below microseconds [141, 154]. 4,4′-Dibromo (Figure 13) and 4,4′-dimethyl groups also led the isomerization taking place in the 10-μs time regime at room temperature

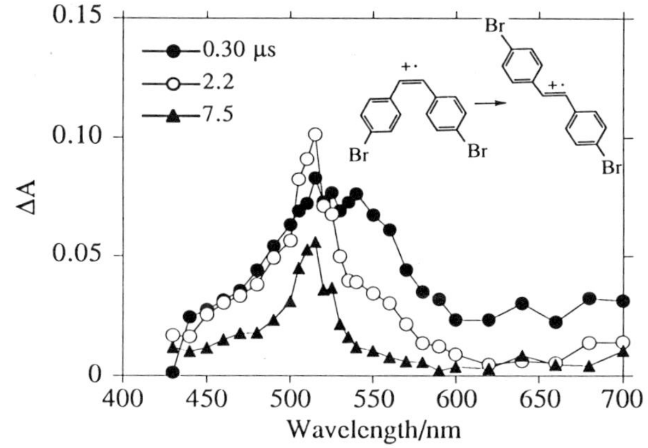

Figure 13. Transient absorption spectra observed on 406-nm laser excitation of DCA in the presence of 4,4′-dibromostilbene and biphenyl as a cosensitizer in acetonitrile [141].

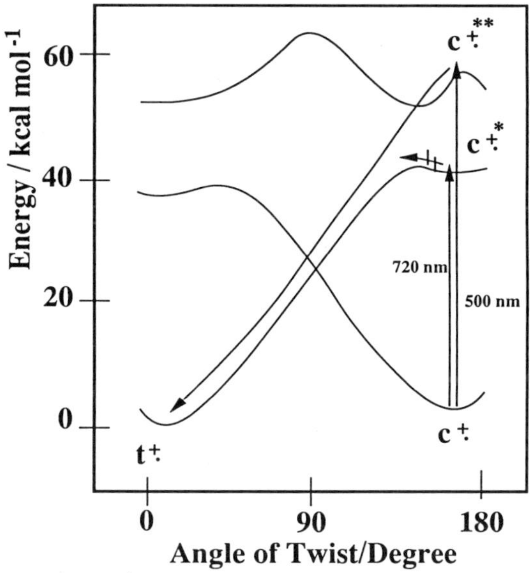

Figure 14. Computed potential energy surfaces of isomerization of stilbene radical cations [147].

with an activation energy of 8 and 3 kcal mol^{-1}, respectively. Replacement of the 4,4′-dibromo group by either a 4,4′-dichloro or a 4,4′-difluoro group reduced the rate constants for cis$^{+\cdot}$ → trans$^{+\cdot}$; however, further excitation of the resulting cis-4,4′-dichloro- or 4,4′-difluorostilbene radical cations with a 500-nm dye laser accelerated isomerization. According to the computed potential energy surface of stilbene radical cations along the twisting of the double bond (Figure 14) [147], the higher excited state resulting from 500-nm excitation of cis$^{+\cdot}$ easily undergoes twisting of the double bond, in contrast to the lowest excited state, produced from 720-nm excitation of cis$^{+\cdot}$, which could isomerize if it could overcome an activation barrier for the twisting.

As to acceleration of the isomerization, addition of salts such as lithium perchlorate in acetonitrile solution and tetrabutylammonium tetrafluoroborate in dichloromethane in DCA-sensitized irradiation of *cis*-stilbene enhanced the isomerization. A salt effect is probably responsible for retardation of recombination between the resulting stilbene radical cations and counteranions (DCA-·, etc.) [143, 146].

ACKNOWLEDGMENTS

The authors thank the Ministry of Education, Science and Culture for their Grant-in-Aid for Scientific Research (Nos. 63104001 and 02453019 for K.T.; Nos. 61740242 and 01740255 for T.A.) and for Specially Promoted Research (No. 03101004) for the partial support of this work. Thanks are also due the University of Tsukuba for partial support by the President's Special Grant for Education and Research, 1990 (K.T.).

REFERENCES

1. J. Saltiel, J. D'Agostino, E. D. Megarity, L. Metts, K. R. Neuberger, M. Wrighton, and O. C. Zafiriou, in *Organic Photochemistry*, Vol. 3, O. L. Chapman, Ed., Marcel Dekker, New York, 1973, p. 1.
2. J. Saltiel and J. L. Charlton, in *Rearrangement in Ground and Excited States*, Vol. 3, P. de Mayo, Ed., Academic Press, New York, 1980, p. 25.
3. J. Saltiel and Y.-P. Sun, in *Photochromism: Molecules and Systems*, H. Durr and H. Bouas-Laurent, Eds., Elsevier, Amsterdam, 1990, p. 64.
4. T. Arai and K. Tokumaru, *Chem. Rev.* **93**, 23 (1993).
5. G. S. Hammond, J. Saltiel, A. A. Lamola, N. J. Turro, J. S. Bradshaw, D. O. Cowan, R. C. Counsell, V. Vogt, and C. Dalton, *J. Am. Chem. Soc. Jpn.* **86**, 3197 (1964).
6. (a) W. G. Herkstroeter and G. S. Hammond, *J. Am. Chem. Soc.* **88**, 4769 (1966). (b) G. S. Hammond, D. E. Demeyer, and J. L. R. Williams, *J. Am. Chem. Soc.* **91**, 5180 (1969).

7. (a) S. Malkin and E. Fischer, *J. Phys. Chem.* **68**, 1153 (1964). (b) K. A. Muszkat and E. Fischer, *J. Chem. Soc. B*, 662 (1967). (c) T. Wismonski-Knittel, G. Fischer, and E. Fischer, *J. Chem. Soc. Perkin Trans.* **2**, 1930 (1974).
8. M. Sumitani and K. Yoshihara, *J. Chem. Phys.* **76**, 738 (1982).
9. (a) R. J. Sension, S. T. Repinec, and R. M. Hochstrasser, *J. Chem. Phys.* **93**, 9185 (1990). (b) S. Abrash, S. Repinec, and R. M. Hochstrasser, *J. Chem. Phys.* **93**, 1041 (1990). (c) S. T. Roseanne, R. J. Sension, A. Z. Szarka, and R. M. Hochstrasser, *J. Phys. Chem.* **95**, 10380 (1991).
10. D. C. Todd, J. M. Jean, S. J. Rosenthal, A. J. Ruggiero, D. Yang, and G. R. Fleming, *J. Chem. Phys.* **93**, 8658 (1990).
11. (a) H. Petek, K. Yoshihara, Y. Fujiwara, Z. Lin, J. H. Penn, and J. H. Frederick, *J. Phys. Chem.* **94**, 7539 (1990). (b) J. H. Frederick, Y. Fujiwara, J. H. Penn, K. Yoshihara, and H. Petek, *J. Phys. Chem.* **85**, 2845 (1991).
12. (a) J. Saltiel, A. Waller, Y.-P. Sun, and D. F. Sears, Jr., *J. Am. Chem. Soc.* **112**, 4580 (1990). (b) J. Saltiel, A. S. Waller, and D. F. Sears, Jr., *J. Photochem. Photobiol. A Chem.* **65**, 29 (1992). (c) J. Saltiel, A. S. Waller, and D. F. Sears, Jr., *J. Am. Chem. Soc.* **115**, 2453 (1993).
13. D. H. Waldeck, *Chem. Rev.* **91**, 415 (1991).
14. H. Goerner and D. Schulte-Froehlinde, *J. Phys. Chem.* **82**, 2653 (1978).
15. H. Goerner and D. Schulte-Froehlinde, *J. Phys. Chem.* **85**, 1835 (1981).
16. (a) H. Goerner, *Ber. Bunsenges. Phys. Chem.* **88**, 1199 (1984). (b) H. Goerner and D. Schulte-Froehlinde, *Ber. Bunsenges. Phys. Chem.* **88**, 1208 (1987).
17. (a) R. S. Mulliken and C. C. Roothaan, *J. Chem. Rev.* **41**, 219 (1947). (b) A. J. Meyer and R. S. Mulliken, *Chem. Rev.* **63**, 639 (1969). (c) R. S. Mulliken, *J. Phys. Chem.* **66**, 2448 (1977).
18. T. Arai, T. Karatsu, H. Sakuragi, and K. Tokumaru, *Tetrahedron Lett.* **24**, 2873 (1983).
19. T. Arai, T. Karatsu, H. Misawa, Y. Kuriyama, H. Okamoto, T. Hiresaki, H. Furuuchi, H. Zeng, H. Sakuragi, and K. Tokumaru, *Pure Appl. Chem.* **60**, 989 (1988), and references therein.
20. (a) T. Karatsu, T. Arai, H. Sakuragi, and K. Tokumaru, *Chem. Phys. Lett.* **115**, 9 (1985). (b) T. W. Ebbesen and K. Tokumaru, *Appl. Opt.* **25**, 4621 (1989).
21. (a) T. Karatsu, M. Tsuhiya, T. Arai, H. Sakuragi, and K. Tokumaru, *Chem. Phys. Lett.* **169**, 36 (1990). (b) T. Karatsu, T. Arai, H. Sakuragi, K. Tokumaru, and J. Wirz, *Bull. Chem. Soc. Jpn.*, **67**, 891 (1994).
22. H. Hamaguchi, M. Tasumi, T. Karatsu, T. Arai, and K. Tokumaru, *J. Am. Chem. Soc.* **108**, 1698 (1986).
23. T. Arai, T. Karatsu, M. Tsuchiya, H. Sakuragi, and K. Tokumaru, *Chem. Phys. Lett.* **149**, 161 (1988).
24. J. Saltiel and B. W. Atwater, *Adv. Photochem.* **14**, 1 (1988).
25. (a) T. Arai, H. Okamoto, H. Sakuragi, and K. Tokumaru, *Chem. Phys. Lett.* **157**, 46 (1989). (b) H. Okamoto, T. Arai, H. Sakuragi, and K. Tokumaru, *Bull. Chem. Soc. Jpn.* **63**, 2881 (1990).

26. H. Okamoto, T. Arai, H. Sakuragi, K. Tokumaru, and Y. Kawanishi, *Bull. Chem. Soc. Jpn.* **64**, 216 (1991).
27. T. Arai, Y. Kuriyama, T. Karatsu, H. Sakuragi, K. Tokumaru, and S. Oishi, *J. Photochem.* **36**, 125 (1987).
28. H. Furuuchi, T. Arai, Y. Kuriyama, H. Sakuragi, and K. Tokumaru, *Chem. Phys. Lett.* **162**, 211 (1989).
29. H. Furuuchi, Y. Kuriyama, T. Arai, H. Sakuragi, and K. Tokumaru, *Bull. Chem. Soc. Jpn.* **64**, 1601 (1991).
30. T. Karatsu, T. Hiresaki, T. Arai, H. Sakuragi, K. Tokumaru, and J. Wirz, *Bull. Chem. Soc. Jpn.* **65**, 3355 (1992).
31. K. Tokumaru and T. Arai, *J. Photochem. Photobiol. A Chem.* **65**, 1 (1992).
32. S. L. Murov, Ed., *Handbook of Photochemistry*, Marcel Dekker, New York, 1973.
33. J. Saltiel, D. W.-L. Chang, E. D. Megarity, A. D. Rousseau, P. T. Shannon, B. Thomas, and A. K. Uriarte, *Pure Appl. Chem.* **41**, 559 (1975).
34. J. Saltiel, A. D. Rousseau, and B. Thomas, *J. Am. Chem. Soc.* **105**, 7631 (1983).
35. J. Saltiel, G. R. Marchand, E. Kirkor-Kaminska, W. K. Smothers, W. B. Mueller, and J. L. Charlton, *J. Am. Chem. Soc.* **106**, 3144 (1984).
36. J. Saltiel, S. Ganapathy, and C. Werking, *J. Phys. Chem.* **91**, 2755 (1987).
37. T. Ni, R. A. Caldwell, and L. A. Melton, *J. Am. Chem. Soc.* **111**, 457 (1989).
38. G. Galiazzo, A. Spalleti, F. Elisei, and G. Gennari, *Gazz. Chim. Ital.* **119**, 277 (1989).
39. V. Kronganz, N. Castel, and E. Fischer, *J. Photochem.* **39**, 285 (1987).
40. (a) T. Wismontski-Knittel and P. K. Das, *J. Phys. Chem.* **88**, 1168 (1984). (b) K. Bhattacharyya, S. K. Chattopadhyay, S. Baral-Tosh, and P. K. Das, *J. Phys. Chem.* **90**, 2646 (1986).
41. G. Orlandi, F. Negri, U. Mazzucato, and G. Bartocci, *J. Photochem. Photobiol. A Chem.* **55**, 37 (1990).
42. H. Misawa, T. Karatsu, T. Arai, H. Sakuragi, and K. Tokumaru, *Chem. Phys. Lett.* **146**, 40 (1988).
43. T. Karatsu, A. Kitamura, H. Zeng, T. Arai, H. Sakuragi, and K. Tokumaru, *Chem. Lett.*, 2193 (1992).
44. O. Kikuchi, K. Segawa, O. Takahashi, T. Arai, and K. Tokumaru, *Bull. Chem. Soc. Jpn.* **65**, 1463 (1992).
45. F. Momicchioli, I. Baraldi, and E. Fischer, *J. Photochem. Photobiol. A Chem.* **48**, 95 (1989).
46. (a) G. Bartocci, F. Masetti, U. Mazzucato, A. Spalletti, G. Orlandi, and G. Poggi, *J. Chem. Soc. Faraday Trans. 2* **84**, 385 (1988). (b) H. Goerner, F. Elisei, and G. G. Sloisi, *J. Chem. Soc. Faraday Trans.* **88**, 29 (1992).
47. T. Arai, H. Sakuragi, and K. Tokumaru, *Chem. Lett.*, 1335 (1980).
48. T. Arai, T. Karatsu, H. Sakuragi, and K. Tokumaru, *Chem. Lett.*, 1377 (1981).

49. T. Arai, H. Sakuragi, K. Tokumaru, Y. Sakaguchi, J. Nakamura, and H. Hayashi, *Chem. Phys. Lett.* **98**, 40 (1983).
50. (a) T. Arai, H. Sakuragi, and K. Tokumaru, *Chem. Lett.*, 261 (1980). (b) T. Arai, H. Sakuragi, and K. Tokumaru, *Bull. Chem. Soc. Jpn.* **55**, 2204 (1982).
51. R. A. Caldwell, G. W. Sovocool, and R. J. Peresie, *J. Am. Chem. Soc.* **95**, 1496 (1973).
52. R. A. Caldwell and C. V. Cao, *J. Am. Chem. Soc.* **104**, 6174 (1982).
53. R. Bonneau, *J. Photochem.* **10**, 439 (1979).
54. R. Bonneau, *J. Am. Chem. Soc.* **104**, 2921 (1982).
55. M. G. Rockley and K. Salisbury, *J. Chem. Soc. Perkin Trans.* 2, 1582 (1973). P. M. Crosby, J. M. Dyke, J. Metcalfe, A. J. Rest, K. Salisbury, and J. R. Sodeau, *J. Chem. Soc. Perkin Trans.* 2, 182 (1977). K. P. Ghiggino, K. Hara, G. R. Mant, D. Phillips, K. Salisbury, R. P. Steer, and M. D. Swords, *J. Chem. Soc. Perkin Trans.* 2, 88 (1978).
56. H. Goerner, D. W. Eaker, and J. Saltiel, *J. Am. Chem. Soc.* **103**, 7164 (1981).
57. J. Saltiel and D. W. Eaker, *Chem. Phys. Lett.* **75**, 209 (1980).
58. T. Arai, H. Okamoto, K. Tokumaru, and K. Ueno, *J. Photochem. Photobiol. A Chem.* **75**, 85 (1993).
59. G. Bartocci, F. Masetti, U. Mazzucato, A. Spalletti, I. Baraldi, and F. Momicchioli, *J. Phys. Chem.* **91**, 4733 (1987).
60. U. Mazzucato, *Gazz. Chim. Ital.* **117**, 661 (1987).
61. F. Elisei, G. G. Aloisi, and U. Mazzucato, *J. Phys. Chem.* **94**, 5818 (1990).
62. J. H. Richards and N. Pisker-Trifunac, *J. Paint Technol.* **41**, 363 (1969).
63. T. Arai, Y. Ogawa, H. Sakuragi, and K. Tokumaru, *Chem. Phys. Lett.* **196**, 145 (1992).
64. T. Arai, O. Takahashi, T. Asano, and K. Tokumaru, *Chem. Lett.*, 205 (1994).
65. N. Castel, E. Fischer, M. Strauch, M. Niemeyer, and W. Luttke, *J. Photochem. Photobiol. A Chem.* **57**, 301 (1991).
66. A. Padwa, *Chem. Rev.* **77**, 37 (1977).
67. (a) A. Padwa and F. Albrecht, *J. Am. Chem. Soc.* **96**, 4849 (1974). (b) A. Padwa and F. Albrecht, *J. Org. Chem.* **39**, 2361 (1974).
68. (a) H. Furuuchi, T. Arai, Y. Kuriyama, H. Sakuragi, and K. Tokumaru, *Chem. Lett.*, 847 (1990). (b) T. Arai, Y. Furuya, H. Furuuchi, and K. Tokumaru, *Chem. Phys. Lett.* **212** 597 (1993).
69. (a) Y. Furuya, H. Furuuchi, T. Arai, and K. Tokumaru, *Symposium on Photochemistry* (*Japan*), Sapporo, 1993, Abstr., p. 445. (b) T. Arai, Y. Furuya, and K. Tokumaru, *J. Phys. Chem.*, **98**, 9945 (1994).
70. (a) H. Goerner and E. Fischer, *J. Photochem. Photobiol. A Chem.* **57**, 235 (1991). (b) H. Goerner, E. Fischer, and N. Castel, *J. Photochem. Photobiol. A Chem.* **52**, 123 (1990). (c) K. Maeda and E. Fischer, *Mol. Photochem.* **9**, 309 (1979).
71. K. Sandros and H.-D. Becker, *J. Photochem.* **39**, 301 (1987).
72. H. Goerner, *J. Photochem. Photobiol. A Chem.* **43**, 263 (1988).

73. K. Sandros and H.-D. Becker, *J. Photochem. Photobiol. A Chem.* **43**, 291 (1988).

74. (a) E. Fischer, *J. Photochem. Photobiol. A Chem.* **43**, 291 (1988). (b) W. H. Laarhoven, Th. J. H. M. Cuppen, N. Castel, and E. Fischer, *J. Photochem. Photobiol. A Chem.* **49**, 137 (1989).

75. (a) H.-D. Becker, *Chem. Rev.* **93**, 145 (1993). (b) H.-D. Becker, *Adv. Photochem.* **15**, 139 (1989).

76. L. Sun and H. Goerner, *Chem. Phys. Lett.* **208**, 43 (1993).

77. (a) A. Spalleti, G. Bartocci, and U. Mazzucato, *Chem. Phys. Lett.* **186**, 297 (1991). Y. Kikuchi, H. Okamoto, T. Arai, and K. Tokumaru, *Chem. Phys. Lett.* **229**, 564 (1994).

78. Y. Kikuchi, H. Okamoto, T. Arai, and K. Tokumaru, *Chem. Lett.*, 833 (1993).

79. Y. Kikuchi, H. Okamoto, T. Arai, and K. Tokumaru, *Chem. Lett.*, 1811 (1993).

80. (a) H. L. Hyndman, B. M. Monroe, and G. S. Hammond, *J. Am. Chem. Soc.* **91**, 2852 (1969). (b) R. Hurley and A. C. Testa, *J. Am. Chem. Soc.* **92**, 211 (1970).

81. R. A. Caldwell, *J. Am. Chem. Soc.* **92**, 3229 (1970).

82. J. Saltiel, D. E. Townsend, and A. Sykes, *J. Am. Chem. Soc.* **95**, 5968 (1973).

83. (a) Y. C. C. Butt, A. K. Sigh, B. H. Baretz, and R. S. H. Liu, *J. Phys. Chem.* **85**, 2091 (1981). (b) F. W. Langkilde, R. Wilbrandt, S. Moller, A. M. Brouwer, F. Negri, and G. Orlandi, *J. Phys. Chem.* **95**, 6884 (1991).

84. (a) H. Hashimoto, Y. Koyama, K. Ichimura, and T. Kobayashi, *Chem. Phys. Lett.* **162**, 517 (1989). (b) H. Hashimoto and Y. Koyama, *J. Phys. Chem.* **92**, 2101 (1988).

85. V. Ramamurthy and R. S. H. Liu, *J. Am. Chem. Soc.* **98**, 2935 (1976).

86. A. Harriman and R. S. H. Liu, *Photochem. Photobiol.* **26**, 29 (1977).

87. B. Veyret, S. G. Davis, M. Yoshida, and K. Weiss, *J. Am. Chem. Soc.* **100**, 3283 (1978).

88. H. Hamaguchi, H. Okamoto, Y. Mukai, and Y. Koyama, *Chem. Phys. Lett.* **107**, 355 (1984).

89. H. Hamaguchi, *J. Mol. Struct.* **126**, 125 (1985).

90. Y. Mukai, H. Hashimoto, and Y. Koyama, *J. Phys. Chem.* **94**, 4042 (1990).

91. N.-H. Jensen, R. Wilbrandt, and R. V. Bensasson, *J. Am. Chem. Soc.* **111**, 7877 (1989).

92. S. Ganapathy and R. S. H. Liu, *J. Am. Chem. Soc.* **114**, 345 (1992).

93. S. Ganapathy and R. S. H. Liu, *Photochem. Photobiol.* **56**, 959 (1992).

94. W. A. Yee, S. J. Hug, and D. S. Kliger, *J. Am. Chem. Soc.* **110**, 2164 (1988).

95. H. Goerner, *J. Photochem.* **19**, 343 (1982).

96. Y. Gong, T. Arai, and K. Tokumaru, *Chem. Lett.*, 753 (1993).

97. K. Sandros, M. Sundahl, O. Wennerstroem, and U. Norinder, *J. Am. Chem. Soc.* **112**, 3082 (1990).

98. M. Sundahl, O. Wennerstroem, K. Sandros, T. Arai, and K. Tokumaru, *J. Phys. Chem.* **94**, 6731 (1990).

99. I. Anger, K. Sandros, M. Sundahl, and O. Wennerstroem, *J. Phys. Chem.* **97**, 1920 (1993).

100. K. Sandros, M. Sundahl, and O. Wennerstroem, *J. Phys. Chem.* **97**, 5291 (1993).

101. M. Sundahl, K. Sandros, O. Wennerstroem, T. Arai, H. Okamoto, and K. Tokumaru, *Chem. Phys. Lett.* **168**, 395 (1990).

102. F. D. Lewis, C. L. Stern, and B. A. Yoon, *J. Am. Chem. Soc.* **114**, 3131 (1992).

103. T. Arai, T. Iwasaki, and K. Tokumaru, *Chem. Lett.*, 691 (1993).

104. J. A. Eenkhoorn, S. Osmund de Silva, and V. Snieckus, *Can. J. Chem.* **51**, 792 (1973).

105. T. Arai, M. Moriyama, and K. Tokumaru, *J. Am. Chem. Soc.* **116**, 3171 (1994).

106. (a) D. A. Lightner and A. F. McDonagh, *Acc. Chem. Res.* **17**, 417 (1984). (b) G. Agati and F. Fusi, *J. Photochem. Photobiol. B Biol.* **7**, 1 (1990). (c) S. Onishi, S. Itoh, K. Isobe, M. Ochi, T. Kunitake, and T. Imai, *Biochem. J.* **257**, 711 (1989).

107. (a) Y. Kanna, T. Arai, H. Sakuragi, and K. Tokumaru, *Chem. Lett.*, 613 (1990). (b) Y. Kanna, T. Arai, and K. Tokumaru, *Bull. Chem. Soc. Jpn.* **66**, 1482 (1993). (c) Y. Kanna, T. Arai, and K. Tokumaru, *Bull. Chem. Soc. Jpn.* **66**, 1586 (1993).

108. R. A. Caldwell, S. J. Riley, A. A. Gorman, S. P. McNeeney, and D. J. Unett, *J. Am. Chem. Soc.* **114**, 4424 (1992).

109. (a) A. A. Gorman, C. Lambert, and A. Prescott, *Photochem. Photobiol.* **51**, 29 (1990). (b) A. A. Gorman, R. Reddoes, I. Hamblett, S. P. McNeeney, A. L. Prescott, and D. J. Unett, *J. Chem. Soc. Chem. Commun.*, 963 (1991).

110. R. A. Caldwell, L. Carlacci, C. E. Doubleday, Jr., T. R. Furlani, H. F. King, and J. W. McIver, Jr., *J. Am. Chem. Soc.* **110**, 6901 (1988).

111. (a) J. L. Goodman, K. S. Peters, H. Misawa, and R. A. Caldwell, *J. Am. Chem. Soc.* **108**, 6803 (1986). (b) R. Bonneau and B. Herran, *Laser Chem.* **4**, 151 (1984).

112. R. A. Caldwell, L. D. Jacobs, T. R. Furlani, E. A. Nalley, and J. Jaboy, *J. Am. Chem. Soc.* **114**, 1623 (1992).

113. M. Irie, *J. Synth. Org. Chem.* **49**, 373 (1991).

114. (a) M. Irie and K. Sayo, *J. Phys. Chem.* **96**, 7671 (1992). (b) Y. Nakayama, K. Hayashi, and M. Irie, *Bull. Chem. Soc. Jpn.* **64**, 202 (1991). (c) Y. Nakayama, K. Hayashi, and M. Irie, *Bull. Chem. Soc. Jpn.* **64**, 789 (1990). (d) K. Uchida, Y. Nakayama, and M. Irie, *Bull. Chem. Soc. Jpn.* **63**, 1311 (1990). (e) Y. Nakayama, K. Hayashi, and M. Irie, *J. Org. Chem.* **55**, 2592 (1990). (f) M. Irie and M. Mohri, *J. Org. Chem.* **53**, 803 (1988). (g) S. Nakayama and M. Irie, *J. Org. Chem.* **53**, 6136 (1988).

115. (a) H. G. Heller, in *Fine Chemicals for Electronic Industry*, P. Bamfields, Ed., Royal Society of Chemistry, London, 1986, p. 120. (b) H. G. Heller, *Chem. Ind. (London)*, 193 (1978). (c) Y. Yokoyama, T. Iwai, Y. Yokoyama, and Y. Kurita, *Chem. Lett.* 225 (1994). (d) S. L. Gilat, S. H. Kawai, and J.-M. Lehn, *J. Chem. Soc. Chem. Commun.*, 1439 (1993).

116. Y. Inoue, S. Kosaka, K. Matsumoto, H. Tsuneishi, T. Hakushi, A. Tai, K. Nakagawa, and L.-H. Tong, *J. Photochem. Photobiol. A Chem.* **71**, 61 (1993).

117. (a) Y. Inoue, N. Yamasaki, T. Yokoyama, and A. Tai, *J. Org. Chem.* **58**, 1011 (1993). (b) Y. Inoue, T. Yokoyama, N. Yamasaki, and A. Tai, *J. Am. Chem. Soc.* **111**, 6480 (1989).

118. U. Mazzucato and F. Momicchioli, *Chem. Rev.* **91**, 1679 (1991).

119. (a) P. J. Vroegop, J. Lugtenburg, and E. Havinga, *Tetrahedron* **29**, 1393 (1973). (b) H. J. C. Jacobs and E. Havinga, *Adv. Photochem.* **11**, 305 (1979).

120. E. Haas, G. Fischer, and E. Fischer, *J. Phys. Chem.* **82**, 1638 (1978). E. Fischer, *J. Phys. Chem.* **85**, 1770 (1981).

121. J. B. Birks, G. Bartocci, G. G. Aloisi, S. Dellonte, and F. Barigalleti, *Chem. Phys.* **51**, 113 (1980).

122. E. Fischer, *J. Photochem.* **17**, 331 (1981).

123. U. Mazzucato, *Pure Appl. Chem.* **54**, 1705 (1982).

124. G. Barocci, U. Mazzucato, I. Baraldi, and E. Fischer, *J. Mol. Struct.* **193**, 173 (1989).

125. U. Mazzucato and F. Momicchioli, *EPA Newsl.* **44**, 31 (1992).

126. T. Wismontski-Knittel, P. K. Das, and E. Fischer, *J. Phys. Chem.* **88**, 1163 (1984).

127. K. P. Ghiggino, P. F. Skilton, and E. Fischer, *J. Am. Chem. Soc.* **108**, 1146 (1986).

128. Y.-P. Sun, D. F. Sears, J. Saltiel, F. B. Mallory, W. M. Mallory, and C. A. Buster, *J. Am. Chem. Soc.* **110**, 6974 (1988).

129. (a) G. Bartocci, U. Mazzucato, F. Masetti, and G. G. Aloisi, *Chem. Phys.* **101**, 461 (1986). (b) G. Bartocci, F. Masetti, U. Mazzucato, A. Spaletti, I. Baraldi, and F. Momicchioli, *J. Phys. Chem.* **91**, 4773 (1987).

130. M. Lamotte, F. J. Morgan, K. A. Muszkat, and T. Wismontski-Knittel, *J. Phys. Chem.* **94**, 1302 (1990).

131. J. Drew, F. Zerbetto, A. G. Szabo, and P. Morand, *J. Phys. Chem.* **94**, 4439 (1990).

132. N. S. Park and D. H. Waldeck, *Chem. Phys. Lett.* **168**, 379 (1990).

133. (a) A. M. Brearley, A. J. G. Stanjord, S. R. Flom, and P. F. Barbara, *Chem. Phys. Lett.* **113**, 43 (1985). (b) S. R. Flom, V. Nagarajan, and P. F. Barbara, *J. Phys. Chem.* **90**, 2092 (1986). (c) A. M. Brearly, S. R. Flom, V. Nagarajan, and P. F. Barbara, *J. Phys. Chem.* **90**, 2092 (1986).

134. P. F. Barbara and W. Jarzeba, *Acc. Chem. Res.* **21**, 195 (1988).

135. (a) A. S. Cherkasov, *Dokl. Acad. Sci. USSR* (*Engl. Transl.*) **146**, 852 (1962). (b) Yu. T. Mazurenko, V. S. Udaltsov, and A. S. Cherkasov, *Opt. Spektrosk.* **46**, 389 (1979).

136. (a) T. Arai, T. Karatsu, H. Sakuragi, K. Tokumaru, N. Tamai, and I. Yamazaki, *Chem. Phys. Lett.* **158**, 429 (1989). (b) T. Arai, T. Karatsu, H. Sakuragi, K. Tokumaru, N. Tamai, and I. Yamazaki, *J. Photochem. Photobiol. A Chem.* **65**, 41 (1992).

137. H. Furuuchi, T. Arai, H. Sakuragi, K. Tokumaru, Y. Nishimura, and I. Yamazaki, *J. Phys. Chem.* **95**, 10322 (1991).

138. (a) T. Karatsu, N. Yoshikawa, A. Kitamura, and K. Tokumaru, *Chem. Lett.*, 381 (1994). (b) Y. Yuda, T. Arai, and K. Tokumaru, Symposium on Photochemistry (Japan), Sapporo, 1993, Abstr., p. 443. (c) T. Arai, Y. Yuda, K. Sano, *Res. Chem. Intermed.* in press.
139. F. D. Lewis, J. R. Petisce, J. D. Oxman, and M. J. Nepras, *J. Am. Chem. Soc.* **107**, 203 (1985).
140. F. D. Lewis, A. M. Bedell, R. E. Dykstra, J. E. Elbert, I. R. Gould, and S. Farid, *J. Am. Chem. Soc.* **112**, 8055 (1990).
141. Y. Kuriyama, T. Arai, H. Sakuragi, and K. Tokumaru, *Chem. Phys. Lett.* **173**, 253 (1990).
142. Y. Kuriyama, T. Arai, H. Sakuragi, and K. Tokumaru, *Chem. Lett.*, 879 (1992).
143. K. Tokumaru, Y. Kuriyama, T. Arai, I. K. Lednev, R. Akaba, and H. Sakuragi, in *Photochemical Processes in Organized Molecular Systems*, K. Honda, Ed., Elsevier, Amsterdam, 1991, p. 199.
144. I. K. Lednev, M. V. Alfimov, Y. Kuriyama, T. Arai, H. Sakuragi, and K. Tokumaru, *J. Photochem. Photobiol. A Chem.* **63**, 201 (1992).
145. Y. Kuriyama, H. Sakuragi, K. Tokumaru, Y. Yoshida, and S. Tagawa, *Bull. Chem. Soc. Jpn.* **66**, 1852 (1993).
146. B. W. Zhang, Y. Kuriyama, Y. Cao, and K. Tokumaru, *Bull. Chem. Soc. Jpn.* **66**, 1859 (1993).
147. (a) T. Oshiyama, O. Takahashi, K. Morihashi, O. Kikuchi, and K. Tokumaru, *Bull. Chem. Soc. Jpn.* **66**, 1622 (1993). (b) O. Kikuchi, T. Oshiyama, O. Takahashi, and K. Tokumaru, *Bull. Chem. Soc. Jpn.* **65**, 2267 (1992).
148. T. Shida and W. H. Hamill, *J. Chem. Phys.* **44**, 2375 (1966).
149. H. Suzuki, K. Ogawa, T. Shida, and A. Kira, *Bull. Chem. Soc. Jpn.* **56**, 66 (1983).
150. H. C. Wang, G. Levin, and M. Szwarc, *J. Am. Chem. Soc.* **99**, 2642 (1977).
151. V. A. Sazhnikov, M. Rakhmatov, and M. V. Alfimov, *Chem. Phys. Lett.* **71**, 33 (1990).
152. T. W. Ebbesen, *J. Phys. Chem.* **92**, 4581 (1988).
153. (a) F. D. Lewis, R. E. Dykstra, I. R. Gould, and S. Farid, *J. Phys. Chem.* **92**, 7042 (1988). (b) F. D. Lewis and R. E. Dykstra, *J. Photochem. Photobiol. A Chem.* **49**, 109 (1989).
154. A. Ishida, K. Tabata, S. Toki, and S. Takamuku, presented at *Annual Meeting of the Chemical Society of Japan*, Kyoto, Apr. 1989, Abstr., 2IIIH07.
155. (a) K. Morishima, S. Tojo, A. Ishida, and S. Takamuku, presented at *Symposium on Radiation Chemistry*, Osaka, Oct. 1991, Abstr., p. 61. (b) S. Takamuku, Y. Okamoto, A. Ishida, and S. Tojo, in *Annual Report 1992*, Radiation Laboratory, Ed., The Institute of Scientific and Industrial Research, Osaka University, Osaka, 1992, p. 43.
156. Y. Kuriyama, F. Hashimoto, M. Tsuchiya, H. Sakuragi, and K. Tokumaru, *Chem. Lett.* 1371 (1994).

COOLING OF A DYE SOLUTION BY ANTI-STOKES FLUORESCENCE

Christoph Zander and Karl Heinz Drexhage
Fachbereich Chemie, Universität Siegen, 57068 Siegen, Germany

CONTENTS

I. INTRODUCTION

When a photon feeds energy to a body through an absorption process, everyday experience appears to teach us that the temperature of the irradiated body will rise invariably. The biosphere may serve as an example:

Advances in Photochemistry, Volume 20, Edited by Douglas C. Neckers, David H. Volman, and Günther von Bünau
ISBN 0-471-11469-3

Life on earth would not exist without heating by sunlight. However, in 1975, cooling of a gas by laser irradiation was proposed by Hänsch and Schawlow [1]. According to their proposal, an atomic gas is excited by a laser whose frequencies are confined to the lower half of the Doppler-broadened absorption line. Thus only atoms moving toward the laser beam are excited. Their momentum is reduced by the absorbed photons: The temperature of the gas is lowered. Wineland et al. [2] reported the first observation of radiation-pressure-induced cooling by means of the Doppler effect. They irradiated Mg^{2+} ions in a Penning trap with a single-mode dye laser and achieved a cooling rate of $5\,\mathrm{K\,s^{-1}}$. Nowadays the cooling of atoms by resonant fields of radiation is a commonly used technique to manipulate atoms in the phase space. The applications reach from spectroscopy of transitions with strongly reduced Doppler broadening [3] to high-precision clocks [4, 5].

Apart from these more recent developments there has been a related discussion that began many years earlier. It dealt with the question of whether cooling of a dye solution by anti-Stokes fluorescence would be possible [6–12]. Vavilov [10] argued that an energy yield of photoluminescence exceeding unity would violate the second law of thermodynamics. Furthermore, Ketskeméty and Farkas [11] derived an analytical expression for an upper boundary of the quantum yield in the anti-Stokes region—presumably confirmed by their experiments—which would exclude any possibility of cooling. On the other hand, Landau [6] showed that an energy efficiency in excess of unity would not violate fundamental laws of thermodynamics, yet he considered this excess, if occurring at all, to be negligible and its observation unlikely. Experimentally, such cooling by anti-Stokes fluorescence has not been observed until recently [13]. In this chapter we describe in detail the experiment that made a quantitative investigation of anti-Stokes cooling possible.

II. CONDITIONS FOR COOLING

As a rule supported by a wealth of experimental evidence, the fluorescence spectrum of organic dye solutions is independent of the wavelength of excitation (Vavilov's law). Therefore, on excitation in the short-wavelength (Stokes) region, the energy difference between a photon absorbed and a photon emitted will cause a corresponding increase in the temperature of the dye solution. On the other hand, when a dye solution is excited in the long-wavelength (anti-Stokes) region, the photons of fluorescence have a higher energy than the photons absorbed. It then follows from energy conservation that the additional energy of the emitted photons is taken out of the heat reservoir of the solution. Consequently, the temperature of the solution should decrease.

To determine the conditions for such cooling by anti-Stokes fluorescence, the net power P_{net} emitted by fluorescence may be considered. This is the difference between the power emitted P_{em} and the power absorbed P_{abs} and can be written as

$$P_{net} = \left(\eta_{fl} \frac{\lambda_0}{\lambda^*} A_d - A_{sol}\right) P_L \tag{1}$$

where P_L is the power of the incident laser beam, λ_0 the wavelength of excitation, and η_{fl} the fluorescence quantum yield of the dye. λ^* is the mean wavelength of fluorescence, given by

$$\lambda^* = \frac{\int F(\lambda)\, d\lambda}{\int (1/\lambda) F(\lambda)\, d\lambda} \tag{1a}$$

Here $F(\lambda)\, d\lambda$ is the energy spectrum of fluorescence. A_d is the absorption by the dye itself and A_{sol} is the absorption of the dye solution, both at wavelength λ_0. These quantities are in general not identical, because spurious absorption due to impurities as well as intrinsic solvent absorption must be taken into account. The absorption due to such nonfluorescent components is described by a quantity A_{nf}. For small values of A_d and A_{nf} it follows $A_{sol} = A_d + A_{nf}$, and we obtain with Eq. (1) for the net power emitted:

$$P_{net} = \left[\eta_{fl} \frac{\lambda_0}{\lambda^*} - \left(1 + \frac{A_{nf}}{A_d}\right)\right] P_L A_d \tag{2}$$

Cooling by anti-Stokes fluorescence will occur for $P_{net} > 0$.

The fluorescence quantum yield of a dye is less than unity ($\eta_{fl} < 1$), due to nonradiative processes in all practical cases. These processes cause a temperature increase that must be overcome if cooling is to be observed. Thus a sufficiently large difference between the wavelength of excitation λ_0 and the wavelength of fluorescence λ^* is necessary to achieve a positive value of P_{net}. On the other hand, in the long-wavelength region the absorbance E of the dye decreases exponentially with decreasing frequency [14, 15], according to

$$\ln E = \frac{hc}{kT} \frac{1}{\lambda_0} + \ln E_0 \tag{3}$$

Here h is Planck's constant, c the velocity of light, k Boltzmann's constant, T the ambient temperature, and E_0 an empirical constant depending on dye concentration and path length. Using this expression we can describe the absorption A_d due to the dye itself in the long-wavelength region by

$$A_d = 1 - 10^{-\exp[(hc/kT)(1/\lambda_0) + \ln E_0]} \tag{4}$$

Because the absorption decreases strongly with increasing wavelength, a marked cooling effect is to be expected only in a limited-wavelength region.

A threshold quantum yield η_{th} above which cooling is possible can be derived from Eq. (2) with $P_{\text{net}} = 0$:

$$\eta_{\text{th}} = \frac{\lambda^*}{\lambda_0}\left(1 + \frac{A_{\text{nf}}}{A_d}\right) \tag{5}$$

In Figure 1 the quantity η_{th} is plotted for an ethanolic solution of the dye Rhodamine 6G (Figure 2). Absorption and fluorescence spectrum of this dye are shown in Figure 3. A value $\lambda^* = 563$ nm was calculated from the fluorescence spectrum according to Eq. (1*a*). The absorbance of this dye can be described in the long-wavelength region by Eq. (3) with $\ln E_0 = -90.9$ for a 10^{-5} *M* solution (path 1 cm). Then the dye absorption A_d is calculated with Eq. (4). For $A_{\text{nf}} = 0$ the dashed line of Figure 1 is obtained. In practice

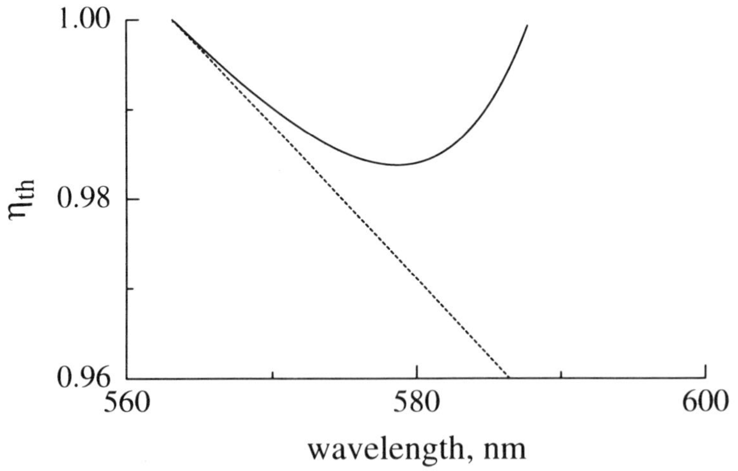

Figure 1. Threshold quantum yield η_{th} vs. exciting wavelength λ_0 for 10^{-5} *M* Rhodamine 6G in ethanol. Parameters: $\lambda^* = 563$ nm, $T = 293$ K, $\ln E_0 = -90.9$. —, $A_{\text{nf}} = 5 \times 10^{-5}$ for a path length of 1 cm; ---, $A_{\text{nf}} = 0$.

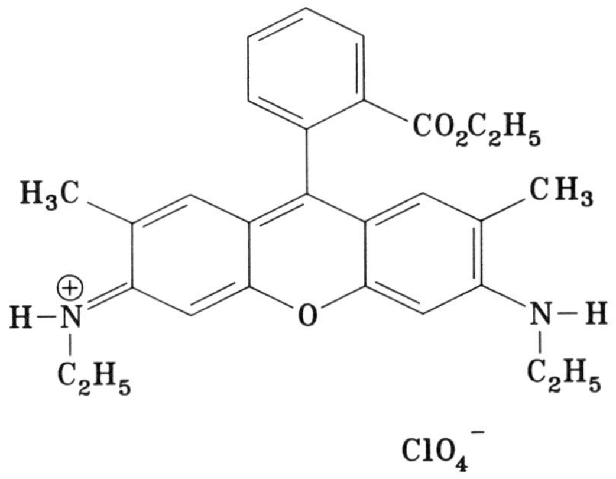

Figure 2. Molecular structure of Rhodamine 6G.

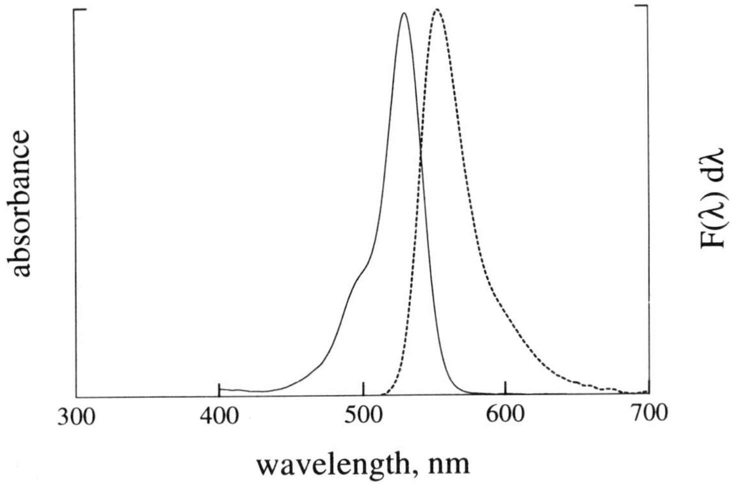

Figure 3. Absorption and fluorescence spectrum of $10^{-5}\,M$ Rhodamine 6G in ethanol. —, Absorbance; – – –, fluorescence.

the absorption of the solvent cannot be neglected. Since the very small absorption of monodeuterated ethanol (C_2H_5OD) is approximately 5×10^{-5} (path 1 cm) in the region of dye fluorescence (Figure 4) [13], the value $A_{nf} = 5 \times 10^{-5}$ independent of wavelength was taken for calculating the solid line of Figure 1. All experiments were performed at temperature $T = 293$ K.

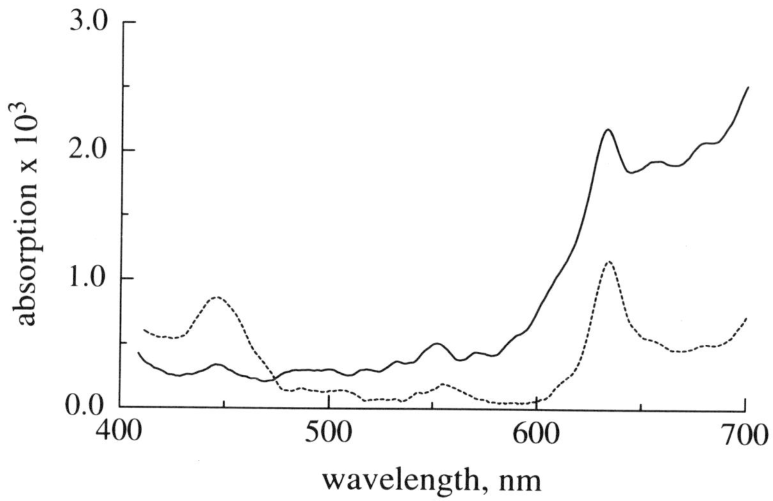

Figure 4. Absorption of ethyl alcohol vs. wavelength. Path length 1 cm. —, Ethyl alcohol (C_2H_5OH); – – –, monodeuterated ethyl alcohol (C_2H_5OD).

As can be seen in Figure 1, at short wavelengths the quantum yield required for cooling decreases if the wavelength λ_0 and thus the spread between λ_0 and λ^* increases. At about 580 nm the influence of the intrinsic solvent absorption A_{nf} together with a decreasing dye absorption A_d begins to dominate the effect of the λ_0–λ^* spread. Therefore, the quantum yield required for cooling rises again. A minimum value of $\eta_{th} = 0.984$ is obtained for $\lambda_0 = 579$ nm.

This consideration shows that cooling by anti-Stokes fluorescence might be possible. However, it would be restricted to a narrow-wavelength region and would require a very high quantum yield of fluorescence. In practice the latter requirement imposes severe restrictions: a dye solution with a fluorescence quantum yield higher than 0.984 has to be found. This seems to be very difficult given the fact that only a few compounds are known with quantum yields above 0.9. Even so the reported quantum yield values are frequently less accurate than assumed by authors and readers alike. In addition, the published values of the quantum yield were measured mostly by excitation in the main absorption band of the dye (i.e., not in the wavelength region relevant for anti-Stokes excitation). In fact, for this region some authors [16, 17] found a decay of the quantum yield with increasing wavelength. To develop a dye solution with the high quantum yield required for anti-Stokes cooling, we need a method that permits very precise measurements of the fluorescence quantum yield.

III. DETERMINATION OF THE FLUORESCENCE QUANTUM YIELD

Conventional methods to measure the fluorescence quantum yield of solutions attain an accuracy of about 5% at best. It is obvious from Figure 1 that this accuracy is by far not sufficient for the problem at hand. Hence a calorimetric method based on the thermal lensing effect was taken into consideration, because in this way the quantum yield can be determined absolutely (i.e., without recurrence to a luminescence standard) and with high confidence. Such a method appears indicated here in particular, because it measures directly the nonradiative deactivation. In case of very high values of the fluorescence quantum yield, this should enhance the accuracy considerably.

In the thermal lensing method, first proposed by Hu and Whinnery [18], a laser beam of appropriate frequency is focused onto the sample, by which the light is absorbed partially (Figure 5). If there is less energy emitted than absorbed by the sample, the radial intensity distribution of the laser beam generates a temperature gradient and therefore a gradient of the refractive index within the sample. Because the refractive index decreases with increasing temperature ($dn/dT < 0$) for most liquids, a divergent lens starts to develop as soon as a shutter S (Figure 5) opens. Consequently, the laser beam starts to spread behind the sample, and its intensity at the detector D drops until a stationary state is reached (Figure 6). However, on prolonged irradiation, the signal at D tends to become erratic due to convection in the liquid.

To obtain a maximum lensing effect the sample should be located one Rayleigh length behind the focus [18]. After opening the shutter S at time $t = 0$ the intensity $I(t)$ at the detector D is expected to obey the expression [18]

$$I(t) = I_0 \left[1 - \Theta \left(1 + \frac{t_c}{2t} \right)^{-1} + \frac{1}{2} \Theta^2 \left(1 + \frac{t_c}{2t} \right)^{-2} \right]^{-1} \tag{6}$$

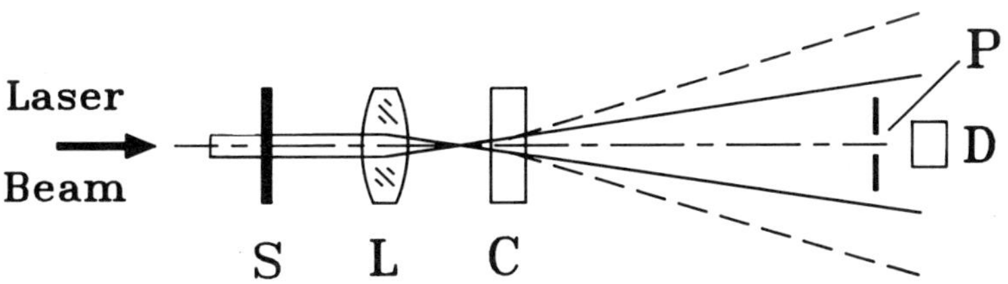

Figure 5. Thermal lensing method. S, shutter; L, lens; C, sample; D, detector; P, pinhole. —, Laser beam without thermal lens; ---, laser beam with thermal lens.

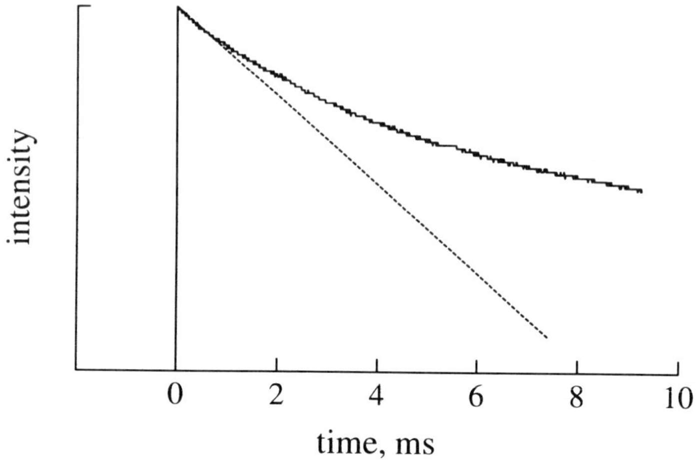

Figure 6. Typical thermal lensing signal of a dye solution. —, Signal; ---, initial slope.

The parameter t_c is a characteristic time of thermal diffusion and is given by

$$t_c = \frac{\omega^2 \rho C}{4\kappa} \tag{6a}$$

where ω is the waist radius (e^{-2} value of the intensity) of the laser beam, ρ the density, C the heat capacity, and κ the heat conductivity of the solution. The quantity Θ is directly proportional to the thermal power P_{th} deposited in the sample:

$$\Theta = P_{\text{th}} \left(\frac{dn}{dT}\right) \frac{1}{\lambda_0 \kappa} \tag{6b}$$

For small times ($t \ll t_c$) Eq. (6) can be approximated by

$$I(t) = I_0 + mt \tag{7}$$

with

$$m = \frac{2I_0}{t_c} \Theta \tag{7a}$$

Thus the initial slope m of the detector signal can be used to determine Θ as well as, with Eq. (6*b*), the quantity $P_{th}(P_{th} = P_{abs} - P_{em})$.

The fluorescence quantum yield of a solution is defined as the number ratio of the photons emitted to the photons absorbed:

$$\eta_{sol} = \frac{P_{em}\lambda^*}{P_{abs}\lambda_0} \tag{8}$$

Therefore, η_{sol} can be expressed by P_{th} and the power P_{abs} absorbed by the sample:

$$\eta_{sol} = \frac{\lambda^*}{\lambda_0}\left(1 - \frac{P_{th}}{P_{abs}}\right) \tag{9}$$

If absorption by impurities and solvent can be neglected, the quantum yield of the solution can be taken as that of the dye ($\eta_{fl} = \eta_{sol}$). Unfortunately, the precision with which thermooptic coefficients are known is inadequate for precise measurements of the quantum yield via Eqs. (6)–(9). To circumvent this problem, Brannon and Magde [19] suggested using a nonfluorescent compound as reference. Thus in addition to the sample, designated below by superscript *s*, the reference compound, denoted by a superscript *r*, is measured under the same conditions (solvent, temperature, etc.). For this reference P_{abs} equals P_{th}, and by combining Eqs. (6*b*), (7*a*) and (9), the following expression for the quantum yield is obtained:

$$\eta_{sol} = \frac{\lambda^*}{\lambda_0}\left(1 - \frac{m^s}{P^s_{abs}I^s_0}\,\frac{P^r_{abs}I^r_0}{m^r}\right) \tag{10}$$

Thus thermooptical constants of the solvent are no longer required for determination of the fluorescence quantum yield.

In contrast to the setup shown in Figure 5, our experiment uses two coaxial laser beams focused onto the sample C by a lens L (Figure 7). The dye laser beam (power 1 to 10 mW) creates the thermal lens in the sample, whereas the helium–neon laser is used only for monitoring development of the thermal lens. To avoid a thermal lens being induced by the helium–neon laser, a neutral density filter F_2 reduces the power of its beam to 6 to 7 μW in the sample. In front of the detector D_1 an interference filter F_1 blocks the beam of the dye laser, and a pinhole P is placed such that only light near the optical axis reaches the detector. To monitor the wavelength of excitation λ_0, part of the dye laser beam is deflected by a glass plate G_1 onto a

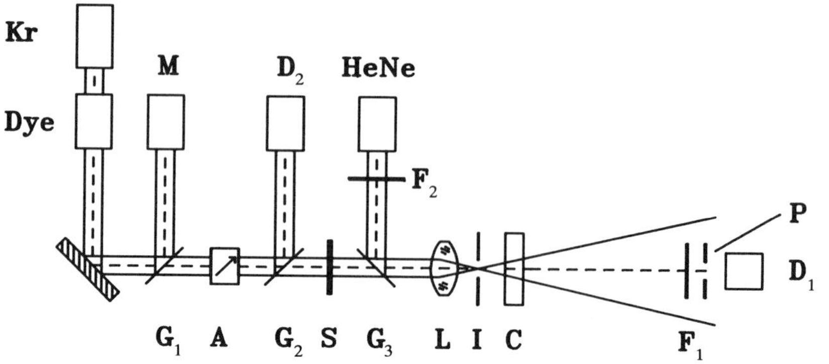

Figure 7. Thermal lensing experiment. Kr, krypton ion laser; Dye, dye laser; M, monochromator; G_1, G_2, G_3, glass plate; A, attenuator; S, shutter, L, lens (f = 200 mm); I, iris diaphragm; C, dye cell (path 1 cm); F_1, interference filter ($\lambda = 633$ nm); F_2, neutral density filter; P, pinhole (diameter $d = 1$ mm); D_1, D_2, detector; HeNe, helium–neon laser.

monochromator M. The intensity of the exciting laser beam can be adjusted by an attenuator A and is measured by the detector D_2. An iris diaphragm I is inserted at the focal plane of the lens to prevent parasitic reflections. In principle, the experiment can be done with a single laser setup (Figure 5). However, the beam quality and amplitude noise of the dye laser would limit the accuracy attainable. The helium–neon laser used for monitoring has the advantage of very small amplitude noise. To determine the mean wavelength of emission λ^* the fluorescence spectrum of the sample was measured with a commercial fluorescence spectrometer. For solutions with a quantum yield higher than 0.9, the accuracy of the method described here is estimated to be $\pm 0.3\%$. This should be sufficient to aid in a search for dye solutions that might be suitable for anti-Stokes cooling.

Using the method described above, the fluorescence quantum yield of several highly fluorescent dye solutions was determined. Here we present only some results for Rhodamine 6G. The fluorescence quantum yield of this dye in ethanolic solution is independent of temperature, and a value of 0.95 for aerated solutions has been reported [20]. It has been known for a long time that the quantum yield of dyes with unsubstituted amino groups increases on deuteration (replacement of H by D at the amino groups) [20]. Therefore, such an effect is also to be expected for Rhodamine 6G (see Figure 2), when dissolved in deuterated ethanol C_2H_5OD. Using the thermal lensing method described above we now determined the values 0.955 and 0.980 for air-equilibrated solutions of Rhodamine 6G–perchlorate in C_2H_5OH and C_2H_5OD, respectively. Furthermore, we found that the

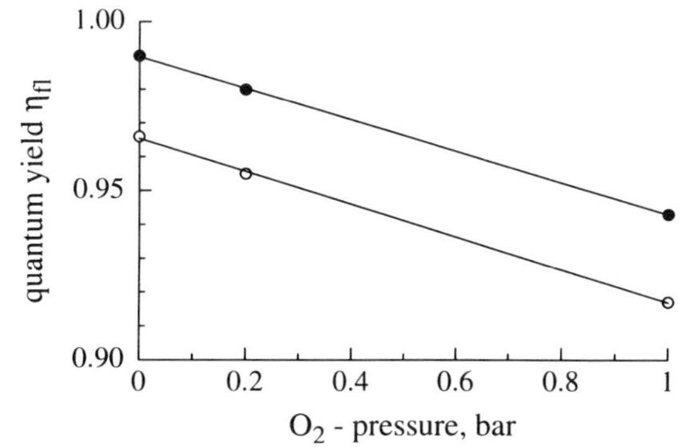

Figure 8. Fluorescence quantum yield of 3×10^{-6} *M* Rhodamine 6G perchlorate in ethanol versus partial pressure of dissolved oxygen. ○, C_2H_5OH; ●, C_2H_5OD.

fluorescence is markedly quenched by dissolved oxygen: On deaeration the quantum yield rises to values of 0.966 and 0.990 for C_2H_5OH and C_2H_5OD, respectively (Figure 8). Hence a solution of Rhodamine 6G (the anion has no influence in dilute solution) in oxygen-free monodeuterated ethanol appears to meet the quantum yield requirement ($\eta_{fl} \geqslant 0.984$) for cooling by anti-Stokes fluorescence. However, the wavelength of excitation in the quantum yield determination above was 531 nm (i.e., near the maximum of the main absorption band). It had to be investigated further if the quantum yield is independent of excitation wavelength even on the low-energy side of the absorption band, where anti-Stokes fluorescence would be excited.

IV. WAVELENGTH DEPENDENCE OF THE FLUORESCENCE QUANTUM YIELD

To investigate the influence of the wavelength of excitation, the fluorescence quantum yield of Rhodamine 6G in ethanol was measured in the anti-Stokes region with the setup of Figure 7. For the initial experiments we used a commercial dye sample of Rhodamine 6G perchlorate (Aldrich, laser grade). The quantum yield values obtained on this material are given in Table 1 (η_{sol}). It appears that the quantum yield of the solution begins to decrease at wavelengths larger than 570 nm. A similar phenomenon was also found by Ketskeméty and Farkas in the case of some other compounds [11]. If the quantum yield η_{fl} of the dye really did decrease in the anti-Stokes region, cooling would not be observable. Therefore, it was very important to

TABLE 1 Fluorescence Quantum Yield of 4.3×10^{-4} *M* Rhodamine 6G Perchlorate (Aldrich, Laser Grade) in Ethanol

λ_0 (nm)	η_{sol}[a]	A_d[b]	A_{nf}[c]$/A_d$	η_{cor}[d]
530	0.955	1	0	0.955
570	0.956	0.6	0	0.956
580	0.947	0.1	0.01	0.956
590	0.920	0.02	0.04	0.957
600	0.820	0.004	0.2	0.984

[a]η_{sol}, measured; [b]A_d, dye absorption according to Figure 9, curve (2); [c]A_{nf}, absorption of nonfluorescent components ($A_{nf} = 7.3 \times 10^{-4}$); [d]$\eta_{cor}$, corrected with Eq. (11).

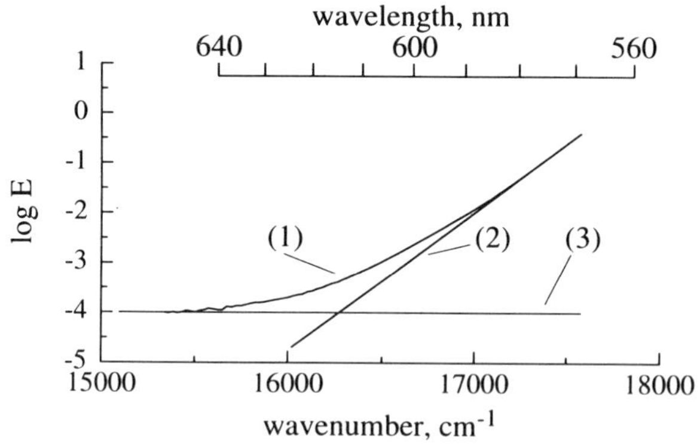

Figure 9. Rhodamine 6G perchlorate (Aldrich, laser grade) in ethanol. Absorbance *E* vs. wavenumber; dye concentration 4.3×10^{-4} *M*, path length 1 cm: (1) measured; (2) expected according to Eq. (3) with $\ln E_0 = -87.1$; (3) assumed for impurities ($E_{imp} = 10^{-4}$). Sum of absorbances (2) and (3) matches experimental absorbance (1).

determine if the decrease observed was a true property of the dye or perhaps due to an artifact.

From Eq. (3) a linear dependence between the logarithm of the dye absorbance and the reciprocal wavelength λ_0 is expected. However, this was not observed in case of the dye solution used for the experiments described above (Figure 9). A deviation from Eq. (3) becomes apparent at wavelengths larger than 580 nm, and it was suspected that this is caused by an additional absorption due to impurities. If a wavelength-independent impurity absorbance $E_{imp} = 10^{-4}$ is assumed and this is added to the dye absorbance expected from Eq. (3), the experimental absorption curve can be reproduced closely (Figure 9). The impurities (absorption 2.3×10^{-4}) and the solvent C_2H_5OH (absorption 5×10^{-4}; see Figure 4) absorb energy but do not contribute to the fluorescence of the solution. Thus it is reasonable to assume for the absorption due to nonfluorescent components the value $A_{nf} = 7.3 \times 10^{-4}$. We then may calculate corrected values of the quantum yield with

$$\eta_{cor} = \eta_{sol} \left(1 + \frac{A_{nf}}{A_d}\right) \tag{11}$$

Here the dye absorption A_d is to be taken from Figure 9, curve (2). Interestingly, the corrected quantum yield η_{cor} remains practically constant throughout the anti-Stokes region. Thus it can be expected that the value $\eta_{fl} = 0.990$ found for Rhodamine 6G in deaerated C_2H_5OD is also independent of wavelength of excitation. It then follows from Figure 1 that cooling should occur on irradiation in the wavelength region 570 to 585 nm.

V. SOLVENT AND DYE

Earlier we noted that intrinsic solvent absorption cannot be neglected, as it induces thermal power in the dye solution. This would have to be overcome by the cooling effect of the dye if net cooling were to be observed. The solvent absorption in the wavelength region of interest (500 to 700 nm) is due to overtone vibrational excitation. Most effective in this respect are O—H vibrations, whereas a much smaller absorption is caused by O—D and C—H vibrations. Ideally, the solvent would not contain any H or D atoms [13]. However, a solvent such as carbon tetrachloride (CCl_4) is not suitable, because Rhodamine 6G perchlorate is insoluble in such a nonpolar solvent. As a compromise, monodeuterated ethanol (C_2H_5OD) was chosen. It has sufficiently low absorption in the region around 580 nm (Figure 4) and is a good solvent for the dye. In addition, the amino groups of the dye (Figure 2) are deuterated via H—D exchange with the hydroxyl groups of the solvent. This is important, as it increases the fluorescence quantum yield (Section III). The absorption as well as the fluorescence spectra of Rhodamine 6G are identical in C_2H_5OH and C_2H_5OD.

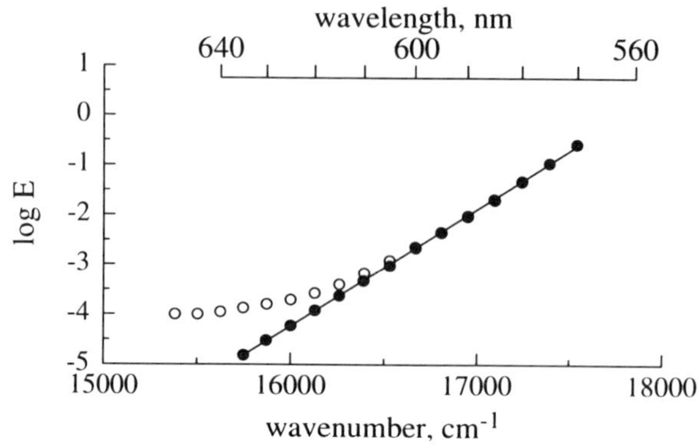

Figure 10. Rhodamine 6G perchlorate in ethanol (4.3×10^{-4} *M*). Absorbance *E* vs. wavenumber. ○, Dye before purification; ●, purified dye; —, expected according to Eq. (3).

The fluorescence quantum yield of Rhodamine 6G is quenched by dissolved oxygen (Section III). Therefore, it is important to remove this gas from the solution. This was done by bubbling with nitrogen for 15 min. The quantum yield of Rhodamine 6G perchlorate in deaerated monodeuterated ethanol has the value 0.990 when excited at 531 nm (Section III). Furthermore, it was shown that the quantum yield of this dye appears to be independent of excitation wavelength, even in the anti-Stokes region (Section IV). However, an extremely high purity is required. This could readily be achieved by column chromatography on silica gel followed by repeated precipitation from ethanolic solution with perchloric acid and water. The result of the purification becomes apparent in the absorption spectrum: The long-wavelength tail of the absorption band follows Eq. (3) exactly up to 640 nm (Figure 10). We estimate the absorption due to impurities in a 3.5×10^{-6} *M* solution of the purified dye to be smaller than 2×10^{-8} (path 1 cm).

For the cooling experiments described in the following section we used a deaerated solution of highly pure Rhodamine 6G perchlorate in monodeuterated ethanol (C_2H_5OD). To prevent as much reabsorption of the fluorescence as possible, a low dye concentration of 3.5×10^{-6} *M* was chosen. The absorption of this solution in the wavelength region of interest is shown in Figure 11. For the absorption measurements as well as for the quantum yield experiments, commercial spectrometer cells have been used. To prevent dye adsorption at the window surfaces the cells were rendered hydrophobic with dichlorodimethylsilane [13, 21].

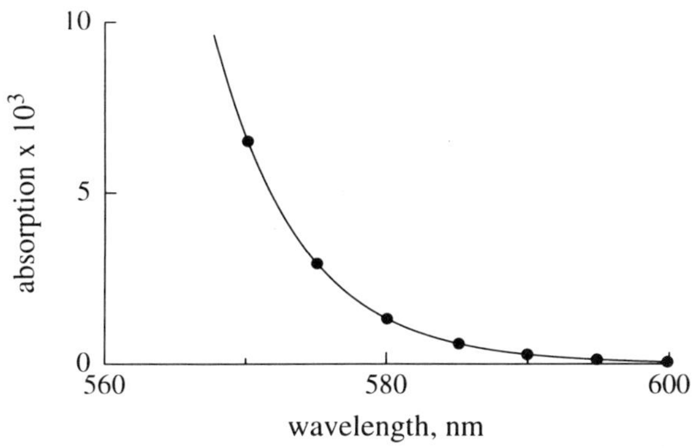

Figure 11. Rhodamine 6G perchlorate (purified) in ethanol. Dye concentration $3.5 \times 10^{-6}\,M$, path length 1 cm. ●, Measured absorption; —, expected according to Eqs. (3) and (4) with $\ln E_0 = -92.1$.

VI. COOLING BY IRRADIATION

To investigate the heat production on anti-Stokes excitation the setup shown in Figure 7 was used. The sample was prepared as described in Section V. It was excited by a Kr-ion pumped dye laser in the region 560 to 600 nm. The excitation power P_L in these experiments varied with wavelength according to the tuning curve of the dye laser (Table 2).

A typical thermal lensing signal $I(t)$ recorded at the detector D_1 (Figure 7) is shown in Figure 12. After opening of the shutter S (Figure 7) at time $t = 0$ the signal (633 nm) increases until the shutter is closed again ($t = 85$ ms). This indicates the formation of a converging lens in the dye solution when it is irradiated by the dye laser. This is opposite to the thermal lens obtained on excitation at 531 nm (Section III) and clearly indicates cooling. After closing of the shutter the signal decreases again due to thermal diffusion (Figure 12).

Because the thermal lensing effect discussed here is very small, it is not practical to use the initial slope of the signal for a quantitative evaluation (cf. Section III). Instead, we determined the change in intensity at the detector D_1 after the stationary state had been reached. This was generally the case at time $t_\infty = 2$ s after opening of the shutter. As the temperature change on irradiation is very small, convective currents in the liquid do not obfuscate the signal in these experiments. It should be noted that all experiments were done at a room temperature of 20°C.

TABLE 2 Thermal Lens Due to Cooling

λ_0 (nm)	$\Delta I/I_0$ (a.u.)	A_d[a] ($\times 10^3$)	P_L (mW)	$-P_{th}$ (μW)	$-P_{th}/P_L$ ($\times 10^{-5}$)
569	1	7.0	13	0.05	0.4
571	3	5.2	23	0.40	2.1
573	6	3.9	32	0.94	2.9
575	8	2.9	38	1.21	3.2
577	10	2.1	45	1.40	3.1
579	9	1.6	51	1.47	2.9
581	8	1.2	55	1.41	2.6
583	7	0.9	57	1.27	2.2
585	6	0.7	58	1.11	1.9
587	5	0.5	56	0.89	1.6

[a]Dye absorption A_d calculated with Eq. (4); concentration $3.5 \times 10^{-6}\,M$, path 1 cm ($\ln E_0 = -92.1$); $\lambda^* = 563$ nm; $\eta_{fl} = 0.990$.

To determine the cooling produced on irradiation, we expand Eq. (6) for $t \to \infty$ and obtain

$$I_\infty = I_0(1 + \Theta) \tag{12}$$

With $I_\infty - I_0 = \Delta I$ it follows for the normalized signal that

$$\frac{\Delta I}{I_0} = \Theta \tag{13}$$

Thus the normalized signal $\Delta I/I_0$ is proportional to the thermal power P_{th} [Eq. (6*b*)]. On the other hand, the thermal power P_{th} can be calculated from [cf. Eq. (9)]

$$P_{th} = P_{abs}\left(1 - \eta_{sol}\frac{\lambda_0}{\lambda^*}\right) \tag{14}$$

If we neglect absorption due to solvent impurities, we may replace P_{abs} by the product $P_L A_d$ and η_{sol} by the quantum yield η_{fl} of the dye. The dye absorption A_d can be calculated with Eq. (4) using $\ln E_0 = -92.1$ for a

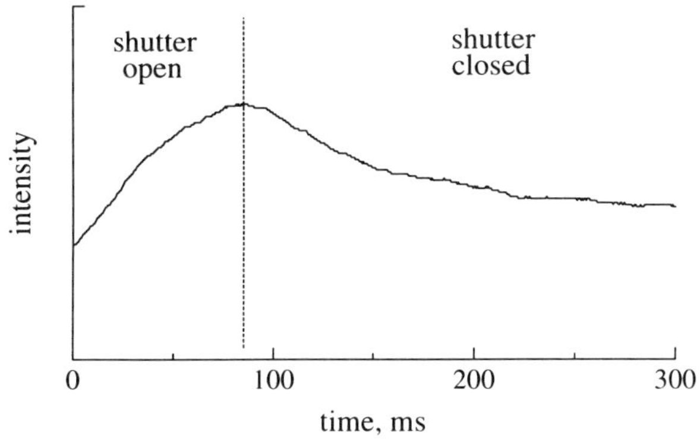

Figure 12. Thermal lensing signal of Rhodamine 6G perchlorate in deaerated monodeuterated ethanol. Intensity increases due to converging lens formed on irradiation at 573 nm.

$3.5 \times 10^{-6}\,M$ solution. For the quantum yield we assume the value $\eta_{fl} = 0.990$, determined for Stokes excitation (Section III). The mean wavelength of fluorescence [Eq. (1*a*)] was obtained as $\lambda^* = 563$ nm.

The thermal power P_{th} calculated with Eq. (14) is given in Table 2 and Figure 13. These data are to be compared with the observed values of the normalized lensing signal $\Delta I/I_0$. The quantity $\Delta I/I_0$ is expected to be proportional to the thermal power P_{th} [Eqs. (6*b*) and (13)]. As can be seen

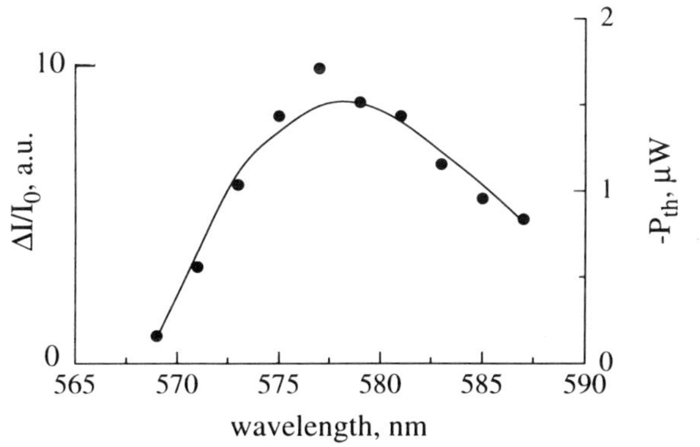

Figure 13. Thermal lens due to cooling. Normalized signal and thermal power expected vs. wavelength of excitation λ_0. ●, $\Delta I/I_0$; —, $-P_{th}$ calculated with Eq. (14).

from Figure 13, there is good agreement. In Table 2 the cooling efficiency $(-P_{th}/P_L)$ is also given. This has a maximum value of 3.2×10^{-5} at the wavelength $\lambda_0 = 575$ nm. With our dye laser we obtained a cooling power of about 1 μW.

Whereas the quantum yield of fluorescence cannot exceed unity ($\eta_{fl} \leqslant 1$), there is no such limitation for the energy yield of fluorescence. This is defined as the ratio of energy emitted to energy absorbed by the dye. The energy yield Y_{fl} is connected with the quantum yield η_{fl} [cf. Eq. (8)] by

$$Y_{fl} = \eta_{fl} \frac{\lambda_0}{\lambda^*} \tag{15}$$

With $\eta_{fl} = 0.990$ and $\lambda^* = 563$ nm we may calculate $Y_{fl}(\lambda_0)$ for the wavelength region of interest and obtain the straight line of Figure 14. For wavelengths of excitation larger than 569 nm an energy yield above unity is obtained.

It is instructive to interpret the experimental results in terms of energy yield Y_{fl}. By combining Eqs. (6*b*), (13), and (14), we obtain

$$\frac{\Delta I}{I_0} = P_{abs}\left(1 - \eta_{fl}\frac{\lambda_0}{\lambda^*}\right) K \tag{16}$$

where the factor K contains thermooptical data of the solvent and is approximately constant ($K < 0$). It then follows from Eq. (15) that

$$\frac{\Delta I}{I_0}\frac{1}{P_{abs}} = -K(Y_{fl} - 1) \tag{17}$$

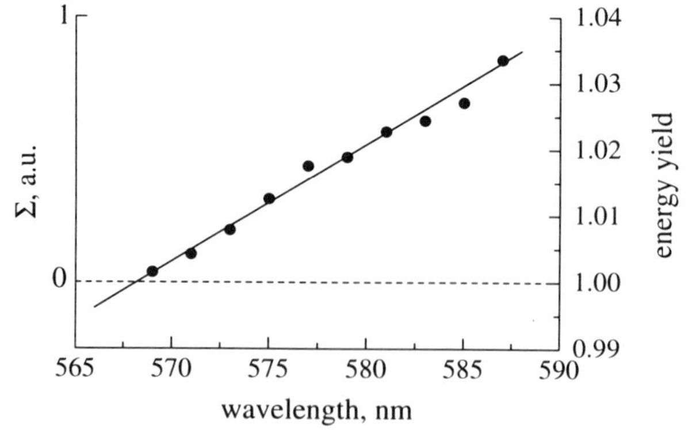

Figure 14. Energy yield vs. wavelength λ_0. ●, Experimental values of Σ; —, Y_{fl} calculated according to Eq. (15) with $\eta_{fl} = 0.990$ and $\lambda^* = 563$ nm.

Therefore, the quantity $\Sigma = f(\Delta I/I_0)(1/P_{\text{abs}})$ is expected to be proportional to $(Y_{\text{fl}} - 1)$, where the factor f may be adjusted such that the slope of Σ matches that of $Y_{\text{fl}}(\lambda_0)$ calculated with Eq. (15). It can be seen from Figure 14 that the values of Σ, obtained from thermal lensing experiments, obey the linear relationship expected theoretically [Eq. (15)] if the quantities η_{fl} and λ^* are independent of wavelength.

VII. CONCLUSIONS

Several conclusions can be drawn from the experiments. Cooling by anti-Stokes fluorescence of a dye solution has been observed apparently for the first time. In contradiction to theoretical considerations by Vavilov [10] an energy yield exceeding unity has been observed and thus does not violate fundamental laws of thermodynamics. However, the cooling effect is very small indeed and therefore difficult to measure, as Landau suspected correctly [6]. To observe anti-Stokes cooling a dye solution with a fluorescence quantum yield very close to unity is required. Furthermore, the dye has to be extremely pure, and the intrinsic solvent absorption needs to be very small. Even so, for 100 mW of incident radiation the cooling power is only on the order of some microwatts.

It follows from the linear dependence between the energy yield and the exciting wavelength that the quantum yield of fluorescence as well as the fluorescence spectrum of the dye Rhodamine 6G is independent of the exciting wavelength even in the anti-Stokes region. At least for this dye there does not exist a reduction of the quantum yield in the anti-Stokes region as postulated by Ketskeméty and Farkas.

Due to the very small dye absorption in the anti-Stokes region the fluorescence quantum yield of dye solutions is markedly reduced by absorption processes not leading to fluorescence. Especially in the far anti-Stokes region the dye absorption becomes comparable to the intrinsic absorption of most solvents. On a particular sample of Rhodamine 6G a decay in the quantum yield with increasing wavelength of excitation was observed. It could be traced to absorption by nonfluorescent impurities. This suggests that the decrease in the quantum yield in the anti-Stokes region found by Ketskeméty and Farkas [11] might be caused by spurious absorption.

Because of the good agreement between the predicted and measured cooling effects, the fluorescence quantum yield of Rhodamine 6G in monodeuterated ethanol has to be very close to the value 0.99 measured by the thermal lensing method. This indicates the high accuracy of this method, which does not require a luminescence standard. In fact, using the thermal lensing method, it should be possible to provide highly accurate luminescence standards for any wavelength region.

ACKNOWLEDGMENTS

We are indebted to Gerhard Deltau for the skillful purification of Rhodamine 6G and thank Dr. Erwin Thiel for many fruitful discussions. Financial support by the Deutsche Forschungsgemeinschaft and the Fonds der Chemischen Industrie is gratefully acknowledged.

REFERENCES

1. T. W. Hänsch and A. L. Schawlow, *Opt. Commun.* **13**, 68 (1975).
2. D. J. Wineland, R. E. Drullinger, and F. L. Walls, *Phys. Rev. Lett.* **40**, 1639 (1978).
3. W. Yu-Zhu, *Chi Kuang* **8**, 10 (1981); *Chem. Abstr.* **95**, 212577e.
4. F. Riehle, T. Kisters, A. Witte, J. Ishikawa, L. L. Ling, and J. Helmcke, *Verhandl. Deut. Phys. Ges.* **25**, 194 (1990).
5. M. A. Kasevich, E. Riis, S. Chu, and R. G. De Voe, *Phys. Rev. Lett.* **63**, 612 (1989).
6. L. Landau, *J. Phys.* **10**, 503 (1946).
7. Y. T. Mazurenko, *Opt. Spectrosc.* **18**, 24 (1965).
8. Y. P. Chukova, *Zh. ETF Pis. Red.* **10**, 458 (1969).
9. Y. P. Chukova, *Zh. Prikl. Spektrosk.* **20**, 412 (1974).
10. S. Vavilov, *J. Phys.* **10**, 499 (1946).
11. I. Ketskeméty and É. Farkas, *Acta Phys. Chem. Szeged* **16**, 77 (1970).
12. P. Pringsheim, *J. Phys.* **10**, 495 (1946).
13. C. Zander, Ph.D. thesis, Universität Siegen, 1991.
14. L. E. Erikson, *J. Lumin.* **5**, 1 (1972).
15. R. Sens, Ph.D. thesis, Universität Siegen, 1984.
16. L. Kozma and Ya. Kheveshi, *Opt. Spectrosc.* **21**, 356 (1966).
17. I. Ketskeméty, L. Kozma, and É. Farkas, *Zh. Prikl. Spektrosk.* **17**, 59 (1972).
18. C. Hu and J. R. Whinnery, *Appl. Opt.* **12**, 72 (1973).
19. J. H. Brannon and D. Magde, *J. Phys. Chem.* **82**, 705 (1978).
20. K. H. Drexhage, *Chapter 4 in Topics in Applied Physics*, Vol. 1, F. P. Schäfer, Ed., Springer-Verlag, Berlin, 1973.
21. H. Hausmann and K. H. Drexhage, unpublished.

ATMOSPHERIC PHOTOCHEMISTRY OF ALTERNATIVE HALOCARBONS

Joseph S. Francisco
Department of Chemistry, Wayne State University, Detroit, Michigan

M. Matti Maricq
Research Laboratory, Ford Motor Company, Dearborn, Michigan

CONTENTS

Advances in Photochemistry, Volume 20, Edited by Douglas C. Neckers, David H. Volman, and Günther von Bünau
ISBN 0-471-11469-3

I. INTRODUCTION

A. Problems Posed by Chlorofluorocarbons

There is a layer of the earth's upper atmosphere, known as the stratosphere, extending from about 10 to 15 km to about 50 km above the earth's surface, in which ozone forms. The peak abundances of this molecule occur between about 20 to 25 km above the surface. Why is ozone so significant? The production of ozone is believed to have been important in protecting living organisms that evolved from the sea onto land. This has allowed life on earth to evolve as we now know it. Ozone shields life from short-wavelength ultraviolet radiation that is damaging to life, a principal manifestation of which in human beings is skin cancer [1]. It has been estimated that a 5% reduction in ozone would amount to about 100,000 additional skin cancer cases a year in the United States [1].

The amount of ozone produced in the stratosphere is maintained by a dynamic balance of production and loss processes. Ozone is produced by the photodissociation of molecular oxygen into oxygen atoms, which then combine with molecular oxygen to form ozone. The ozone produced is stabilized by collisions with a third body. The process by which ozone is destroyed photolytically is the one that shields life from deleterious solar ultraviolet radiation. These steps are illustrated in the following mechanism, known as the *Chapman mechanism*:

$$O_2 + h\nu \rightarrow 2O \tag{1}$$

$$O + O_2 + M \rightarrow O_3 + M \tag{2}$$

$$O_3 + h\nu \rightarrow O + O_2 \tag{3}$$

$$O + O_3 \rightarrow 2O_2 \tag{4}$$

The first step in the production of ozone, the photolysis of molecular oxygen [reaction (1)], is rate limiting. While ozone production is slow, there are chemical reactions that can rapidly destroy it. One of the major species that is efficient in the removal of ozone is chlorine. The role of chlorine species in the depletion of ozone has been investigated actively since 1974, when Rowland and Molina [2] drew attention to the potential impact of human-made materials known as chlorofluorocarbons (CFCs) on ozone produced in the stratosphere. Chlorofluorocarbons are widely used in our daily life as refrigerants, aerosol propellants, cleaning solvents, and in fire-extinguishing applications. CFCs are stable, chemically inert, and have low toxicity. These properties make CFCs ideal for many applications and account for their wide use. However, the release of chlorine from the photodissociation of chlorofluorocarbons poses a central threat to ozone produced in the stratosphere:

$$CX_3Cl + h\nu\,(\lambda \approx 200\text{ nm}) \rightarrow CX_3 + Cl \tag{5}$$

where X represents a combination of fluorine and chlorine substituents. Each chlorine released has been estimated to remove about 10^5 odd-oxygen species (i.e., O atom and ozone) [3, 4]. The reason chlorine is so effective is that it participates in catalytic cycles for the removal of ozone:

$$Cl + O_3 \rightarrow ClO + O_2 \tag{6}$$

$$ClO + O \rightarrow Cl + O_2 \tag{7}$$

$$\text{net:} \quad O + O_3 \rightarrow 2O_2 \tag{8}$$

The chlorine not only destroys the ozone that is produced from reaction (2) in the Chapman mechanism, but it also inhibits the formation of ozone by sequestering oxygen atoms in the form of chlorine monoxide (ClO). Under certain meterological conditions the ClO is itself stored in the form of chlorine nitrate, to be re-released when the meterology changes. This is what makes the ozone depletion process so effective in forming the antarctic ozone hole, for example.

What breaks the cycle? Chlorine atoms released from chlorofluorocarbons can be removed by the reaction with hydrocarbons, principally methane (CH_4),

$$Cl + CH_4 \rightarrow CH_3 + HCl \qquad (9)$$

In the process this reaction produces hydrochloric acid (HCl). The HCl is hygroscopic, and so is soluble in cloud droplets and is usually rained out. The cycle is also broken by the reaction of chlorine oxide with nitrogen dioxide (NO_2) in the atmosphere to make chlorine nitrate ($ClONO_2$):

$$ClO + NO_2 \rightarrow ClONO_2 \qquad (10)$$

However, the removal of chlorine and its reservoir species (HCl and $ClONO_2$) by physical processes is slow. Consequently, chlorine is a serious threat to stratospheric ozone. To amplify this concern, the natural level of chlorine is about 0.6 part per billion (ppb) and results from methyl chloride (CH_3Cl) released from the oceans [3]. As of 1989, the total chlorine level from all known sources was 3.7 ppb [3], nearly six times the natural level of chlorine. Prior to 1987 there were no controls over the production of chlorofluorocarbons from industrial sources; however, in 1987, international agreements were drawn up to limit the production and use of CFCs, with the expected complete phaseout of these materials by the year 2000 [4–6]. Now the question that must be addressed is the following: What are the atmospheric implications of the new materials found to replace the chlorofluorocarbons?

B. Alternative Halocarbons and Scope of the Review

A number of substitute compounds have been identified to have physical and thermophysical properties comparable to the CFCs that would allow their use as CFC replacements. There are two classes of such compounds under consideration: (1) hydrofluorocarbons (HFCs) and (2) hydrochlorofluorocarbons (HCFCs). In the HFCs and HCFCs one or more halogen atoms are replaced with hydrogen atoms; this in principle should make these materials susceptible to degradation in the troposphere. Furthermore, an important consideration regarding the HFCs is that even if they are not completely degraded in the troposphere, they do not provide a chlorine source in the stratosphere.

The partially hydrogenated materials react much more quickly in the troposphere than do their fully halogenated counterparts, because the hydrogen on the halocarbon is readily extracted by reaction with OH radicals. Consequently, the alternative halocarbon could be degraded in the troposphere, presumably before they have an opportunity to reach the stratosphere. This strategy, in principal, should minimize the loss of ozone resulting from alternative halocarbons emitted into the atmosphere. However, this strategy is effective if the lifetimes of HFCs and HCFCs are short compared to their transport into the stratosphere; that is, if the OH hydrogen abstraction rate,

$$R\text{—}H + OH \rightarrow R + H_2O \tag{11}$$

is sufficiently rapid for their removal. However, this is not the only issue. The next issue concerns the acceptability of these materials, in particular: (1) what are the photooxidation products that result from the degradation of the alternative halocarbons; and (2) what role will photooxidation products have on the chemistry of the atmosphere? To evaluate the environmental impact of alternative halocarbons on the chemistry of the atmosphere, these questions need to be addressed.

In this chapter we review recent studies on the atmospheric photooxidation processes of alternative halocarbons and the relevance to the gas-phase chemistry of both the troposphere and stratosphere. Emphasis is placed on kinetic and thermochemical data that relate directly to the atmospheric roles played by HFCs and HCFCs and to the identification of potentially new reactive intermediates and products that may affect current chemistry in the atmosphere. We also review the relationship between CX_3CX_2H and CX_3H degradation chemistry. From the review of the energetics and reaction pathways resulting from alternative halocarbon degradation, we elucidate the role that substituents and their placement in the CX_3CX_2H framework play in controlling the type of degradation by-products resulting from the oxidation process.

II. NOMENCLATURE FOR CFCs, HFCs, AND HCFCs

One of the confusions that the novice first encounters in this area is the association of the names of halocarbon compounds and their chemical formula. The standard nomenclature used is quite different from the IUPAC convention.

Chlorofluorocarbons (CFCs) is the name traditionally given to the group of fully halogenated methanes. Two of the most widely used compounds are CFC-11 ($CFCl_3$) and CFC-12 (CF_2Cl_2). The chemical formula can be obtained from the numerical designation by adding 90 to it and equating the number of fluorine atoms with the units digit, the number of hydrogen atoms with the tens digit, the number of carbon atoms with the hundreds digit, and placing chlorine atoms at the remaining bond sites. Thus, for CFC-12, adding 90 gives 102, whereby number of F atoms = 2, number of H atoms = 0, and number of C atoms = 1, implying the chemical formula CF_2Cl_2.

This nomenclature extends to the alternative compounds termed hydrochlorofluorocarbons (HCFCs), or more simply hydrofluorocarbons (HFCs) for the subset of compounds containing no chlorine atoms. For example, adding 90 to HFC-134a gives 224; thus the compound has four fluorine atoms, two hydrogen atoms, and two carbon atoms, with no sites remaining for chlorine atoms. The four fluorine atoms can be distributed on the two carbons in either a symmetric or an asymmetric fashion. Various isomers are distinguished by attaching a letter (none, a, b, or c) to the numerical designation; thus HFC-134a corresponds to the asymmetric isomer CF_3CFH_2, and HFC-134 labels the compound CF_2HCF_2H.

III. EXPERIMENTAL INVESTIGATION OF THE ATMOSPHERIC DEGRADATION OF HCFCs

A. Overview of Degradation Pathways

1. Atmospheric Chemistry of Hydrocarbons. Hydrocarbon compounds that are released into the troposphere are oxidized by a series of chemical reactions initiated by one or more of a variety of free radicals, such as OH, O_3, and NO_3. The principal reaction pathways are illustrated in Figure 1. An excellent review of the chemistry summarized in this figure has been

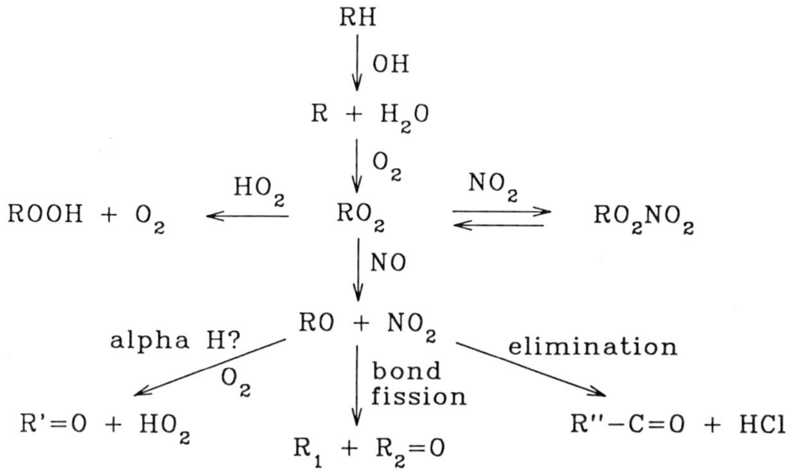

Figure 1. Overview of the initial steps in hydrocarbon atmospheric degradation.

given by Atkinson [7]. For saturated hydrocarbons, including the HCFCs, hydrogen abstraction by hydroxyl radicals [8],

$$RH + OH \rightarrow R + H_2O \tag{12}$$

is the primary initial oxidation step. As is expected for an abstraction reaction, the reaction rates exhibit a sizable positive temperature dependence, due to the activation energy for the hydrogen atom transfer. They are in general relatively slow, with room-temperature rate constants on the order of 10^{-14} to $10^{-15}\,cm^3\,s^{-1}$. An accurate knowledge of these rate constants is very important because, along with physical removal mechanisms such as transport into the stratosphere, they determine the tropospheric lifetimes of the hydrocarbons. Consequently, significant effort has gone into measurements of HCFC reaction rate constants with hydroxyl radicals. For the small but significant fraction of HCFCs that are transported into the stratosphere, photolysis and reaction with $O(^1D)$ also become important removal mechanisms. Besides producing an alkyl radical, the photolysis liberates a chlorine or fluorine atom, the former contributing to the catalytic destruction of stratospheric ozone.

In the presence of molecular oxygen and at a high total pressure ($> \sim 10$ torr), alkyl radicals are rapidly converted into peroxy radicals via the three-body addition reaction [9]

$$R + O_2 + M \rightarrow RO_2 + M \tag{13}$$

At low pressures the addition reaction is in competition with a second, bimolecular, channel which may become increasingly significant at higher temperatures; for example [10],

$$C_2H_5 + O_2 \rightarrow C_2H_4 + HO_2 \tag{14}$$

in the case of the ethyl radical. Under atmospheric conditions (high total pressure and low temperature), however, the second channel is generally of negligible importance and there have been no indications to the contrary for the halogenated alkyl radicals relevant to HCFC degradation. Although alkyl radicals can react rapidly with other radicals, in recombination reactions for example, the overwhelming concentration of O_2 relative to any other reactive species in the atmosphere makes all other alkyl radical reactions of negligible importance. The O_2 concentration is sufficient to guarantee the conversion of an alkyl radical into the corresponding peroxy radical on a submicrosecond time scale; thus, although the temperature and pressure dependencies of the addition reaction are of physical chemical interest, they are of limited importance in terms of atmospheric chemistry.

Peroxy radicals are stable with respect to reaction with the major constituents of the atmosphere. Their chemistry is dominated by reactions with peroxy radicals, HO_2 and RO_2, and with the nitrogen oxides, NO, NO_2, and NO_3. Two recent articles review the spectroscopy, kinetics, and reaction mechanisms of organic peroxy radicals [11, 12]. Three of the more important reaction channels are illustrated in Figure 1. The reaction with NO_2 proceeds exclusively by addition to form the corresponding alkylperoxy nitrate. This reaction is generally not important in determining the end products of the degradation process due to the usually rapid dissociation rate of the peroxy nitrate; however, it can be relevant to dynamical issues, because it sequesters NO_2.

Reaction with HO_2 converts alkyl peroxy radicals to the corresponding hydroperoxides. These reactions are relatively fast, with rate constants on the order of $10^{-12}\,cm^3\,s^{-1}$. Where temperature-dependent studies have been reported, a negative temperature dependence has been found, suggesting that the reaction proceeds via complex formation [12]. There is only a limited number of literature references to studies of $HO_2 + RO_2$ reactions in general, whereas only a few preliminary reports of rate-constant measurements are available for HCFC-based peroxy radicals. There are two loss mechanisms expected for the hydroperoxides [7]. The first is via ultraviolet

(UV) photolysis to produce the corresponding alkoxy radical and OH, and the second is reaction with OH radicals, the major channel of which regenerates the peroxy radical.

In general, the most important peroxy radical reaction is with NO. This reaction proceeds via two channels:

$$RO_2 + NO \rightarrow RO + NO_2 \tag{15}$$

$$RO_2 + NO + M \rightarrow RONO_2 + M \tag{16}$$

Formation of the alkyl nitrate, however, represents a minor channel, which becomes increasingly important for larger parent alkanes. Reaction (15) is typically fast, with rate constants on the order of $10^{-11}\,cm^3\,s^{-1}$. By providing a route for oxidizing NO to NO_2, it is a key step in photochemical smog formation [13]. There is, therefore, a price for substituting HCFCs in place of CFCs. Tropospheric degradation of the replacements reduces the amount transported into the stratosphere, thereby decreasing stratospheric ozone depletion; however, this degradation will contribute an increase (small relative to that from other hydrocarbon emissions) in tropospheric ozone generation.

Another path for peroxy radical removal, not shown in Figure 1, is by reaction with other peroxy radicals (except HO_2, which is considered separately). Due to the low concentrations of atmospheric peroxy radicals relative to NO or HO_2, this pathway is expected to be of minor importance. Reactions between peroxy radicals are usually dominated by two channels [11, 12]:

$$RO_2 + R'O_2 \rightarrow RO + R'O + O_2 \tag{17}$$

$$RO_2 + R'O_2 \rightarrow R_{-H}O + R'OH + O_2 \tag{18}$$

Aside from molecular oxygen, the first channel gives products identical to those obtained from the RO_2 + NO reaction; thus this channel contributes no new qualitative features to the degradation process. When there is an α-hydrogen available, the second channel produces stable molecular products of a ketone, or aldehyde, and an alcohol. These, in turn, react with OH radicals and begin the oxidation process again.

The alkoxy radicals produced from peroxy radical reactions with NO or with each other have, in general, three atmospheric fates [7]: reaction with O_2, dissociation, or isomerization. The reaction with oxygen,

$$RO + O_2 \rightarrow R_{-H}O + HO_2 \tag{19}$$

occurs when an α-hydrogen is available and produces a ketone and a hydroperoxy radical. Reaction of the ketone with OH radicals initiates its further oxidation and, among other possibilities, the HO_2 radical reacts with NO, producing NO_2 and replenishing the OH population.

There often exists a competitive dissociation channel,

$$RO + M \rightarrow R_1 + R_2 = O + M \qquad (20)$$

which creates a smaller alkyl radical and a ketone. The ketone will undergo further oxidation, whereas the alkyl radical will immediately add oxygen to form a new peroxy radical. The competition between reactions (19) and (20) depends on temperature and pressure; therefore, the relative rate k_{dis}/k_{O_2} should decrease with increasing altitude. For larger alkoxy radicals, isomerization may become more important than dissociation, but this has not been found to be the case for HCFCs.

Recently, a fourth reaction pathway has been discovered for some alkoxy radicals [14, 15]. The 1-chloroethoxy radical undergoes three-centered HCl elimination,

$$CH_3CHClO + M \rightarrow CH_3CO + HCl + M \qquad (21)$$

releasing hydrogen chloride and an acetyl radical. Preliminary data show the same elimination process to occur for the chloromethoxy radical, and it seems likely that other alkoxy radicals containing hydrogen and chlorine atoms at the α position will experience a similar fate.

From the preceding discussion it is apparent that the main result of the initial degradation steps is to produce from the parent hydrocarbon a variety of ketones and secondary alkyl radicals. When the parent compound is fully hydrogenated, this process will repeat until finally producing carbon monoxide at the end of the chain, which is then itself oxidized by OH to CO_2. The situation is somewhat different for halogenated alkanes. Substitution by F and/or Cl atoms implies that at some point in the degradation process the fragments will no longer contain a hydrogen atom with which an OH radical can react. As a result, the degradation process outlined above stops at species such as F_2CO, $CF_3C(O)F$, CF_3O, and related compounds. The first two examples are stable molecules and are removed from the troposphere by absorption into cloud water or by transport to the stratosphere. In the first instance, the compounds are hydrolyzed and rained out. In the second case, photolysis can initiate radical chemistry with potentially important implications for ozone depletion, especially for the versions of these molecules that contain chlorine atoms. The CF_3O radical is currently thought to react with H_2O or hydrocarbons to produce CF_3OH, which is hydrolyzed upon uptake by cloud water or undergoes heterogeneous decomposition into HF and F_2CO [16].

2. Introduction to HFC and HCFC Atmospheric Chemistry. Even before Rowland and Molina [2] postulated the mechanism by which photodissociation of CFCs that are transported into the stratosphere could lead to ozone destruction, it was recognized that CFCs would not undergo the usual atmospheric degradation of hydrocarbons and that they had immense atmospheric lifetimes. Because they lacked hydrogen atoms, CFCs were not susceptible to attack by OH radicals, and thus the oxidation process outlined in Section III.A could not begin. The fact that HCFCs retain one or more hydrogen atoms makes them reactive toward OH. In fact, the possibility of their degradation in the troposphere was one of the reasons for choosing them to replace the CFCs.

Although the HCFCs are oxidized in the troposphere, this does not guarantee a particular replacement to be environmentally safe. Aside from the greenhouse threat posed by the HCFCs and the ozone depletion caused by the fraction of HCFCs released into the atmosphere that reaches the stratosphere, another issue must be addressed: that the degradation products, too, must be environmentally benign. Halogen substitution of hydrocarbons alters quantitatively and qualitatively the degradation pathways outlined in the preceding section. In some cases, only the rate constant for a particular reaction changes; in other cases a particular pathway is forbidden or a new one becomes feasible. Because the C—F bond is much stronger than the C—Cl bond, this is often dictated by the distribution of chlorine and hydrogen atoms on the carbon atoms. As will become evident below, it is the distribution of the halogen atoms on the alkoxy radical that is most important in determining the possible outcomes of the degradation process.

Recognizing the importance of the pattern of halogen substitution provides a useful set of criteria by which to classify a HCFC: (1) Does it contain chlorine atoms? (This distinction is already made by the HFC vs. HCFC designation.) (2) Does the alkoxy radical produced by atmospheric degradation (see Figure 1) contain an α-hydrogen? (3) Does the alkoxy radical contain a β-hydrogen? and (4) Does the alkoxy radical contain an α-chlorine? This classification scheme should provide a useful framework within which to view the detailed discussion of HCFC atmospheric degradation process that is presented below.

The first criterion distinguishes those compounds that can release chlorine into the atmosphere from those that cannot. Furthermore, the absence of chlorine reduces the rates of photodissociation of both the HCFC and its oxidation products. The second criterion distinguishes those HCFCs for which the corresponding alkoxy radical can react with O_2 from those for which they cannot. In the latter case C—C bond fission and Cl or HCl elimination remain the feasible options. Reactions with O_2 or Cl elimination lead to the formation of stable aldehydes; C—C bond fission and HCl elimination cause fragmentation of the original two-carbon skeleton into two one-carbon species. In the former case, the third criterion separates those HCFCs that produce trihaloaldehydes from those that do not. In the latter case, a lack of β-hydrogens implies the formation of a fully halogenated radical fragment. In both instances the absence of β-hydrogens leads to species substantially more resistant to further atmospheric degradation than the original HCFC. The last criterion separates those HCFCs for which Cl elimination from the corresponding alkoxy radical is possible from those for which it is not. Using the classification above as a guide, we consider next the atmospheric degradation process in detail.

B. Reactions and Spectroscopy of HCFCs

1. OH + RH. HCFCs are partially halogenated, saturated, alkanes. The initial step in their atmospheric degradation is hydrogen abstraction by OH radicals, reaction (12). Knowledge of the rate constants of these reactions and their temperature dependence is crucial to determining their tropospheric lifetimes and hence the fraction that survives transport into the stratosphere. To this end, there have been numerous investigations of the RH + OH rate constant. The results are collected in Table 1 along with rate constants for some other halogenated ethanes not generally considered HCFCs.

TABLE 1 HFC and HCFC Reactions with OH

		$k_T = A \exp(-E/RT)$		
Reactant	$k_{298} \times 10^{15}$ (cm^3 molecule^{-1} s^{-1})	$A \times 10^{12}$ (cm^3 s^{-1})	E/R (K)	Reference
$CHCl_2CF_2Cl$ (HCFC-122)	53.0 ± 4.1	$1.13^{+0.21}_{-0.16}$	918 ± 52	17
$CHCl_2CF_3$ (HCFC-123)	36.9 ± 3.7	$0.65^{+0.11}_{-0.10}$	840 ± 40	18
	59 ± 6 (303 K)	1.18	900 ± 150	19
	35.2 ± 2.8	1.1 ± 0.2	1040 ± 140	20
	42 ± 2	1.12 ± 0.05	1000 ± 100	21
	35.4 ± 3.4	1.4 ± 0.4	1102^{+157}_{-105}	22
	28.4 ± 4.3	—	—	23

TABLE 1 (Continued)

		$k_T = A\exp(-E/RT)$		
Reactant	$k_{298} \times 10^{15}$ (cm^3 molecule^{-1} s^{-1})	$A \times 10^{12}$ (cm^3 s^{-1})	E/R (K)	Reference
$CHClFCClF_2$ (HCFC-123a)	12.3 ± 1.0	$0.92^{+0.25}_{-0.20}$	1281 ± 85	17
$CHClFCF_3$	9.44 ± 0.75	$0.445^{+0.110}_{-0.088}$	1150 ± 60	18
(HCFC-124)	9.37 ± 0.7	0.613 ± 0.04	1244 ± 90	22
	12.4 ± 1.9	—	—	23
CHF_2CF_3	1.90 ± 0.27	0.541 ± 0.083	1700 ± 100	24
(HFC-125)	2.9 ± 1.0 (303 K)	0.28	1350 ± 100	19
	2.49 ± 0.28	—	—	25
	5.00 ± 1.0	0.17 ± 0.015	1100 ± 100	21
$CH_2ClCClF_2$ (HCFC-132b)	24.2 ± 1.6	2.02 ± 0.24	1263 ± 35	26
CH_2ClCF_3	15 ± 2	$38.9^{+46.2}_{-21.1}$	3200 ± 500	21
(HCFC-133a)	16	1.1 ± 0.3	1260 ± 60	27
	10.5 ± 2.3	—	—	23
CHF_2CHF_2	5.7	1.6	1680	28
(HFC-134)	6.7	2.8	1800	21
CH_2FCF_3	3.8	1.3	1740	28
(HFC-134a)	5.0 ± 0.44	$1.03^{+0.18}_{-0.15}$	1588 ± 52	17
	2.38 ± 0.22 (270 K)	—	—	29
	4.34 ± 0.35	$0.57^{+0.14}_{-0.11}$	1430 ± 60	18
	6.9 ± 1.5 (301 K)	0.58	1350 ± 100	19
	5.18 ± 0.7	3.7 ± 1.5	1990 ± 280	20
	8.44 ± 0.73	1.10 ± 0.11	1424 ± 35	26
	5.15 ± 0.58	—	—	25
	6.9 ± 0.9	$3.2^{+2.3}_{-1.3}$	1800 ± 200	21
CH_3CCl_3	10.6 ± 1.1 (293 K)	5.04 ± 0.96	1797 ± 65	26
$CH_2ClCHCl_2$	318 ± 20 (295 K)	1.65 ± 0.27	483 ± 55	26
CH_3CCl_2F	5.92 ± 0.54	1.47 ± 0.32	1640 ± 100	24
(HCFC-141b)	6.12	1.42 ± 0.60	1623 ± 293	29
	13.6 ± 2.7 (303 K)	0.58	1100 ± 250	19
	7.01 ± 1.2	0.36 ± 0.11	1140 ± 210	20
CH_3CF_2Cl	2.45 ± 0.31 (270 K)	—	—	29
(HCFC-142b)	2.95 ± 0.25	$1.14^{+0.35}_{-0.26}$	1750 ± 75	18
	3.7 ± 1.4 (303 K)	0.26	1230 ± 250	19
	4.02 ± 1.0	0.98 ± 0.28	1660 ± 200	20
	4.6 ± 1.7	—	—	30
	6.7 ± 1.3	$3.31^{+4.3}_{-1.9}$	1800 ± 300	21
	3.4	1.8 ± 0.5	1790 ± 150	27
	3.27 ± 0.58	1.15 ± 0.15	1748 ± 30	31
	2.83 ± 0.42	—	—	23
CH_2FCHF_2	18.3 ± 1.8	—	—	25
(HFC-143)	49.0 ± 5.0	—	—	21
CH_3CF_3	1.35 ± 0.25	2.12 ± 0.65	2200 ± 200	24
(HFC-143a)	1.71 ± 0.43	—	—	25
	1.49 ± 0.10	69.2 ± 12.0	3200 ± 500	21

TABLE 1 (Continued)

		$k_T = A \exp(-E/RT)$		
Reactant	$k_{298} \times 10^{15}$ (cm^3 molecule^{-1} s^{-1})	$A \times 10^{12}$ (cm^3 s^{-1})	E/R (K)	Reference
CH_2FCH_2F (HFC-152)	112 ± 12	—	—	25
CH_3CHF_2 (HFC-152a)	37.6 ± 6.0	$1^{+0.18}_{-0.15}$	980 ± 50	18
	56 ± 9 (303 K)	1.42	1050 ± 250	19
	42.2 ± 4.5	0.96 ± 0.20	940 ± 130	20
	51 ± 5	$2.95^{+1.03}_{-0.76}$	1200 ± 100	21
	37 ± 4	—	—	32
	35 ± 6	—	—	27
	31 ± 7	—	—	23
CH_3CH_2F (HFC-161)	37.0 ± 3.7	—	—	32
C_2H_5Cl	393 ± 53	—	—	30
	390 ± 70	—	—	23
$CHCl_2F$ (HCFC-21)	35.2 ± 4.4	1.21 ± 0.13	1060 ± 40	33
	33.9 ± 8.6	—	—	30
	39 ± 2	$4.79^{+0.96}_{-0.81}$	1400 ± 100	21
	28.8 ± 2.4	1.87 ± 0.2	1245 ± 26	31
	26 ± 4	—	—	34
$CHClF_2$ (HCFC-22)	4.9 ± 0.45	$0.81^{+0.15}_{-0.12}$	1516 ± 53	17
	4.75 ± 0.63	1.28 ± 0.33	1670 ± 96	33
	4.58 ± 0.58	—	—	30
	4.1 ± 0.9	$9.55^{+1.65}_{-1.42}$	2300 ± 200	21
	5.3	2.1 ± 0.6	1780 ± 150	27
	4.8 ± 0.46	0.925 ± 0.10	1575 ± 71	31
	3.4 ± 0.7	—	—	34
CHF_3 (HFC-23)	0.17	3.02 ± 0.29	2910 ± 45	33
	0.35 ± 0.17	—	—	32
	$0.2^{+0.2}_{-0.15}$	—	—	34
CH_2ClF (HCFC-31)	49.4 ± 6.0	2.44 ± 0.32	1150 ± 45	33
	44.5 ± 6.6	—	—	30
	37	3.1 ± 0.9	1320 ± 100	27
	42.1 ± 4	2.84 ± 0.3	1259 ± 50	31
	37 ± 6	—	—	34
CH_2F_2 (HFC-32)	11.3 ± 1.0	1.57 ± 0.21	1470 ± 100	24
	11.2 ± 0.75	4.40 ± 0.37	1770 ± 30	33
	7.2 ± 1.0	7.41 ± 0.97	2100 ± 200	21
	11.7 ± 1.4	—	—	32
	7.8 ± 1.2	—	—	34
CH_3F (HFC-41)	14.0 ± 1.8	0.818 ± 0.066	1887 ± 30	33
	21.8 ± 1.8	—	—	32
	16 ± 3.5	—	—	34

The rate constants are rather slow, ranging from 1.7×10^{-16} cm^3 s^{-1} for HFC-23 [33] to $(1.12 \pm 0.12) \times 10^{-13}$ cm^3 s^{-1} for HFC-152 [25]. Rate constants for two of the chloroethanes listed are about three times higher than the upper end of this range. An examination of Table 1 reveals a correlation between increasing fluorine vs. chlorine substitution and decreasing rate constant for OH attack. For example, in the series of compounds CHF_nCl_{3-n} the rate constant decreases by almost two orders of magnitude from about 3.5×10^{-14} cm^3 s^{-1} for $n = 1$ to 2.5×10^{-16} cm^3 s^{-1} for $n = 3$. The trend is the same, but the effect is smaller, when the halogen atoms are beta to the hydrogen atoms; thus the rate constants for $CH_3CF_nCl_{3-n}$ decrease from 1.06×10^{-14} cm^3 s^{-1} for $n = 0$ to about 1.5×10^{-15} cm^3 s^{-1} for $n = 3$. These effects can be traced to the effect that fluorination has on the C—H bond strength and hence the activation energy of the hydrogen atom abstraction.

Where multiple measurements of the same rate constant have been made, they show reasonably good agreement for the faster reactions, but have variations by more than a factor of 2 for some of the slower halogenated ethane reactions. For example, the rate constants for OH + HFC-152a range from $(3.1 \pm 0.7) \times 10^{-14}$ cm^3 s^{-1} [23] to $(4.2 \pm 0.45) \times 10^{-14}$ cm^3 s^{-1} [20], not including the outlying value recorded by Brown et al. [19]. In contrast, the rate constants for OH + HFC-134a and OH + HCFC-142b show considerably more variation; from 3.8×10^{-15} cm^3 s^{-1} [28] to $(8.44 \pm 0.73) \times 10^{-15}$ cm^3 s^{-1} [26] for HFC-134a and $(2.83 \pm 0.42) \times 10^{-15}$ cm^3 s^{-1} [23] to $(6.7 \pm 1.3) \times 10^{-15}$ cm^3 s^{-1} [21] for HCFC-142b. An examination of the references for these rate constants shows that the measurements divide into two groups: those reported between 1976 and 1983, and the more recent studies spurred by discovery of the antarctic ozone hole and published since 1989. Generally, the larger rate constants were reported in the earlier studies. As the recent measurements were made, it became clear that small amounts of impurities, usually halogenated alkenes, could lead to erroneously large rate constants. Consequently, in the studies published since 1989, considerable care was taken to obtain pure HCFC samples and to report a quantitative analysis of the impurities. This resulted in lower values for the measured rate constants and smaller discrepancies between the various research groups. The reasons for the residual discrepancies are not well understood. Recommended rate constant expressions for most of the HCFC + OH reactions listed in Table 1 are available in recent rate-constant compilations and we defer to these [9, 35].

2. Cl, F + RH. Although atmospherically important, the RH + OH reaction is often too slow and the OH radical too difficult to generate for this reaction to be useful for producing alkyl radicals in laboratory studies. Instead, many studies use the photodissociation of Cl_2 or F_2 and generate the radicals via

$$Cl_2 + h\nu, \text{discharge} \rightarrow 2Cl \quad (22)$$

$$Cl + RH \rightarrow R + HCl \quad (23)$$

or the analogous fluorine reactions. In many situations, even a low concentration of the parent, unsubstituted, alkane is sufficient to make the alkyl radical formation nearly instantaneous on the time scale of its further reactions; hence a precise value of the rate constant for reaction (23) is not necessary for kinetic modeling purposes. However, with regard to the HCFCs, chlorine attack, and even the reaction with fluorine atoms, is often relatively slow and needs to be included in the kinetic modeling. More important, the rate constants are required to establish upper limits to hydrocarbon impurities in HCFC sample, which, because of their much faster reaction with chlorine, can be tolerated without interfering with the desired measurement.

Tables 2 and 3 collect the available rate constant data for Cl + HCFC and F + HCFC reactions, respectively. The rate constants for chlorine attack vary over a wide range, from $(2.5 \pm 0.83) \times 10^{-17}\ cm^3\ s^{-1}$ for CF_3CH_3 [40] to $(8.04 \pm 0.54) \times 10^{-12}\ cm^3\ s^{-1}$ for CH_3CH_2Cl [36]. Where multiple measurements of the same rate constants are available they are generally in good agreement. However, a trend is noticeable between the rate constants reported by Tuazon et al. [37] and Wallington and co-workers [38], on one hand, and Tschuikow-Roux and co-workers [40], on the other. All three groups measured the rates relative to the reference reaction $Cl + CH_4 \rightarrow CH_3 + HCl$. While the first two groups are in very good agreement with each other, their rate constants are generally smaller than those from the latter group, possibly indicating a systematic error arising from one or both experimental methods.

TABLE 2 HFC and HCFC Reactions with Cl

		$k_T = A \exp(-E/RT)$		
Reactant	$k_{298} \times 10^{15}$ ($cm^3\ molecule^{-1}\ s^{-1}$)	$A \times 10^{11}$ ($cm^3\ s^{-1}$)	E/R (K)	Reference
C_2H_5Cl	7300 ± 900			14
	8700 ± 1000			15
	8040 ± 570	2.34 ± 0.42	310 ± 56	36
$CHCl_2CF_3$	14 ± 3			37
(HCFC-123)	12.2 ± 1.8			38
$CHClFCF_3$	2.7 ± 0.6			37
HCFC-124)				

TABLE 2 (Continued)

Reactant	$k_{298} \times 10^{15}$ (cm^3 molecule^{-1} s^{-1})	$k_T = A\exp(-E/RT)$ $A \times 10^{11}$ (cm^3 s^{-1})	E/R (K)	Reference
CHF_2CF_3	0.24 ± 0.05			37
(HFC-125)				
CHF_2CHF_2	1.9 ± 0.2			39
(HFC-134)	3.27 ± 1.09	1.39 ± 0.31	2490 ± 73	40
CH_2FCF_3	1.6 ± 0.3			41
(HFC-134a)	1.6 ± 0.3			37
	1.38 ± 0.18			38
CH_3CFCl_2	2.1 ± 0.2			41
(HCFC-141b)	2.4 ± 0.5			37
	2.04 ± 0.22			38
CH_3CF_2Cl	0.56 ± 0.20			41
(HCFC-142b)	0.37 ± 0.08			37
	0.390 ± 0.052			38
	0.53 ± 0.11			42
$CH_2FC\underline{H}F_2$	24 ± 7	0.825 ± 0.16	1736 ± 69	40
$C\underline{H}_2FCHF_2$	23 ± 7	1.15 ± 0.23	1845 ± 69	40
(HFC-143)				
CH_3CF_3	0.0249 ± 0.0083	1.73 ± 0.38	4009 ± 74	40
(HFC-143a)				
CH_2FCH_2F	670 ± 80			43
(HFC-152)	1377 ± 440	5.73 ± 1.24	1111 ± 71	40
CH_3CHF_2	240 ± 50			37
(HFC-152a)	239 ± 7			38
	270 ± 50			44
$CH_3C\underline{H}F_2$	429 ± 133	1.27 ± 0.27	1011 ± 70	40
$C\underline{H}_3CHF_2$	4.13 ± 1.30	1.42 ± 0.29	2427 ± 71	40
$CH_3C\underline{H}_2F$	6510 ± 2020	2.64 ± 0.53	417 ± 70	40
$C\underline{H}_3CH_2F$	690 ± 214	2.08 ± 0.42	1015 ± 70	40
(HFC-161)				
$CHFCl_2$	21 ± 4			37
(HCFC-21)	9.3			45
CHF_2Cl	1.7 ± 0.2			41
(HCFC-22)	2.0 ± 0.4			37
	1.4 ± 0.2			42
CH_2FCl	110 ± 25			37
(HCFC-31)	191 ± 57	3.16 ± 0.61	1522 ± 68	46
CH_2F_2	68.3 ± 20.8	2.49 ± 0.49	1758 ± 69	40
(HFC-32)				
CH_3F	340 ± 70			37
(HFC-41)	324 ± 51			47
	421 ± 127	3.78 ± 0.75	1340 ± 68	40
	470			48
	359	0.479 ± 0.11	772 ± 54	49

TABLE 3 HFC and HCFC Reactions with F

Reactant	$k_{298} \times 10^{13}$ (cm^3 $molecule^{-1}$ s^{-1})	Reference
$CHCl_2CF_3$ (HCFC-123)	12 ± 4	50
CHF_2CF_3	3.5 ± 0.8	50
(HFC-125)	3.3 ± 1.0	51
CHF_2CHF_2 (HFC-134)	10 ± 3	50
CH_2FCF_3	13 ± 3	50
(HFC-134a)	14 ± 3	50
	21 ± 7	52
CH_3CF_2Cl (HCFC-142b)	39 ± 9	50
CH_2FCH_2F (HFC-152)	380 ± 110	43
CH_3CHF_2 (HFC-152a)	170 ± 40	50
CHF_3	1.4 ± 0.4	50
(HCFC-23)	1.2 ± 0.3	53
	1.2 ± 0.2	54
CH_2ClF (HCFC-31)	78 ± 4	55
CH_2F_2	43 ± 9	50
(HFC-32)	100 ± 40	56
	57 ± 3	55
	40 ± 16	57
CH_3Cl	330 ± 70	50
	240 ± 50	58
	310 ± 90	59
CH_3F	370 ± 80	50
(HFC-41)	270 ± 20	57

There are a few trends in the rate constants that become apparent from an examination of Table 2. Successive halogen atom substitution decreases the rate constant for hydrogen atom abstraction by Cl atoms. The decrease is more pronounced at the β than at the α position and it is more pronounced for fluorine than for chlorine substitution. A comparison of C—H bond energies to activation energies shows a clear correlation between the two quantities [40]. For the successive fluorination of methane, the C—H bond energy decreases after the substitution of one fluorine atom, but then increases for subsequent fluorine substitutions. In the case of chlorinated methanes, both the C—H bond energy and the activation energy for hydrogen-atom abstraction decrease with increasing chlorine substitution. A number of plausible explanations for the trends in reaction rate and bond energy have been put forward, the most promising of which is that the trends arise from a combination of inductive and resonance effects [40]. The inductive effect, which increases with the number of halogen substituents, lowers the reactivity of both the α and β hydrogens toward chlorine attack, with the effect larger for the α-hydrogen. The resonance effect,

$$R—CH—X \rightleftarrows R—CH{=}X$$

lowers the energy of the corresponding alkyl radical, thereby increasing the reactivity of the α-hydrogen. Both the inductive and resonance effects are larger for the more electronegative fluorine atom; thus the more pronounced effect of fluorine vs. chlorine substitution. The combination of effects explains both the overall decrease in reactivity with increasing halogen substitution and the higher reactivity of the α vs. β hydrogens.

Compared to the reactions with chlorine atoms, considerably fewer data are available for the F + HCFC reaction. Most of the rate constants were derived by the relative rate method by Wallington et al. [50]. Where other measurements are available, they are usually in good agreement, with the exception of the overly large rate constant reported by Nielsen et al. [56] for CH_2F_2 and the comparatively low values reported by Wickramaaratchi et al. [58] and Manocha et al. [57] for CH_3Cl and CH_3F, respectively. Owing to the higher reactivity of fluorine vs. chlorine atoms, the hydrogen-atom abstraction reactions listed in Table 3 for fluorine attack are much faster than their counterparts in Table 2 for chlorine-atom attack. However, these rate constants also show the trend that the rate constant decreases as halogen substitution increases.

3. $O(^1D)$ + RH. Once HCFCs are transported into the stratosphere, as a fraction of them will be, reaction with $O(^1D)$ atoms,

$$O(^1D) + RH \rightarrow \text{products} \tag{24}$$

provides another mechanism for their removal besides OH attack. In addition to a direct abstraction channel, two other processes are possible:

$$O(^1D) + RH \rightarrow O(^3P) + RH \tag{25}$$

$$O(^1D) + RH \rightarrow O(^3P) + [RH]^* \rightarrow O(^3P) + \text{products} \tag{26}$$

Reaction (25) represents collisional relaxation of the excited oxygen atom. While halogenated hydrocarbons form too small a fraction of the available collision partners to be of atmospheric consequence for the relaxation channel, this pathway must be considered for laboratory experiments relying on $O(^1D)$ or $O(^3P)$ vs. time profiles to deduce the rate constant of reaction (24). In reaction (26), a sizable fraction of the 190 kJ mol^{-1} excess energy of $O(^1D)$ is transferred to internal excitation of the HCFC, which then can dissociate to products. Between reactions (24) and (26) a range of products is possible, including OH + R, ClO + R′, and $O(^3P)$ + HCl + chlorofluoroalkene.

Data for the reaction of $O(^1D)$ with HCFCs are scarce. We are unaware of any published product studies. Rate constants for reaction (24), provided primarily by two studies, are available for a variety of HCFCs, but only at 298 K. Table 4 collects the measured rate constants, giving values based on the removal of $O(^1D)$, which represents the sum of the rate constants for collisional relaxation and reaction, and giving rate constants for the removal of the HCFC. The rate constants reported by Green and Wayne [61] were measured relative to the reaction $O(^1D) + N_2O \rightarrow$ products, using a rate constant of 2.2×10^{-10} cm^3 s^{-1} for the latter reaction. These were subsequently corrected for the currently accepted value of 1.16×10^{-10} cm^3 s^{-1} by Warren et al. [60]. The corrected values are listed in Table 4.

Unlike the reactions with OH, the $O(^1D)$ + RH reaction is fast, with rate constants on the order of 10^{-10} cm^3 s^{-1} for most of the HCFCs. With the exception of CH_3CF_2Cl, the rate constants measured by Green and Wayne [61] are consistently higher than those determined by Warren et al. [60], although the difference is often near the limit of the combined error bars. The rate constant determined by Pitts et al. [64] for CHF_2Cl appears to be inordinately high. In contrast to the OH + RH reactions, which are slow, the $O(^1D)$ reactions should be insensitive to minor impurities; thus these cannot explain differences in the reported rate constants, and their origin remains unclear.

The rate constants for HCFC removal show a correlation with respect to the chlorine content of the molecule. For molecules of the form $RCX_{3-n}Cl_n$, where R = H, CH_3 or CF_3 and X = H, F or a combination of the two, the reaction rate increases with n. For example, the reaction of $O(^1D)$ with HCFC-141b is faster than the reaction with HCFC-142b, and the reaction with HCFC-21 is faster than that with HCFC-22. This is consistent with an abstraction mechanism in which chlorine atom abstraction is preferred because of the weaker C—Cl bond.

TABLE 4 HFC and HCFC Reactions with O(^{1}D)

Reactant	$k_{total} \times 10^{10}$ (cm^3 molecule^{-1} s^{-1})	$k_{HCFC\ rem} \times 10^{10}$ (cm^3 molecule^{-1} s^{-1})	Reference
$CHCl_2CF_3$	1.98 ± 0.15	1.56 ± 0.19	60
(HCFC-123)		2.2 ± 0.35^a	61
$CHClFCF_3$	0.86 ± 0.04	0.59 ± 0.10	60
(HCFC-124)			
CHF_2CF_3	1.23 ± 0.06	$0.18^{+0.15}_{-0.22}$	60
(HFC-125)		0.46 ± 0.12^a	61
$CH_2ClCClF_2$		1.62 ± 0.35^a	61
(HCFC-132)			
CH_2ClCF_3	1.20 ± 0.06	$0.96^{+0.04}_{-0.08}$	60
(HFC-133a)		1.51 ± 0.35^a	61
CH_2FCF_3	0.485 ± 0.025	0.03 ± 0.03	60
(HFC-134a)			
CH_3CFCl_2	2.58 ± 0.20	1.78 ± 0.16	60
(HCFC-141b)			
CH_3CF_2Cl	2.15 ± 0.20	1.59 ± 0.18	60
(HCFC-142b)		1.39 ± 0.35^a	61
CH_3CF_3		0.58 ± 0.12^a	61
(HFC-143a)			
CH_3CHF_2	2.02 ± 0.15	0.93 ± 0.16	60
(HFC-152a)			
$CHFCl_2$	1.9 ± 0.6		62
(HCFC-21)	2.0 ± 0.1^b		63
CHF_2Cl	1.08 ± 0.20	0.78 ± 0.16	60
(HCFC-22)	0.95 ± 0.30		62
		0.93 ± 0.23^a	61
	0.98 ± 0.06^b		63
		1.9 ± 0.2^a	64
CHF_3	0.40 ± 0.03^b		63
(HFC-23)			
CH_2F_2		0.46 ± 0.23^a	61
(HFC-32)			

[a]Ref. 61 using the currently recommended value of the rate coefficient for O(^{1}D) + N_2O of 1.16×10^{-10} cm^3 molecule^{-1} s^{-1}.
[b]Corrected for $\gamma = 1$ as referenced in Ref. 63.

4. HCFC Ultraviolet Absorption Spectra. While the presence of hydrogen atoms makes HCFCs susceptible to OH attack in the troposphere, this reaction is sufficiently slow that some transport of the HCFCs into the stratosphere will occur. For example, the tropospheric lifetime, with respect to reaction with OH, of CF_3CCl_2H is approximately 1.6 years, whereas that of CF_2ClCH_3 is about 25 years [65]. Once the HCFCs have reached the stratosphere, their photolysis by ultraviolet (UV) radiation in the range 190 to 230 nm releases active chlorine to participate in ozone destruction. The photolysis is in competition with chemical reaction of the HCFCs with stratospheric OH radicals and $O(^1D)$ atoms; therefore, measurements of the UV absorption cross sections are required to ascertain the ozone-depletion potentials of the HCFCs.

Figure 2 illustrates the UV spectra of HCFC-123, HCFC-124, and HCFC-142b, which are taken from the recent work of Orlando et al. [65] and Gillotay and Simon [66]. While spectra recorded by these two groups are in good agreement, they differ significantly from the earlier measurements of Green and Wayne [67] and Hubrich and Stuhl [68]. The disagreement is due, at least in part, to sample impurities. The spectra in Figure 2 are consistent with those of other chlorinated alkanes. The absorption is due to the long-wavelength tail of the $n(C) \rightarrow \sigma^*(C—Cl)$ transition, which for methyl chloride [69] peaks at 173 nm. The absorption cross sections on the long-wavelength side of the band decrease with decreasing temperature, as is expected from a lowering of the populations in rotationally and vibrationally excited states of the ground electronic state. Orlando et al. [65] also examined the UV spectra of HFC-134a and HFC-152a and found these species to have absorption cross sections $<3 \times 10^{-24}$ cm^2 between 190 and 230 nm. This is expected from the fact that the $n(F) \rightarrow \sigma^*(C—F)$ transition lies at a shorter wavelength than that of the analogous transition in chloroalkanes.

The photolysis rate of HCFC-123 is about two orders of magnitude larger than those of HCFC-124 and HCFC-142b and is comparable to that of CFC-12. For HCFC-123, photolysis represents the major removal pathway in the stratosphere. In the case of HCFC-124 and HCFC-123, photolysis is less important than loss by reaction with OH and $O(^1D)$.

C. O_2 Addition to Halogenated Alkyl Radicals

Once formed, an alkyl radical reacts rapidly with O_2 to form the corresponding peroxy radical. With high-pressure limiting rate constants in the range of 10^{-12} cm^3 s^{-1}, the atmospheric lifetime of an alkyl radical is expected to be on the order of 1 μs or less, and no other removal mechanism

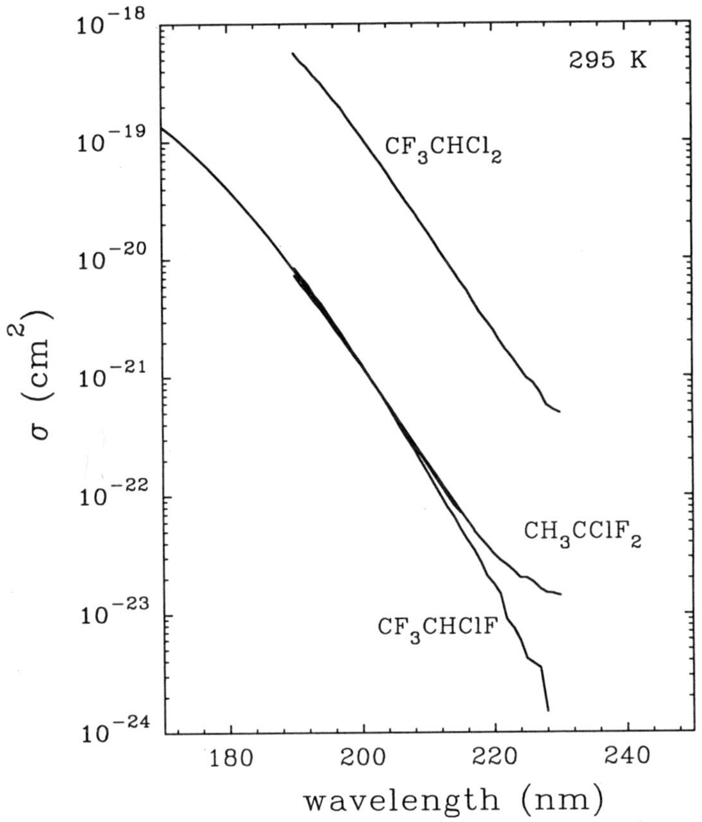

Figure 2. UV absorption cross sections for some HCFCs. (Data from Refs. 65 and 66.)

is expected to compete with the O_2 addition. Despite fundamental interest from a physical chemistry standpoint, details of the temperature and pressure dependence of the R + O_2 reaction is, therefore, of little atmospheric significance. This is probably the explanation for the paucity of data concerning O_2 addition to HCFC-based alkyl radicals.

The oxygen addition reaction rate depends on the amount and nature of a third body, which collisionally stabilizes the adduct. Rate constants for such reactions are conventionally expressed within the Troe formalism [70] by

$$k = \frac{k_0[M]k_\infty}{k_0[M] + k_\infty} F_c^{(1 + [\log_{10}(k_0[M]/k_\infty)/N(F_c)]^2)^{-1}} \quad (27)$$

where k_0 represents the low-pressure third-order rate constant, k_∞ is the high-pressure limiting bimolecular rate constant, $N(F_c) = 0.75 - 1.27 \log F_c$, and F_c controls the broadening of the falloff curve. Where possible, we adopt here the approximation made in the evaluation of DeMore et al. [35] that $F_c = 0.6$ and $N(F_c) = 1$.

To our knowledge rate constants have been measured for O_2 addition to CCl_3, $CFCl_2$, CF_3, and CF_3CFH. Ryan and Plumb [71] measured a three-body rate constant of $k_0 = 5.8 \times 10^{-31}$ cm^6 s^{-1} for the reaction between CCl_3 and O_2 in He and extrapolated this to a high-pressure limit of $k_\infty = 2.5 \times 10^{-12}$ cm^3 s^{-1}. As expected, the rate constant increases when the more efficient N_2 is the bath gas. Danis et al. [72] determined a third-order rate constant of $k_0 = (1.6 \pm 0.3) \times 10^{-30} (T/298)^{-6.3}$ cm^6 s^{-1} in nitrogen and a high-pressure limit of $k_\infty = (3.2 \pm 0.7) \times 10^{-12} (T/298)^{-1.2}$ cm^3 s^{-1}. The latter value is intermediate between the rate constant of $1.44 \times 10^{-12} (T/298)^{-3.6}$ cm^3 s^{-1} found by Russel et al. [73] in 760 torr N_2 and the value $(5.13 \pm 0.51) \times 10^{-12}$ cm^3 s^{-1} reported by Cooper et al. [74] in 700 torr Ar.

The oxygen addition rate increases when one or more of the chlorine atoms are replaced by fluorine. The three-body and high-pressure limiting rate constants for $CFCl_2 + O_2$ are $k_0 = (5.0 \pm 0.8) \times 10^{-30}$ cm^6 s^{-1} and $k_\infty = (6 \pm 1) \times 10^{-12}$ cm^3 s^{-1}, respectively [75]. For $CF_3 + O_2$, the rate increases further. The third-order and limiting rate constants are $k_0 = (1.8 \pm 0.4) \times 10^{-29}$ cm^6 s^{-1} and $k_\infty = 8 \times 10^{-12}$ cm^3 s^{-1} for a He buffer [76] and $k_0 = (1.9 \pm 0.4) \times 10^{-29} (T/298)^{-4.7}$ cm^6 s^{-1} and $k_\infty = (9 \pm 2) \times 10^{-12}$ cm^3 s^{-1} in a N_2 bath gas [77]. The effect of replacing chlorine by fluorine is rather marked, with a threefold increase in the third-order rate constant in N_2 for the substitution of one fluorine atom and a factor of 12 increase upon replacing all three chlorine atoms. This can be rationalized on the basis that fluorine substitution increases the C—O bond strength; for example, the bond energy decreases from 155 kJ mol^{-1} for CF_3—O_2 to 128 kJ mol^{-1} for CH_3—O_2 [78].

A high-pressure rate constant of $(2.1 \pm 0.5) \times 10^{-12}$ cm^3 s^{-1} was measured for the $CF_3CFH + O_2$ reaction at total pressures between 200 and 600 torr N_2 [79], although this limit is probably reached at a considerably lower total pressure (for C_2H_5 the high-pressure limit of 9×10^{-12} cm^3 s^{-1} is reached at a few torr [80]). Although smaller than the corresponding rate constants for the halogenated methyl radicals discussed above and that of the ethyl radical, this value is expected to be typical of the halogenated ethanes. For comparison, a solution-phase rate constant of 2.2×10^{-12} cm^3 s^{-1} has been reported [81] for O_2 addition to CF_3CClH.

Although not an alkyl radical, FCO is an important intermediate species in the degradation of certain HCFCs, which adds O_2 to form $FC(O)O_2$. The addition reaction has a rate constant of $(8.0 \pm 0.8) \times 10^{-13}\ cm^3\ s^{-1}$ in 300 torr of N_2, which is independent of temperature in the range 213 to 358 K [82]. Given the small size of the FCO radical, it is unlikely that this rate constant represents the high-pressure limit. In 760 torr of SF_6 the rate constant is somewhat larger, $(1.2 \pm 0.2) \times 10^{-12}\ cm^3\ s^{-1}$, as expected at the higher total pressure and given that SF_6 is a more efficient collision partner than N_2 [83].

D. Haloalkylperoxy Radical Spectra and Reactions

The peroxy radical represents the first point along the atmospheric degradation pathway at which qualitative differences in the oxidation chemistry appear. The relative stability of peroxy radicals makes them unreactive toward the major constituents of the atmosphere; thus they are removed principally by reactions with NO, NO_2, HO_2, and other peroxy radicals, of which CH_3O_2 is the major component. Peroxy radicals exhibit a strong characteristic UV absorption band, typically in the range 200 to 250 nm, which makes them amenable to experimental investigation. Because they represent important intermediates in tropospheric chemistry and since they are relatively easy to study, peroxy radicals have received a good deal of attention. Reviews of their chemistry and spectroscopy have recently been compiled by Lightfoot et al. [12] and Wallington et al. [11]. Owing to their importance in the issue of CFC replacements, however, a significant amount of work on halogenated peroxy radicals has appeared since these reviews.

1. UV Spectroscopy. When an alkyl radical, with an unpaired electron in a sp_3 orbital, approaches an oxygen molecule, which has two unpaired π^* electrons, the overlap of the alkyl sp_3 orbital with the in-plane oxygen π and π^* orbitals forms σ_{CO} bonding, σ^*_{CO} antibonding, and oxygen lone-pair orbitals, as shown by the energy-level diagram in Figure 3. This overlap produces a σ C—O bond at the expense of reducing the O—O bond from double to 1.5 in bond order. The resulting peroxy radical has overall C_s or no symmetry, depending on the symmetry of the alkyl group itself. The unpaired electron occupies an out-of-plane oxygen π^* orbital, giving the ground electronic state $^2A''$ symmetry.

Calculations have been performed by Jafri and Phillips [84] to examine the low-lying electronic states of CH_3O_2. These reveal that the $1^2A'$ excited state, obtained by promotion of an in-plane oxygen lone-pair electron to the out-of-plane π^* orbital, is bound and lies in the near infrared, about 87 kJ mol^{-1} above the ground state. The $2^2A''$ state, which is reached by an oxygen $\pi \rightarrow \pi^*$ transition, lies in the UV and has been assigned to the

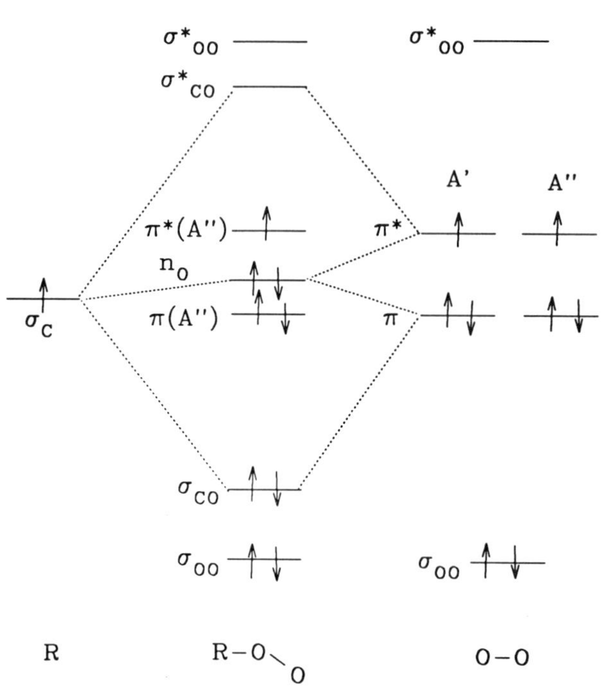

Figure 3. Energy-level correlation diagram for the reaction between an alkyl radical and O_2. For simplicity, we have retained the symmetry designations of the oxygen orbitals that do not participate in the R—O bond.

CH_3O_2 absorption band observed at 238 nm. The state has only a weak minimum, if any, in the O—O coordinate, implying rapid dissociation of the excited state, which is consistent with the broad featureless appearance of the absorption band.

The UV spectra of peroxy radicals are very similar to each other, as evident from the HCFC-derived peroxy radicals illustrated in Figures 4 to 11. All the spectra are broad and featureless with $\lambda_{max} \cong 200$ to 250 nm and $\sigma_{max} \cong 4 \times 10^{-18}$ cm^2. The similarity exists because the $X^2A'' \rightarrow 2^2A''$ transition involves orbitals localized on the O—O moiety and because the upper state is dissociative. The variations that occur in the position of λ_{max} arise from changes in the structure of the alkyl group, such as the substitution of halogen atoms, that effect secondary shifts in the energies of the oxygen orbitals.

The majority of peroxy radical UV line shapes show a Gaussian dependence of absorption cross section on the logarithm of the wavelength. As

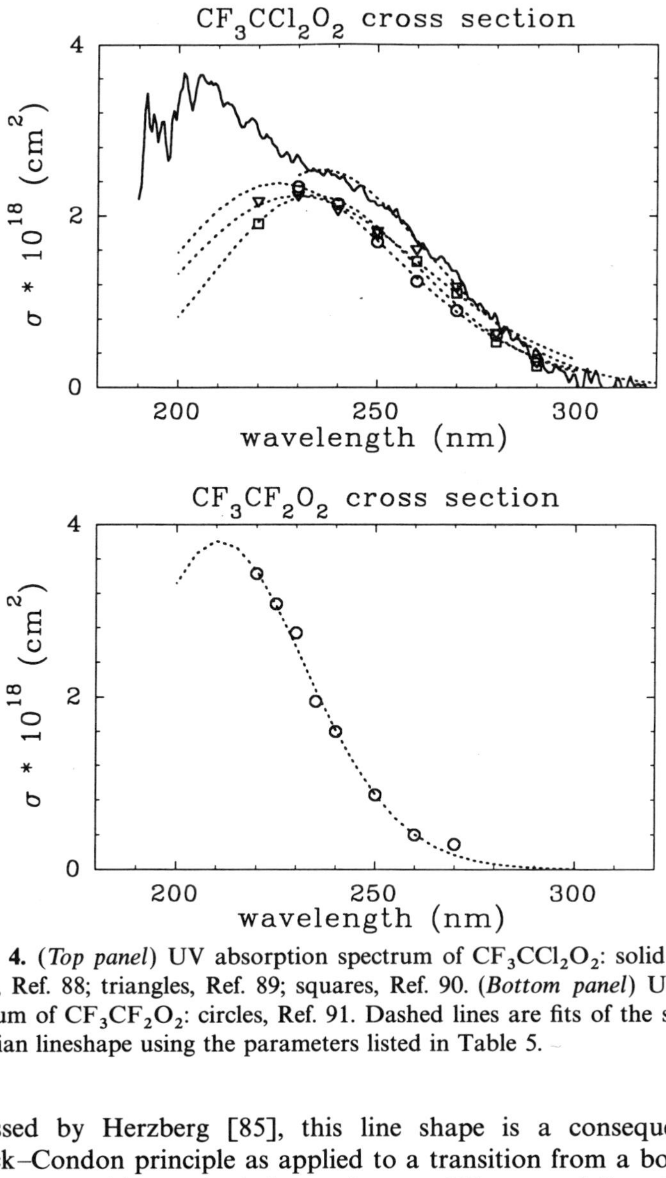

Figure 4. (*Top panel*) UV absorption spectrum of $CF_3CCl_2O_2$: solid line, Ref. 87; circles, Ref. 88; triangles, Ref. 89; squares, Ref. 90. (*Bottom panel*) UV absorption spectrum of $CF_3CF_2O_2$: circles, Ref. 91. Dashed lines are fits of the spectra to the Gaussian lineshape using the parameters listed in Table 5.

discussed by Herzberg [85], this line shape is a consequence of the Franck–Condon principle as applied to a transition from a bound ground state to a repulsive excited electronic state. Winans and Stueckelberg [86] suggested replacing the upper repulsive state wavefunction by a delta function at the classical turning point, in which case the Franck–Condon factor is $\langle \psi'(E) | \psi''(v'' = 0) \rangle \sim \exp[-c(r_t - r_{eq})^2]$, where r_t indicates the

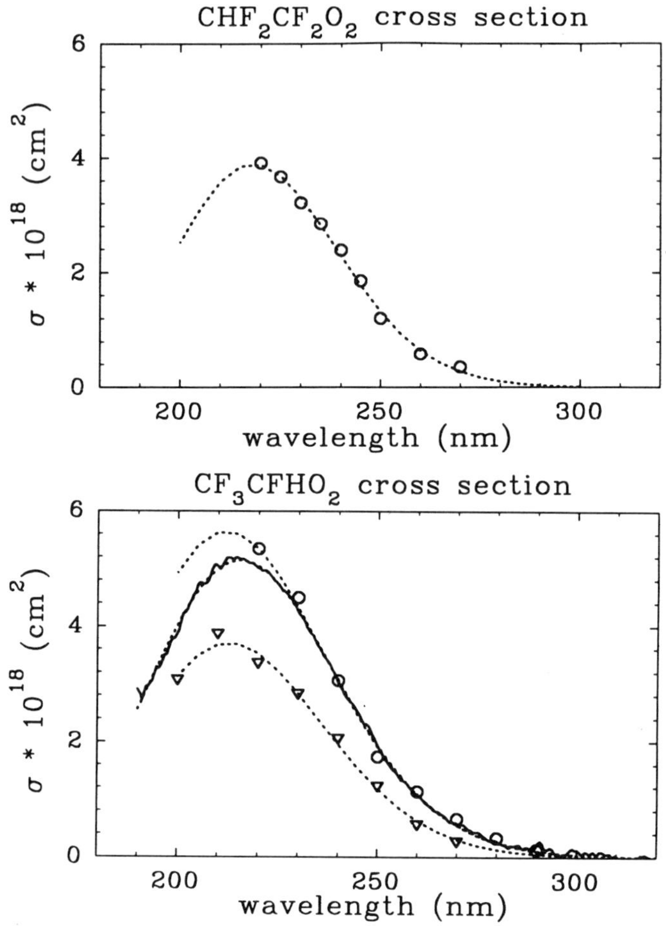

Figure 5. (*Top panel*) UV absorption spectrum of $CHF_2CF_2O_2$: circles, Ref. 39. (*Bottom panel*) UV absorption spectrum of CF_3CFHO_2: solid line, Ref. 92; circles, Ref. 52; triangles, Ref. 89. Dashed lines are fits of the spectra to the Gaussian line shape using the parameters listed in Table 5.

classical turning point of the upper state potential surface at a total energy of E, r_{eq} represents the equilibrium O—O bond length of the ground state, and c is a constant. The delta function is a severe approximation for the repulsive wavefunction; however, the same result is obtained by the following argument. The continuum wavefunction exhibits a broad, nearly Gaussian shape at the classical turning point and a sinusoidal dependence on internuclear distance as r increases. Owing

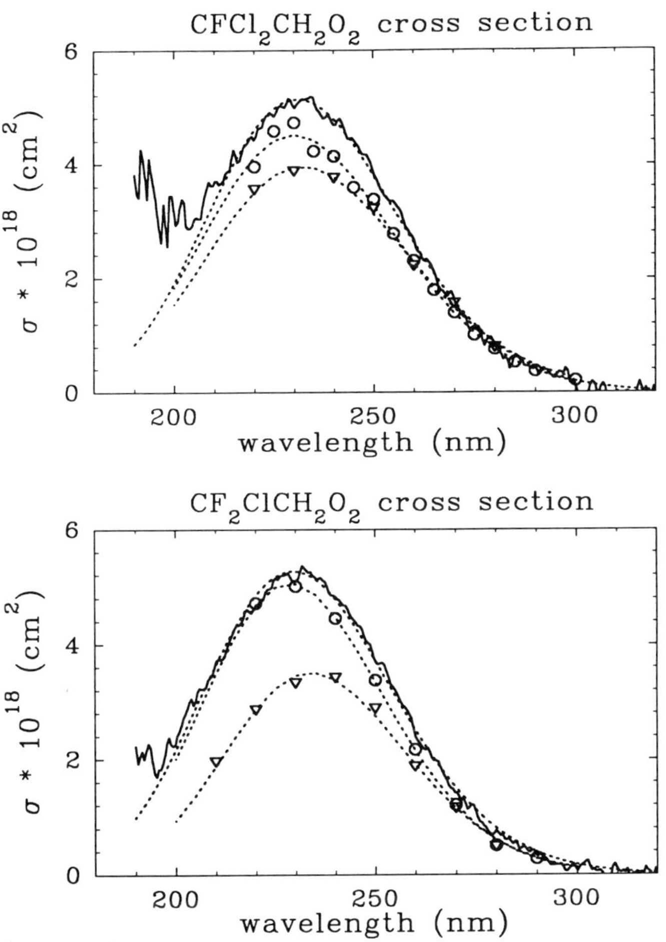

Figure 6. (*Top panel*) UV absorption spectrum of $CFCl_2CH_2O_2$: solid line, Ref. 87; circles, Ref. 93; triangles, Ref. 89. (*Bottom panel*) UV absorption spectrum of $CF_2ClCH_2O_2$: solid line, Ref. 87; circles, Ref. 93; triangles, Ref. 89. Dashed lines are fits of the spectra to the Gaussian line shape using the parameters listed in Table 5.

to the rapid oscillations of the sinusoidal portion of the repulsive wavefunction, its overlap with the ground-state wavefunction is negligible. The Franck–Condon factor is, therefore, dominated by the overlap between the approximately Gaussian-shaped portion of the repulsive wavefunction at the classical turning point and the ground-state vibrational wavefunction. Because the latter, too, has a Gaussian shape, the overlap integral has the same Gaussian dependence on $r_t - r_{eq}$ as obtained above with the

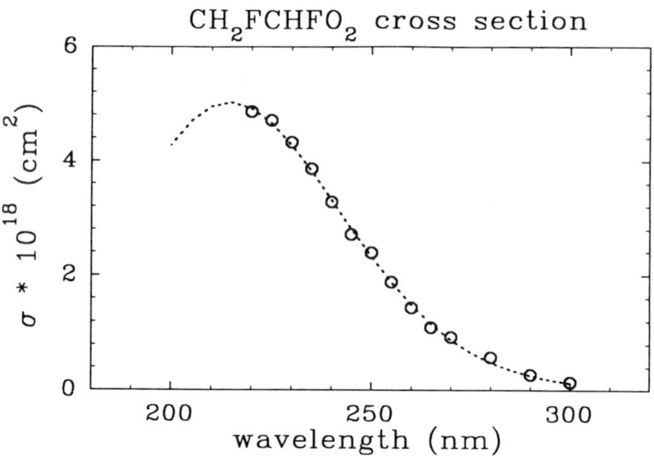

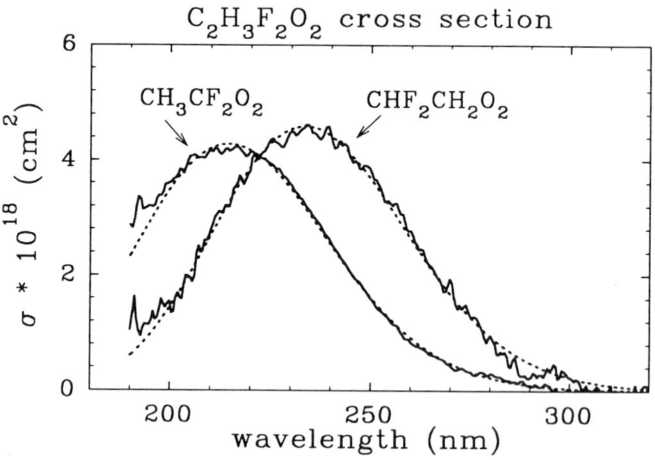

Figure 7. (*Top panel*) UV absorption spectrum of CH_2FCHFO_2: circles, Ref. 43. (*Bottom panel*) UV absorption spectra of $CH_3CF_2O_2$ and $CHF_2CH_2O_2$: solid lines, Ref. 87. Dashed lines are fits of the spectra to the Gaussian line shape using the parameters listed in Table 5.

cruder delta function approximation. If, as is reasonable, the potential energy of the repulsive curve depends exponentially on r, the classical turning point also varies exponentially with the energy of the upper-state wavefunction, implying that

$$\sigma(\lambda) = \sigma_{\max} \exp\left\{-\alpha\left[\ln\left(\frac{\lambda_{\max}}{\lambda}\right)\right]^2\right\} \tag{28}$$

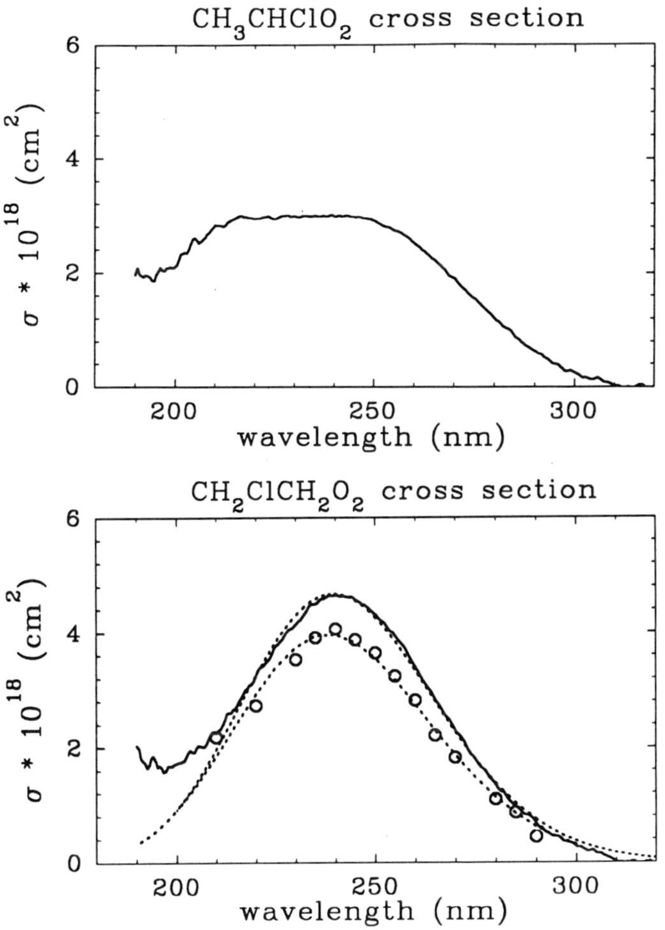

Figure 8. (*Top panel*) UV absorption spectrum of CH_3CHClO_2: solid line, Ref. 14. (*Bottom panel*) UV absorption spectrum of $CH_2ClCH_2O_2$: solid line, Ref. 14; circles, Ref. 94. Dashed lines are fits of the spectra to the Gaussian line shape using the parameters listed in Table 5.

The fits of this line shape to the peroxy radical UV spectra are shown by the dashed lines in Figs. 4 to 11. Besides providing an excellent representation of the absorption profile, Eq. (28) conveniently summarizes the spectrum by three physically meaningful parameters. σ_{max} and λ_{max} give, respectively, the intensity and position of the absorption maximum, whereas α provides a measure of the width of the absorption profile. Values of these parameters for the peroxy radical spectra relevant to HCFCs are collected in Table 5.

TABLE 5 Cross Sections of HFC- and HCFC-Based Peroxy Radicals

	$\sigma_{fit} = \sigma_{max} \exp\{-\alpha[\ln(\lambda_{max}/\lambda)]^2\}$			
Reactant	$\sigma_{max} \times 10^{18}$ (cm^2)	λ_{max} (nm)	α	Reference
$CF_3CCl_2O_2$	2.54	236	41.2	87
	2.39	225	29.9	88
	2.24	230	27.2	89
	2.25	234	40.9	90
$CF_3CF_2O_2$	3.81	211	50.7	91
$CHF_2CF_2O_2$	3.90	218	57.8	39
CF_3CFHO_2	5.14	215.5	44.1	92
	5.64	212	39.9	52
	3.70	213	44.3	89
$CFCl_2CH_2O_2$	5.13	231	47.6	87
	4.50	230	45.9	93
	3.96	232	42.7	89
$CF_2ClCH_2O_2$	5.27	229	47.1	87
	5.06	229	51.7	93
	3.51	234	54.1	89
CH_2FCHFO_2	5.01	214.5	32.8	43
$CH_3CF_2O_2$	4.29	214.5	41.6	87
$CHF_2CH_2O_2$	4.59	233.5	47.9	87
CH_3CHClO_2	—	—	—	14
$CH_2ClCH_2O_2$	4.67	239.5	50.1	14
	3.97	238	48.1	94
CF_3O_2	4.33	209	60.5	54
	3.66	206	46.8	53
CHF_2O_2	4.53	218	58.0	56
CH_2FO_2	5.29	221	40.5	47
	4.62	226	53.1	95
CH_2ClO_2	4.06	230	34.3	87
	3.75	234	41.8	95
$FC(O)O_2$	3.00	230	50.1	82
	3.07	230	49.6	83
$CF_3C(O)O_2$	—	—	—	96

An examination of the spectra presented in Figures 4 to 11 reveals moderate disagreement between cross sections reported for some peroxy radicals, including CF_3CFHO_2, $CF_2ClCH_2O_2$, and $CFCl_2CH_2O_2$. Cross sections for these species measured by Jemi-Alade et al. [89] are systematically lower by about 30% than those measured by Maricq and Szente

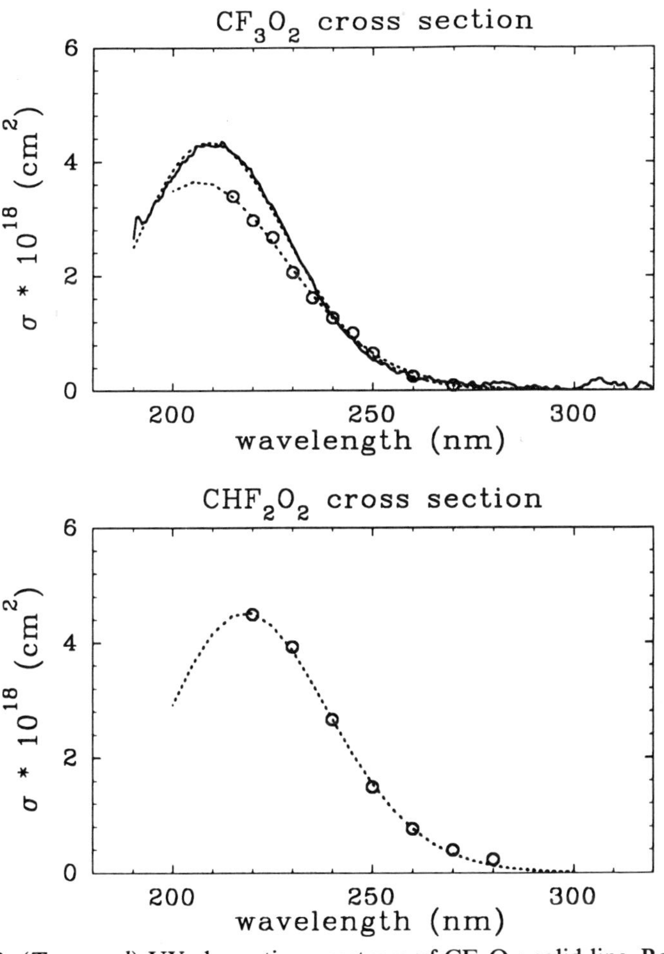

Figure 9. (*Top panel*) UV absorption spectrum of CF_3O_2: solid line, Ref. 54; circles, Ref. 53. (*Bottom panel*) UV absorption spectrum of CHF_2O_2: circles, Ref. 56. Dashed lines are fits of the spectra to the Gaussian line shape using the parameters listed in Table 5.

[87,92] and by Wallington and Nielsen [52,93]. The probable reason for the discrepancy stems from the use, by Jemi-Alade et al., of chlorine atoms, as opposed to fluorine atoms, to initiate peroxy radical formation, which were used by the other two groups. As seen from Table 2, chlorine-atom attack of the haloalkane precursor for these peroxy radicals is very slow, whereas the corresponding fluorine-atom reactions (Table 3) are much faster. It is possible that radical–radical reactions such as the rapid reaction between chlorine atoms and peroxy radicals [97] reduced the peroxy radical

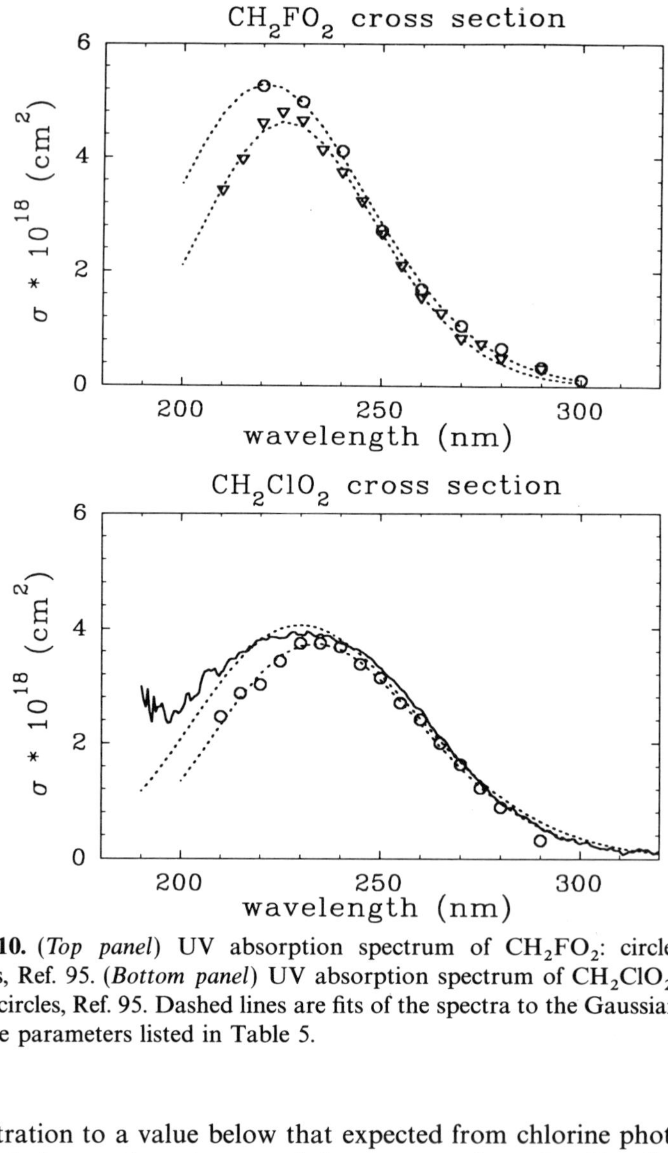

Figure 10. (*Top panel*) UV absorption spectrum of CH_2FO_2: circles, Ref. 46; triangles, Ref. 95. (*Bottom panel*) UV absorption spectrum of CH_2ClO_2: solid line, Ref. 87; circles, Ref. 95. Dashed lines are fits of the spectra to the Gaussian line shape using the parameters listed in Table 5.

concentration to a value below that expected from chlorine photolysis and thereby led to underestimates of the cross sections. Smaller discrepancies (~10%) are noted for the UV spectra of $CH_2ClCH_2O_2$ and CF_3O_2; however, the reasons for this are not clear.

The Gaussian line shape gives an excellent fit to most of the peroxy radical UV spectra. There are, however, a few exceptions, notably $CF_3CCl_2O_2$, CH_3CHClO_2, and $CF_3C(O)O_2$. The deviations from the Gaussian shape

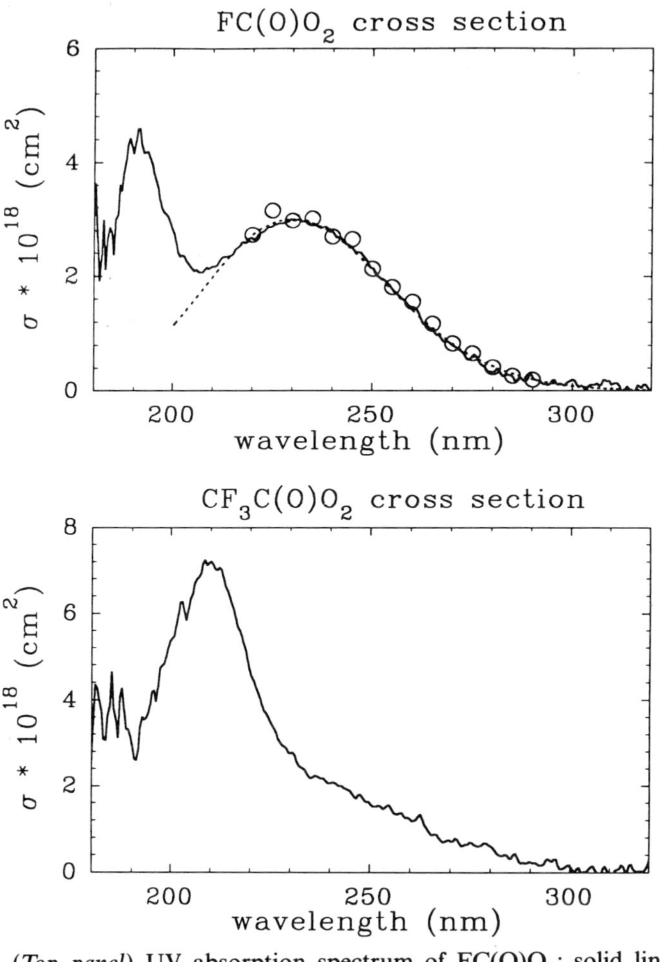

Figure 11. (*Top panel*) UV absorption spectrum of $FC(O)O_2$: solid line, Ref. 82; circles, Ref. 83. (*Bottom panel*) UV absorption spectrum of $CF_3C(O)O_2$: solid line, Ref. 96. Dashed lines are fits of the spectra to the Gaussian line shape using the parameters listed in Table 5.

occur on the short-wavelength side of the absorption band. Fluorinated alkylperoxy radicals show very little deviation, whereas the peroxy radicals with chlorine atoms at the α-position differ the most. A possible explanation is that overlap from long-wavelength tails of transitions to higher electronic states cause the absorption profile to deviate from the Gaussian shape. The onset of the next-shorter wavelength transition is visible in the spectra of $CFCl_2CH_2O_2$, CH_2ClO_2, and $FC(O)O_2$. Apparently, chlorine substitution moves these higher-frequency absorption bands to longer wavelengths, where they cause more interference with the $X^2A'' \rightarrow 2^2A''$ transition. A similar shift of the higher-frequency absorption band to longer wavelengths is probably the reason for the good fit to a Gaussian shape obtained for $FC(O)O_2$ as compared to the clearly non-Gaussian shape of the $CF_3C(O)O_2$ spectrum.

2. Self-Reaction Kinetics. Of all peroxy radical reactions, the self-reaction between two identical peroxy radicals is perhaps the most studied. The measurement of peroxy radical UV absorption cross sections, discussed above, often occurs under the assumption that all the chlorine or fluorine atoms produced by photolysis are converted quantitatively into peroxy radicals; however, this assumption must be corrected for by the loss of peroxy radicals from self-reaction. Furthermore, studies of $RO_2 + NO$ or $RO_2 + HO_2$ reactions usually take place at sufficiently high RO_2 concentrations to require knowledge of the self-reaction rate constant, in order to interpret the results of the kinetics measurements. Both concerns make laboratory studies of peroxy self-reaction kinetics an important issue. In contrast, the steady-state atmospheric concentrations of HCFC-based peroxy radicals are probably too small for their self-reactions to be relevant to atmospheric chemistry. In this context, the most important peroxy–peroxy radical reactions would be between the HCFC-based peroxy radicals and CH_3O_2, but such reactions have not been studied to date.

Table 6 collects the available HCFC-based peroxy radical self-reaction rate constants. There are two important points to keep in mind regarding this table. The first is that the peroxy radical self-reaction proceeds by two major channels:

$$\text{(a)}\quad RO_2 + RO_2 \rightarrow 2RO + O_2 \tag{29}$$

$$\text{(b)}\quad RO_2 + RO_2 \rightarrow R_{-H}O + ROH + O_2 \tag{30}$$

(A third, minor channel also exists in some cases [11]. For the example of methylperoxy radicals the channel $CH_3O_2 + CH_3O_2 \rightarrow CH_3OOCH_3 + O_2$ proceeds with a 0.07 branching ratio.) The branching ratio, where available, is listed in the column marked k_a/k, as the fraction proceeding via the alkoxy channel. The second point is that because O_2 is present to generate the peroxy radical in the first place, the alkoxy radical can react with O_2,

$$RO + O_2 \rightarrow R_{-H}O + HO_2 \tag{31}$$

to form HO_2. The latter radical, in turn, can remove peroxy radicals via

$$RO_2 + HO_2 \rightarrow ROOH + O_2 \tag{32}$$

TABLE 6 HFC- and HCFC-Based Peroxy Radical Self-Reactions

		$k_T = A\exp(-E/RT)$			
Reactant	$k_{298} \times 10^{12}$ (cm^3 molecule^{-1} s^{-1})	$A \times 10^{13}$ (cm^3 s^{-1})	E/R (K)	k_a/k[a]	Reference
CH_3CHClO_2	$k = 5.2 \pm 1.3$				14
$CH_2ClCH_2O_2$	$k_{obs} = 3.6 \pm 0.6$	1.1 ± 0.7	-1020 ± 170		94
	$k_{obs} = 6.0 \pm 0.8$				14
				0.69	98
$CF_3CCl_2O_2$	$k_{obs} = 3.3 \pm 0.5$				88
$CF_3CF_2O_2$	$k_{obs} = 2.1 \pm 0.38$				91
$CHF_2CF_2O_2$	$k_{obs} = 2.7 \pm 0.6$				39
CF_3CFHO_2	$k = 7 \pm 1.0$	7.8 ± 1.3	-605 ± 40		92
				0.84	99
$CFCl_2CH_2O_2$	$k_{obs} = 4.36 \pm 0.64$				93
$CF_2ClCH_2O_2$	$k_{obs} = 4.13 \pm 0.58$				93
CHF_2O_2	$k_{obs} = 5.0 \pm 0.7$			1	56
CH_2ClO_2	$k_{obs} = 3.78 \pm 0.45$	3.1 ± 1.1	-735 ± 95		95
				~1	100
					101
CH_2FO_2	$k_{obs} = 4.0 \pm 0.5$			>0.77	47
	3.1 ± 0.7	3.3 ± 1.2	-700 ± 100		95
CF_3O_2	$k_{obs} = 3.1 \pm 0.25$				54
	$k = 1.8$				
	$k_{obs} = 3.6 \pm 0.9$				53
	$k = 1.8$				
$FC(O)O_2$	$k_{obs} = 6.5 \pm 0.5$	25 ± 4	-286 ± 40		82
	$k_{obs} = 6.0 \pm 0.7$				83
$CF_3C(O)O_2$	$k_{obs} \simeq 10$				96

[a] k_a is the rate constant for alkoxy radical formation reaction (17) and k represents the overall rate constant for peroxy radical loss.

thus making the apparent loss of peroxy radicals more rapid than the loss dictated by reactions (29) and (30) alone. Conversely, a prediction of the self-reaction rate constant based on the observed rate of peroxy radical disappearance will generally be an overestimate of the true rate constant. In Table 6 "$k_{obs} =$" is used to indicate "observed" rate constants that have not been corrected for the secondary chemistry, whereas "$k =$" indicates that the listed value represents the sum of the rate constants for reactions (29) and (30).

The data in Table 6 suggest a few general trends. The room-temperature HCFC-based peroxy radical self-reaction rate constants all lie in the rather limited range 2 to 7×10^{-12} cm^3 s^{-1}. Where available, the rate constant exhibits a negative temperature dependence. This is commonplace among peroxy radicals [11, 12] and suggests that the reaction proceeds by a rate-limiting complex formation step followed by dissociation of the complex into products. Finally, the tabulated branching ratios indicate that alkoxy radical formation is the dominant channel for the HCFC-based peroxy radical self-reactions.

Although not truly an HCFC, the 1-chloroethylperoxy radical is included here because of the interesting HCl elimination reaction that the corresponding alkoxy radical undergoes (see Section IIIF). This elimination removes the concern of secondary interference from HO_2 produced via reaction (31); instead, acetylperoxy radicals are formed. Interference from the secondary chemistry that these radicals introduce precludes the use of UV spectroscopy to determine the CH_3CHClO_3 self-reaction rate constant. That value was measured, instead, by monitoring the HCl eliminated from the alkoxy radical product [14]; thus k as opposed to k_{obs} is listed in Table 6. The branching ratio of $k_a/k = 0.95$ was determined from the amount of HCl that was produced by elimination from the alkoxy radical relative to the initial peroxy radical concentration.

Two groups have reported rate constants at 298 K for the self-reaction of $CH_2ClCH_2O_2$ with substantially different results: $k_{obs} = (6.0 \pm 0.8) \times 10^{-12}$ cm^3 s^{-1} by Maricq et al. [14] and $k_{obs} = (3.6 \pm 0.6) \times 10^{-12}$ cm^3 s^{-1} by Dagaut et al. [95]. Unlike the case for pseudo-first-order reactions, second-order reactions monitored by absorption spectroscopy require knowledge of the optical cross section in order to extract the reaction rate constant. The two rate constants were determined using different cross sections, which accounts for approximately 12% of the discrepancy. The origin of the remainder is unclear. Product studies of the $CH_2ClCH_2O_2$ self-reaction indicate a branching ratio [98] of $k_a/k = 0.69$. While interference from HO_2 produced in the reaction between the alkoxy radical and O_2 is possible in principle, Maricq et al. [14] did not observe HO_2 levels sufficiently high to merit significant interference from its secondary reactions. This suggests that $k_{obs} = k$, but further work is needed to confirm the lack of secondary effects.

Neither of the peroxy radicals derived from HCFC-123 and HFC-125 contain α-hydrogen atoms. This has two implications. First, the molecular channel, reaction (30) is not feasible; thus $k_a/k = 1$. Second, reaction (31) between the alkoxy radical and O_2 is not possible, eliminating the interference due to secondary removal of the peroxy radicals by HO_2. However, both alkoxy radicals provide other sources of secondary complications due to their dissociation,

$$CF_3CF_2O \rightarrow CF_3 + F_2CO \tag{33}$$

in the case of HFC-125. Subsequently, CF_3 adds O_2 to form a peroxy radical, which can react with itself or $CF_3CF_2O_2$, increasing the $CF_3CF_2O_2$ loss rate and producing the corresponding alkoxy radicals. Product studies [91] show the appearance of two trioxide compounds, $CF_3O_3CF_3$ and $CF_3CF_2O_3CF_3$, implying that

$$CF_3CF_2O_2 + CF_3O \rightarrow CF_3CF_2O_3CF_3 \tag{34}$$

represents yet another peroxy radical loss channel. Without the rate constants for reaction (34) and the $CF_3O_2 + CF_3CF_2O_2$ reactions, it is not possible to deduce the self-reaction rate constant, and only k_{obs} has been reported.

For HCFC-123, the observations of Jemi-Alade et al. [89] suggest that the alkoxy radical decomposes via

$$CF_3CCl_2O \rightarrow Cl + CF_3C(O)Cl \tag{35}$$

regenerating a chlorine atom that can attack the parent hydrocarbon and thereby replace the peroxy radical. It is also possible that

$$CF_3CCl_2O_2 + CF_3CCl_2O \rightarrow CF_3CCl_2O_3CCl_2CF_3 \tag{36}$$

contributes to peroxy radical disappearance. Finally, peroxy radicals could be lost by

$$CF_3CCl_2O_2 + Cl \rightarrow CF_3CCl_2O + ClO \tag{37}$$

a process that can be very rapid [97]. In the absence of further kinetic information, it is impossible to derive the actual peroxy radical self-reaction rate constant from k_{obs}.

The lack of an α-hydrogen implies that as with the HCFC-123- and HFC-125-based peroxy radicals, the molecular reaction channel for $CF_2HCF_2O_2$ is not possible. The product study of Nielsen et al. [39] shows only the formation of F_2CO and indicates that $k_a/k = 1$. The lack of an α-hydrogen also implies that the alkoxy radical cannot react with O_2; instead, it dissociates via

$$CF_2HCF_2O \rightarrow CF_2H + F_2CO \tag{38}$$

The alkyl fragment adds O_2 to form CF_2HO_2, which can then react with itself or with $CF_2HCF_2O_2$. It is, consequently, not feasible to extract the self-reaction rate constant due to the incompletely understood secondary chemistry.

The product study of Wallington et al. [99] indicates a room-temperature branching ratio of $k_a/k = 0.85$ for the CF_3CFHO_2 self-reaction. Maricq and Szente [92] found this branching ratio to be consistent with their time-resolved UV absorption measurements over the range 211 to 372 K. An extrapolation of Wallington et al.'s data from 273 to 211 K disagrees with this result at low temperatures, giving $k_a/k = 0.22$ at 211 K; however, this is likely due to the large extrapolation that is required. Time-resolved UV spectra [92] reveal that the absorption, belonging initially to CF_3CFHO_2, changes during the course of reaction to the spectrum corresponding to CF_3O_2. This is consistent with the product study of Wallington et al. [99], which showed a competition between

$$CF_3CFHO \rightarrow CF_3 + HFCO \tag{39}$$

$$CF_3CFHO + O_2 \rightarrow CF_3C(O)F + HO_2 \tag{40}$$

for loss of the alkoxy radical. At O_2 concentrations below about 600 torr, the dissociation channel dominates and explains the appearance of CF_3O_2 in the time-resolved study. Because the CF_3CFHO_2 and CF_3O_2 concentrations were both monitored, the actual HFC-134a- based peroxy self-reaction rate constant was determined [92]; hence the designation "$k =$" in Table 6. The rate constants are well fit by an Arrhenius expression, but with a negative temperature dependence. Rate constants ranging from $(12 \pm 12) \times 10^{-12}\,cm^3\,s^{-1}$ at 216 K to $(5 \pm 3) \times 10^{-12}\,cm^3\,s^{-1}$ at 372 K were also reported for the cross reaction

$$CF_3O_2 + CF_3CFHO_2 \rightarrow CF_3O + CF_3CFHO + O_2 \tag{41}$$

The only kinetic data for the self-reactions of $CFCl_2CH_2O_2$ and $CF_2ClCH_2O_2$ are the k_{obs} values measured via pulsed radiolysis by Wallington and Nielsen [93]. The presence of α-hydrogen atoms implies the availability of the molecular channel represented by reaction (30); however, the branching ratio is unknown. The α-hydrogens also imply that there will be a competition between dissociation of the alkoxy radical and its reaction with O_2. Since the extent of this is unknown, the actual self-reaction rate constant cannot be obtained from k_{obs}.

A product study [56] of the CHF_2O_2 self-reaction reveals a single carbon-containing product, F_2CO. This and the lack of CHF_2OH suggest that the branching ratio is $k_a/k = 1$. F_2CO is formed by the alkoxy radical reaction with O_2, which also produces HO_2. While the CHF_2OOH product of the reaction between HO_2 and the peroxy radical was not observed, secondary removal of the CHF_2O_2 could not be ruled out and, hence, only k_{obs} was reported.

Temperature-dependent rate constants for the CH_2ClO_2 self-reaction were measured by Dagaut et al. [95] but not corrected for possible secondary reactions with HO_2. The product studies of Sanhueza and Heicklen [100] suggest that the branching ratio is $k_a/k = 1$. However, two preliminary investigations reveal that the secondary chemistry may be both more complicated and more interesting. Recent experiments [87] indicate that the rate constant for

$$Cl + CH_2ClO_2 \rightarrow \text{products} \qquad (42)$$

is on the order of $10^{-10}\ cm^3\ s^{-1}$. Because the attack of methyl chloride by Cl is slow ($k = 5 \times 10^{-13}\ cm^3\ s^{-1}$), it is possible that, except at high CH_3Cl concentrations, a fraction of the peroxy radicals is lost by reaction (42) and a systematic error is introduced into the observed self-reaction rate constant. A recent product study [102] of the alkoxy radical shows that there is a competition between

$$CH_2ClO \rightarrow HCO + HCl \qquad (43)$$

and the $RO + O_2$ reaction. Both the latter reaction and the reaction between HCO and O_2 produce HO_2, which can remove CH_2ClO_2; thus the observed second-order decay in peroxy radical concentration provides an upper limit to the self-reaction rate constant.

The CH_2FO_2 self-reaction has been studied by two groups [47, 95]. The observed rate constants at 298 K are in agreement within the combined error bars. The reaction shows a negative temperature dependence, similar to that of other halogenated ethylperoxy radicals. The product study of Wallington et al. [47] reveals HFCO to be the major product, but also shows infrared (IR) absorption features attributed to CH_2FOOH under some conditions; as a result, a lower limit of $k_a/k > 0.77$ is reported for the branching ratio.

There have been two investigations of the CF_3O_2 self-reaction [53, 54]. The observed second-order rate constants at 298 K are in good agreement. Due to the absence of hydrogen atoms, the molecular channel, reaction (30), is not possible; thus the alkoxy radical is the only product expected from the self-reaction. The CF_3O radical is very interesting from a chemical standpoint and very important from an atmospheric standpoint and is discussed in more detail in Section IV. A product study of the CF_3O_2 self-reaction shows CF_3OOOCF_3 as the only carbon-containing product. It presumably arises from

$$CF_3O_2 + CF_3O \rightarrow CF_3OOOCF_3 \qquad (44)$$

This has significant implications for the interpretation of the observed peroxy radical decay rate constant. If reaction (44) is very fast (an estimated rate constant of $2.5 \times 10^{-11}\,cm^3\,s^{-1}$ was obtained from modeling the product study), then four, as opposed to two, peroxy radicals are effectively removed for every reactive encounter between two CF_3O_2 molecules, implying that the actual self-reaction rate constant is half the observed value. A more careful analysis by both studies gives the corrected self-reaction rate constant as $k = 1.8 \times 10^{-12}\,cm^3\,s^{-1}$.

FCO radicals arise as intermediates in the oxidation of the halogenated methyl radicals produced by UV photolysis of CFCs in the stratosphere and from the atmospheric degradation products of HCFCs. A more complete discussion of $FC(O)O_x$ chemistry, both experimental and theoretical, is given in Section IV. FCO radicals rapidly add O_2 to form the corresponding fluoroformylperoxy radical. There have been two measurements of the $FC(O)O_2$ self-reaction rate constant at 298 K, which are in very good agreement. As with many other peroxy radicals, the self-reaction shows a negative temperature dependence, although to a lesser degree than the haloalkylperoxy radicals. $CF_3C(O)$ radicals arise from the reaction between OH and $CF_3C(O)H$ which is produced in HCFC degradation. The CF_3CO dissociation rate is on the order of $10^5\,s^{-1}$ at 298 K and considerably slower at lower temperatures [96]; thus its predominant fate in the atmosphere is to add O_2. Preliminary measurements indicate that the $CF_3C(O)O_2$ self-reaction rate constant is $\sim 10^{-11}\,cm^3\,s^{-1}$ [96]. The major product is $CF_3C(O)O$, which falls apart to CF_3 and CO_2 on the submicrosecond time scale.

3. RO_2 + NO Reaction Kinetics. The reaction between peroxy radicals and nitric oxide proceeds, in general, via two pathways:

$$\text{(a)}\quad RO_2 + NO \rightarrow RO + NO_2 \tag{45}$$

$$\text{(b)}\quad RO_2 + NO + M \rightarrow RONO_2 + M \tag{46}$$

Pathway (b) is always the minor channel. For small hydrocarbon peroxy radicals the branching ratio k_b/k increases from <0.014 for $R = C_2H_5$ to approximately 0.2 for *n*-hexane. As the hydrocarbon grows further, a branching ratio of $k_b/k = 0.4$ is approached (e.g., for *n*-octane) [12]. With respect to the HCFC-based peroxy radical plus NO reactions, there have been no reports of either a pressure dependence to the reaction or of observations of alkyl nitrate formation. Most of the studies base the conclusion that channel (b) is insignificant on the observation that the NO_2 produced approximately equals the initial peroxy radical concentration.

In some cases the fact that no $RONO_2$ is observed provides an upper limit for k_b/k.

Rate constants for the $RO_2 + NO$ reaction are tabulated in Table 7. Most of the rate constants, including all of those for the halogenated ethylperoxy radicals, are the result of a single study by Sehested et al. obtained using pulsed radiolysis in 1 atm of SF_6 [103]. The other rate constants were measured at total pressures below 10 torr using discharge flow or laser photolysis to generate the radicals and mass spectrometry to detect them. Only one system, CF_3O_2, has been investigated by more than two groups. In this case, the reported 298 K rate constants are all in very good agreement.

It is striking that there is almost no variation in the reported rate constant with the pattern of halogen substitution. The rate constants range from a low value of $(1.18 \pm 0.10) \times 10^{-11}$ cm^3 s^{-1} for $CF_2ClCH_2O_2$ to a high value of $(1.87 \pm 0.20) \times 10^{-11}$ cm^3 s^{-1} for CH_2ClO_2 [for $FC(O)O_2$ a slightly

TABLE 7 HFC- and HCFC-Based Peroxy Radical Reactions with NO

Reactant	$k_{298} \times 10^{11}$ (cm^3 molecule^{-1} s^{-1})	$k_T = k_{298}(T/298)^{-n}$ n	Reference
$CF_3CF_2O_2$	$>1.07 \pm 0.15$		103
$CHF_2CF_2O_2$	$>0.97 \pm 0.13$		103
CF_3CFHO_2	1.28 ± 0.36		52
$CFCl_2CH_2O_2$	1.28 ± 0.11		103
$CF_2ClCH_2O_2$	1.18 ± 0.10		103
CH_2FCFHO_2	>0.87		43
CH_2ClO_2	1.87 ± 0.20		103
CH_2FO_2	1.25 ± 0.13		103
CHF_2O_2	1.26 ± 0.16		103
$CFCl_2O_2$	1.45 ± 0.2	1.3 ± 0.2	104
	1.60 ± 0.20		105
CF_2ClO_2	1.31 ± 0.12		103
	1.6 ± 0.3	1.5 ± 0.4	104
CCl_3O_2	1.7 ± 0.2	1.0 ± 0.2	104
	1.86 ± 0.28		71
CF_3O_2	1.53 ± 0.20		106
	1.68 ± 0.26		107
	1.54 ± 0.35		108
	1.45 ± 0.2	1.2 ± 0.2	104
	1.78 ± 0.36		109
$FC(O)O_2$	2.5 ± 0.8		83

higher value of $(2.5 \pm 0.8) \times 10^{-11}\ cm^3\ s^{-1}$ is reported]. Furthermore, the few systems for which rate constant vs. temperature data are available show only slight negative temperature dependencies [e.g., CF_2ClO_2 has the largest dependence with $k \sim (T/298)^{-1.5}$]. These observations are consistent with a reaction mechanism [110] in which the peroxy radical and nitric oxide form a collision complex,

$$RO_2 + NO \rightarrow ROONO^*$$

$$ROONO^* \rightarrow RO + NO_2$$

$$ROONO^* + M \rightarrow RONO_2 + M$$

which isomerizes and dissociates to yield the alkoxy radical and NO_2 or isomerizes and is collisionally quenched to produce the alkyl nitrate. While thermodynamic data for RO_2 radicals are sparse, the exothermicity for reaction (16) when $R = CH_3$ is approximately $\Delta H = -220\ kJ/mol^{-1}$, based on the heats of formation of $11\ kJ\ mol^{-1}$ for CH_3O_2 [111] and $-120\ kJ\ mol^{-1}$ and $90\ kJ\ mol^{-1}$ for CH_3ONO_2 and NO, respectively [112]. Although this value will vary somewhat upon halogen substitution, it remains that collisional stabilization of the alkyl nitrate requires the removal of a substantial amount of excess energy. Apparently for the HCFC-based and other relatively small peroxy radicals, this process is inefficient compared to dissociation of the complex into an alkoxy radical and nitrogen dioxide.

4. $RO_2 + NO_2$ Reaction Kinetics. The reaction

$$RO_2 + NO_2 + M \leftrightarrow RO_2NO_2 + M \tag{47}$$

has been studied for the HCFC-based peroxy radicals with $R = CF_3$, $CClF_2$, CCl_2F, and CCl_3. This reaction serves as a temporary reservoir for the RO_2 and NO_2 species; thus both the forward and reverse rate constants for reaction (47) are needed to infer its atmospheric importance. Table 8 lists low-pressure rate constants, limiting high-pressure rate constants and the center broadening factors, F_c, for both the forward and reverse directions of reaction (47).

The forward rate constants in Table 8 were measured by Carlap et al. [113] and Wu and Carr [114] in the pressure range 1 to 10 torr, whereas the dissociation rate constants were measured over the pressure range 7 to 600 torr by Köppenkastrop and Zabel [115]. Because the falloff region extends well above 10 torr, the limiting high-pressure bimolecular rate constants reported for the peroxy plus NO_2 reactions were determined by

TABLE 8 HCFC-Based Peroxy Radical Reactions with NO_2

Reactant[a]	k_0^b	k_∞^c	F_c	Reference
CF_3O_2	$(2.7 \pm 0.8) \times 10^{-29} (T/298)^{-(4.7 \pm 1.0)}$	$8.9 \times 10^{-12} (T/298)^{-(0.72 \pm 0.3)}$	$e^{-T/416}$	113
$CF_3O_2NO_2$		$2 \times 10^{16} e^{-12,200/T}$		12
$CClF_2O_2$	$(5.0 \pm 1.0) \times 10^{-29} (T/298)^{-(6.2 \pm 1.2)}$	$(4.5 \pm 2.3) \times 10^{-12} (T/298)^{-(2.5 \pm 0.6)}$	$0.78 e^{-T/569}$	114
$CClF_2O_2NO_2$	$1.8 \times 10^{-3} e^{-10,500/T}$	$1.6 \times 10^{16} e^{-12,000/T}$	0.28	115
CCl_2FO_2	$(5.5 \pm 1.6) \times 10^{-29} (T/298)^{-(5.5 \pm 1.2)}$	$8.3 \times 10^{-12} (T/298)^{-(0.66 \pm 0.3)}$	$e^{-T/342}$	113
$CCl_2FO_2NO_2$	$1.0 \times 10^{-2} e^{-11,000/T}$	$6.6 \times 10^{16} e^{-12,200/T}$	0.28	115
CCl_3O_2	$(9.2 \pm 3.0) \times 10^{-29} (T/298)^{-(6.0 \pm 1.5)}$	$14.9 \times 10^{-12} (T/298)^{-(0.3 \pm 0.3)}$	$e^{-T/260}$	113
$CCl_3O_2NO_2$	$6.3 \times 10^{-3} e^{-10,200/T}$	$4.8 \times 10^{16} e^{-11,800/T}$	0.22	115

[a] RO_2 implies forward reaction; RO_2NO_2 implies reverse reaction.
[b] Units are $cm^6\ s^{-1}$ for forward reaction and $cm^3\ s^{-1}$ for reverse.
[c] Units are $cm^3\ s^{-1}$ for forward reaction and s^{-1} for reverse.

fitting RRKM falloff curves to the low-pressure experimental data and to high-pressure rate constants inferred from the dissociation reaction. Generally, the RRKM calculations reproduce the low-pressure data within about 25%, but are only within a factor of 2 of the rate constants at atmospheric pressure. Accurate measurements of the forward rate constants at high pressures are thus required to verify the expressions for k_∞ listed in Table 8.

In addition to the RO_2NO_2 dissociation rates listed in Table 8, preliminary data are available for the dissociation of some halogenated ethylperoxy nitrates and halogenated acetylperoxy nitrates. Kirchner et al. [116] measured temperature-dependent dissociation rates at 1000 mbar for $CClF_2CH_2O_2NO_2$, $CCl_2FCH_2O_2NO_2$, and $CClF_2C(O)O_2NO_2$. Because of experimental difficulties and the narrow temperature range of the experiments, they arbitrarily set the preexponential factor to $2 \times 10^{16}\ s^{-1}$. Fits to the data then yielded activation energies of 90.5 ± 4.0, 90.5 ± 4.0, and 116 kJ mol^{-1}, respectively. FTIR product studies of $CCl_3C(O)O_2NO_2$ and $CF_2ClC(O)O_2NO_2$ decomposition by Haahr et al. [117] indicate that these species have thermal stabilities comparable to or greater than peroxyacetyl nitrate. The HCFC alkylperoxy nitrate dissociation lifetimes at 1 atm and 298 K (see Table 8) are about an order of magnitude longer than that of methylperoxy nitrate but are still well short of the lifetime for peroxyacetyl nitrate, a very important reservoir for NO_2. Based on this limited database, it appears that halogenated alkylperoxy nitrates will not play an important atmospheric role, but halogenated acetylperoxy radicals may be sufficiently stable to provide a small pathway for the transport of halogen atoms into the stratosphere.

5. Other RO_2 Reactions. There is a limited number of rate constant measurements for the reactions of HCFC-based peroxy radicals with HO_2 and O_3. Whereas the removal of peroxy radicals is dominated by reaction with NO in NO_x-rich environments, such as the urban atmosphere, the reaction with HO_2 can become important in areas with low NO_x levels. Preliminary rate constants exist for the reaction of HO_2 with two halogenated ethylperoxy radicals. For $CF_3CCl_2O_2$, Hayman et al. [90] report a rate constant of $(1.9 \pm 0.3) \times 10^{-12}\ cm^3\ s^{-1}$ at 298 K. The measurements of Maricq et al. [118] provide a rate constant of $(1.8^{+2.4}_{-1.0}) \times 10^{-13} e^{(910 \pm 220)/T}$ $cm^3\ s^{-1}$ for the reaction between CF_3CFHO_2 and HO_2. The room temperature rate constant of $4.3 \times 10^{-12}\ cm^3\ s^{-1}$ is in good agreement with the value of $4 \times 10^{-12}\ cm^3\ s^{-1}$ reported by Hayman [119] for this reaction.

CF_3O_2 is a peroxy radical formed either from the degradation of HFC-23 (CF_3H) or from the dissociation of CF_3CFHO, which results from the oxidation of HFC-134a, followed by oxygen addition to the CF_3 fragment. As discussed in more detail below, both CF_3O_2 and CF_3O were considered as potential catalysts for ozone depletion because of the following reactions:

$$CF_3O_2 + O_3 \rightarrow CF_3O + 2O_2 \tag{48}$$

$$CF_3O + O_3 \rightarrow CF_3O_2 + O_2 \tag{49}$$

which have the overall effect of converting two ozone molecules into three oxygen molecules. There have been four measurements of the rate constant for reaction (48). All report upper limits that are in good agreement with each other. Nielsen and Sehested [120] find that $k < 5 \times 10^{-15}\ cm^3\ s^{-1}$, Maricq and Szente [121] report $k < 1 \times 10^{-14}\ cm^3\ s^{-1}$, Turnipseed et al. [122] find $k < 7 \times 10^{-15}\ cm^3\ s^{-1}$, and Ravishankara et al. [123] report $k < 3 \times 10^{-15}\ cm^3\ s^{-1}$.

In analogy with reactions (48) and (49) the reactions of $FC(O)O_2$ and FC(O)O with O_3 also form a potential ozone destruction cycle. Preliminary data for the $FC(O)O_2 + O_3$ reaction indicate an upper limit of $k < 1.5 \times 10^{-15}\ cm^3\ s^{-1}$ at 298 K [87]. The same data place an upper limit of $k < 3 \times 10^{-15}\ cm^3\ s^{-1}$ for the $FC(O)O + O_3$ reaction, consistent with the recently reported value of $k < 6 \times 10^{-14}\ cm^3\ s^{-1}$ [83]. These rate constants are roughly four orders of magnitude smaller than those for the reaction between $FC(O)O_x$ and NO. The much slower rate constants for reaction with ozone as opposed to NO imply that $FC(O)O_x$ will be rapidly removed from the atmosphere and contribute little to ozone destruction.

E. Haloalkoxy Radicals

As with the peroxy radicals from which they are formed, alkoxy radicals undergo a variety of reactions. There are, in general, four possible reaction pathways: (1) reaction with O_2 if a hydrogen atom is available alpha to the oxygen atom, (2) C—C or C—Cl bond fission, (3) isomerization, and (4) elimination of HCl if the radical is of the form XCHClO. Isomerization becomes a significant possibility only for long-chain alkoxy radicals and is not an important channel for HCFC-based alkoxy radicals. The extent to which each of the other three channels contributes to the atmospheric degradation of a particular HCFC depends on the pattern of Cl and F substitution of the parent hydrocarbon.

Direct spectroscopic measurements of HCFC-based alkoxy radicals and real-time determinations of their reaction rates are scarce. The laser-induced fluorescence (LIF) spectrum [124, 125] of CF_3O has been measured and a UV absorption feature has tentatively been assigned [92] to CF_3CFHO. Three-center elimination of HCl from CH_3CHClO and CH_2ClO has been observed by real-time measurement of the HCl production [10, 87], but the alkoxy radicals have not been directly detected. In other cases, our understanding of alkoxy radical chemistry comes from product studies and theoretical considerations. Below we summarize the fate of the alkoxy radicals derived from the peroxy radicals listed in Tables 6 and 7.

CF_3CCl_2O lacks hydrogen atoms; thus channels (1) and (4) are unavailable and it must decay via

$$CF_3CCl_2O \rightarrow CF_3 + Cl_2CO \qquad \Delta H = -0.9\,\text{kcal mol}^{-1} \qquad (50)$$

$$CF_3CCl_2O \rightarrow CF_3C(O)Cl + Cl \qquad \Delta H = -14.0\,\text{kcal mol}^{-1} \qquad (51)$$

From *ab initio* calculations [126] of the heats of reaction, both C—C and C—Cl bond fission are thermodynamically allowed. However, since the activation energy for dissociation of the C—Cl bond is predicted to be 2.7 kcal mol^{-1} as compared to 9.8 kcal mol^{-1} for the C—C bond, Cl elimination should be the dominant pathway. This is consistent with the observations of Jemi-Alade et al. [89] and Edney et al. [127]. The latter group found 100% conversion of HCFC-123 to $CF_3C(O)Cl$, but observed none of the Cl_2CO product from C—C bond fission in their product study. Jemi-Alade et al. observed an unusually slow decay of $CF_3CCl_2O_2$ in their flash photolysis experiments, which they explained by a chain reaction consisting of

$$CF_3CCl_2O_2 + CF_3CCl_2O_2 \rightarrow 2CF_3CCl_2O + O_2 \qquad (52)$$

followed by reaction (51), hydrogen abstraction from HCFC-123 by the chlorine atom released by the alkoxy radical, and oxygen addition to CF_3CCl_2 to regenerate the peroxy radical. Trifluoroacetyl chloride generated from reaction (51) decays either by incorporation into cloud water or by photolysis, which releases active chlorine, CF_3 radicals and CO. The CF_3 radical begins a series of reactions involving species of the form CF_3O_x and is discussed in more detail in the following section.

Both CF_3CF_2O and CHF_2CF_2O lack chlorine or hydrogen substituents at the α-position; thus of the four channels listed above, C—C bond fission,

$$CF_3CF_2O \rightarrow CF_3 + F_2CO \qquad \Delta H = 0.1\ \mathrm{kcal\ mol^{-1}} \qquad (53)$$

$$CHF_2CF_2O \rightarrow CHF_2 + F_2CO \qquad (54)$$

is the sole feasible decay path for these alkoxy radicals. C—F bond fission at the α-position is not possible, because of the much larger C—F vs. C—Cl bond energy (fluorine dissociation from CF_3CF_2O is endothermic by 25.5 kcal mol^{-1} [126]). Reaction with NO, in analogy with the $CF_3O + NO$ reaction discussed below, represents another possible loss process. In this case it is unlikely to be important, because the theoretical value for the activation energy of C—C bond fission is 10.4 kcal mol^{-1}, which assuming an A factor of $2 \times 10^{13}\ \mathrm{s^{-1}}$, implies a lifetime for dissociation of about 10 μs at 277 K. This decay process is supported by product studies in which nearly 100% formation of F_2CO was observed based on HFC loss [39, 91, 128, 129]. The fluorinated methyl radicals produced by the C—C bond fission undergo the same atmospheric degradation processes as HFC-23 and HFC-32, eventually yielding CF_3O and CF_2HO radicals. The likely fate of the carbonyl fluoride product is uptake by cloud droplets and subsequent hydrolysis, although in the upper stratosphere UV photolysis will also occur.

CF_3CFHO, the alkoxy radical produced from the degradation of HFC-134a, can decay by two channels:

$$CF_3CFHO \rightarrow CF_3 + HFCO \qquad \Delta H = 4.7\ \mathrm{kcal\ mol^{-1}} \qquad (55)$$

$$CF_3CFHO + O_2 \rightarrow CF_3C(O)F + HO_2 \qquad (56)$$

The activation energy for dissociation has been calculated to be 13.0 kcal mol^{-1} [126]; thus at 277 K the rate of reaction (55) is expected to be on the order $10^3\ \mathrm{s^{-1}}$ (based on an A factor of $2 \times 10^{13}\ \mathrm{s^{-1}}$). This estimate is in accord with the measured reaction rate of $1.5 \times 10^4\ \mathrm{s^{-1}}$ in 150 torr of N_2 [92]. Wallington et al. [99] showed that dissociation and reaction with O_2 are competitive under atmospheric conditions by measuring a relative rate for these reactions of $k_{O_2}/k_{dis} = 1.6 \times 10^{-25} e^{3600/T}$ cm^3 in the high-pressure limit. The relatively steep temperature dependence of the dissociation reaction implies that, under conditions of the U.S. standard atmosphere, the branching fraction for dissociation decreases from 0.81 at sea level of 0.22 at an altitude of 12 km. Following O_2 addition and reaction with NO, the CF_3 dissociation product leads to CF_3O. HFCO reacts only slowly with hydroxyl radicals ($k < 4 \times 10^{-15}$ cm^3 s^{-1} [130]) and will probably be incorporated into cloud water and hydrolyzed to HF and HCOOH. $CF_3C(O)F$ does not undergo any further gas-phase reactions. It is only slowly photolyzed, implying that hydrolysis in cloud water is the probable fate of this species. However, the trifluoroacetic acid that is formed has no currently known removal process.

The alkoxy radicals derived from HCFC-141b and HCFC-142b have the form $CXFClCH_2O$, where X = Cl and F, respectively. They can undergo either reaction with O_2 or C—C bond fission:

$$CXFClCH_2O + O_2 \rightarrow CXFClCHO + HO_2 \quad (57)$$

$$CXFClCH_2O \rightarrow CXFCl + H_2CO \quad (58)$$

Although complicated by secondary chemistry, an indirect analysis of a product study by Edney et al. [127] suggests that the branching fraction for dissociation is about 0.5 in the case of $CFCl_2CH_2O$. The aldehyde formed by reaction (57) can undergo attack by OH to form a trihaloacetyl radical. This species can dissociate directly to CXFCl + CO or react with O_2 and subsequently NO to yield CXFClC(O)O, which will rapidly dissociate to CXFCl + CO_2. O_2 addition followed by reaction with NO will convert the CXFCl fragment, either from degradation of the aldehyde or from reaction (58), to CXFClO.

A product study of the CH_2FCHFO radical showed that HFCO accounts for 91 ± 10% of the loss of HFC-152 precursor at 298 K and 760 torr of air [43] Under these conditions C—C bond fission,

$$CH_2FCHFO \rightarrow CH_2F + HFCO \quad (59)$$

dominates over the reaction with O_2,

$$CH_2FCHFO + O_2 \rightarrow CH_2FC(O)F + HO_2 \quad (60)$$

However, reaction (60) may become competitive at lower temperatures.

The CH_3CHClO radical is unusual since it has been found to undergo three-center elimination of HCl via reaction (21) [14]. Real-time observations of HCl formation indicate that at 298 K, the elimination proceeds with a rate $>2 \times 10^6\,s^{-1}$. This lower limit exceeds the O_2 reaction rate by a factor of 50 in 760 torr air, assuming that the rate constant for O_2 reaction is the same as for ethoxy radicals ($8 \times 10^{-15}\,cm^3\,s^{-1}$) [9]. The acetyl radicals produced by the elimination add O_2 to form $CH_3C(O)O_2$. The acetylperoxy radical can combine with NO_2 to form the reservoir species peroxy-acetyl nitrate, or it can react with NO to yield $CH_3C(O)O$, which rapidly falls apart into CH_3 and CO_2.

Product studies [47, 56] have shown that alkoxy radicals of the form CHFXO, with X = H, F, react with O_2,

$$CHFXO + O_2 \rightarrow XFCO + HO_2 \qquad (61)$$

to form HO_2 and a fluorine-substituted formaldehyde. If the rate constant for reaction (61) is similar to those of methoxy and ethoxy radicals, this reaction will dominate the reaction of CHFXO with trace species such as NO. The HFCO product reacts very slowly with OH radicals [130], leaving heterogeneous chemistry and transport as the principal removal mechanisms for the XFCO species. The fraction that is transported into the stratosphere will eventually be photolyzed to FCO, generating a sequence of $FC(O)O_x$ oxidation products.

Because they lack hydrogen atoms, the reaction between alkoxy radicals CFClXO and O_2, where X = F or Cl, cannot occur. Instead, these radicals eliminate Cl via

$$CFClXO \rightarrow XFCO + Cl \qquad (62)$$

When this process occurs in the stratosphere it releases active chlorine and thereby contributes to the ozone depletion potential of any HCFC from which it is derived. Photolysis of XFCO generates FCO, which undergoes a sequence of atmospheric reactions generating radicals of the form $FC(O)O_x$, as described in detail in the following section. When X = Cl photolysis also releases a second chlorine atom. *Ab initio* calculations show that chlorine atom elimination is exothermic for both CF_2ClO and $CFCl_2O$, with $\Delta H = -16.1$ kcal mol^{-1} and -24 kcal mol^{-1}, respectively [131]. The theoretical activation energies are very low, 2.4 kcal mol^{-1} for CF_2ClO and -3.2 kcal mol^{-1} for $CFCl_2O$; thus the predicted RRKM dissociation lifetimes are less than 10^{-11} s.

Alkoxy radicals of the form CHClXO, where X = H, Cl, F, can eliminate HCl via

$$CHClXO \rightarrow XCO + HCl \qquad (63)$$

in competition with the reaction of the alkoxy radical with O_2 and C—Cl bond fission. For the case X = H, the relative rate of reaction (63) versus the reaction with O_2 has recently been studied by Kaiser and Wallington [102], who find that $k_{O_2}/k_{dis} = 4.6 \times 10^{-18}\ cm^3$ at 298 K and 700 torr total pressure. In the ambient atmosphere this implies a branching fraction of 0.96 for O_2 addition; thus, unlike the 1-chloroethoxy radical, HCl elimination represents a minor pathway for atmospheric CH_2ClO loss.

The CF_3O radical is unique among the alkoxy radicals considered here in that none of the four reaction pathways listed above are available to it. The radical is too small to isomerize, it contains no hydrogen atoms for O_2 to attack, and the C—F bond is too strong for dissociation to occur. Its removal from the atmosphere will therefore occur by reaction with trace compounds. Rapid reactions with NO or hydrocarbons (RH) to yield, respectively, $FNO + F_2CO$ and $CF_3OH + R$ will provide sinks for the alkoxy radical. On the other hand, rapid reaction with O_3 will lead to a catalytic ozone destruction cycle. Recent kinetic measurements have shown that the reaction with ozone is slow and the reactions with NO and hydrocarbons are significantly faster, whereby the CF_3O removal processes will dominate over the catalytic cycle. The details of the CF_3O plus trace gas chemistry is the subject of the Section IV.

F. Summary of the Primary Steps in HFC and HCFC Atmospheric Degradation

Owing to the halogen substituents, HFC and HCFC atmospheric degradation leads to a number of products that do not occur in the oxidation of the unsubstituted hydrocarbons. It is important to understand the chemistry of these secondary species to assess the atmospheric impact of a particular HFC or HCFC. In some cases they continue to undergo gas-phase chemistry. Other species are resistant to gas-phase attack and are removed by photolysis or heterogeneous processes.

Figure 12 summarizes the atmospheric degradation pathways for the example of HFC-134a. The primary steps are those that were discussed in detail above. They begin with the halogenated ethane and end with the products of the alkoxy reactions. At this point either the C—C bond is broken, yielding a one-carbon radical and a halogenated formaldehyde (X_2CO, where X is a combination of F, Cl, or H) or an acid halide [$CX_3C(O)X$] plus HO_2 is formed. In the first case, the radical fragment begins anew the primary degradation process. The second time through the primary degradation sequence produces a halogenated formaldehyde or, from CF_3, the products CF_3OH and FNO.

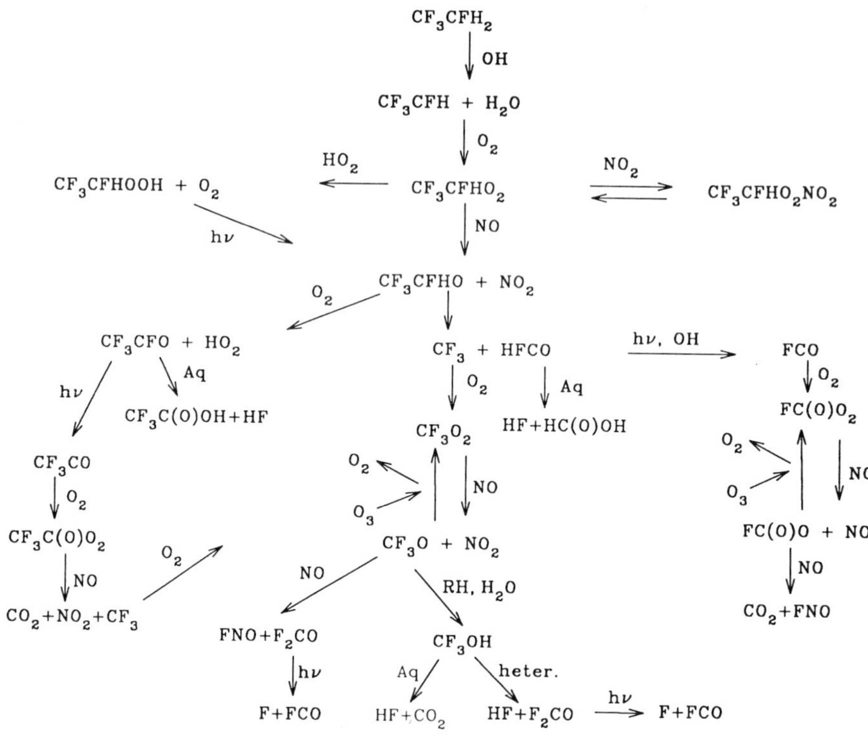

Figure 12. Reaction pathways for the atmospheric degradation of HFC-134a (CF_3CFH_2).

The substituted formaldehyde and acid halide compounds react only slowly with OH, if at all. Uptake into cloud water and hydrolysis yields

$$X_2CO + H_2O \rightarrow XCOOH + HX \tag{64}$$

$$CX_3C(O)X + H_2O \rightarrow CX_3COOH + HX \tag{65}$$

The aqueous chemistry of the hydrolysis products is still under investigation and is beyond the scope of the present chapter. Alternatively, the secondary species can be photolyzed via

$$X_2CO + h\nu \rightarrow XCO + X \tag{66}$$

$$CX_3C(O)X + h\nu \rightarrow CX_3CO + X \tag{67}$$

$$\rightarrow CX_3 + CO + X$$

In general, the absorption cross section is considerably larger when X = Cl vs. X = F and the absorption occurs at longer wavelength. As a result, photolysis of a C—Cl bond is substantially more efficient than of a C—F bond and will compete more effectively with heterogeneous removal. In either case, the haloformyl and haloacetyl radicals continue to undergo interesting gas-phase chemistry.

IV. REVIEW OF EXPERIMENTAL AND THEORETICAL STUDIES OF KEY SECONDARY SPECIES

A. Trifluoromethoxy Radicals (CF_3O)

The potential importance of the trifluoromethoxy radical to atmospheric chemical process was recognized by Francisco and Williams in 1984 [132]. Later, the potential role was amplified by its proposed formation from the degradation of alternative halocarbons [133]. The major production route of the CF_3O radical from alternative halocarbons is from the release of the trifluoromethyl group (CF_3). The CF_3 radical reacts with molecular oxygen to produce the trifluoroperoxy radical (CF_3O_2), which oxidizes NO to NO_2:

$$CF_3 + O_2 \rightarrow CF_3O_2 \tag{68}$$

$$CF_3O_2 + NO \rightarrow CF_3O + NO_2 \tag{69}$$

Trifluoromethoxy radicals have been postulated to be an intermediate in a number of reactions involving the oxidation of trifluoromethyl radicals [12, 92, 104, 109, 134]. The first measurement of the stability of CF_3O radical was by Kennedy and Levy [135], where they measured an activation energy of 31 kcal mol^{-1} for fluorine atom extrusion. The CF_3O radical was produced from the thermal decomposition of bis(trifluoromethyl) peroxide (CF_3OOCF_3). The activation energy was not a direct measurement. Francisco et al. [136] reexamined the decomposition of bis(trifluoromethyl) peroxide using infrared laser–induced dissociation. These authors found that the mechanism for the dissociation involved the formation of the two CF_3O radicals via cleavage of the OO bond,

$$CF_3OOCF_3 \rightarrow 2CF_3O \tag{70}$$

However, there was no direct detection of the CF_3O radical. A series of theoretical studies aimed at spectroscopically characterizing the CF_3O radical led to its direct detection. The first theoretical characterization of the CF_3O radical performed by Francisco and Williams [137] identified geometries, vibrational frequencies, and *ab initio* transition energies for low-lying excited states for CF_3O radicals. This study led to the first experimental identification of the radical in the infrared in an argon matrix. The matrix experiment performed by Andrews et al. [138] reported evidence of CF_3O in the photolysis of CF_3I in the presence of ozone. Two absorption bands in the photolysis product, at 1223.2 and 1222.0 cm^{-1}, were identified as being due to the CF stretching mode of the CF_3O radical. An independent study of CF_3O was reported by Clemitshaw and Sodeau [139], who observed an intense peak at 1221.6 cm^{-1} upon photolysis of a CF_3I/O_2 mixture at 4.2 K. Francisco et al. [140] performed *ab initio* calculations on decomposition pathways of CF_3O radicals and found that extrusion of fluorine has an activation energy 35.2 kcal mol^{-1} using PMP4/6-31G*//UMP2/6-31G* wavefunctions. This is consistent with the earlier experimental results of Kennedy and Levy [135], which reported an activation energy of 31.0 kcal mol^{-1}. This led the authors to suggest that the stability of CF_3O radicals will depend in part on the reactions that produce it. Reactions that produce CF_3O radicals with large excess of internal energies above its dissociation threshold had little likelihood of producing detectable quantities of CF_3O. This led to the first laser-induced fluorescence spectrum of CF_3O produced by infrared multiphoton dissociation of CF_3OOCF_3 [125]. Based on a comparison to *ab initio* calculations [137, 141], the reported origin of the $\tilde{A}^2A_1-\tilde{X}^2E$ electronic transition occurred at 21,000 cm^{-1} above the ground state and revealed a CO progression of 718 cm^{-1} in the excited $^2A_1$ state. This spectrum and assignment was later confirmed independently by Saathoff and Zeller [125]. The laser-induced fluorescence spectrum is shown in Figure 13.

The reactions of CF_3O radicals are critical toward accessing the potential impact of secondary products resulting from alternative halocarbon degradation on ozone perturbations. The potential for ozone removal by CF_3O radicals was recognized by Francisco and Williams [132]. Example cycles originally pointed out are:

Cycle I:

$$CF_3O_2 + O \rightarrow CF_3O + O_2 \tag{71}$$

$$CF_3O + O_3 \rightarrow CF_3O_2 + O_2 \tag{72}$$

$$O + O_3 \rightarrow 2O_2 \tag{73}$$

Cycle II:

$$CF_3 + O_3 \rightarrow CF_3O + O_2 \tag{74}$$

$$CF_3O + O \rightarrow CF_3 + O_2 \tag{75}$$

$$O + O_3 \rightarrow 2O_2 \tag{76}$$

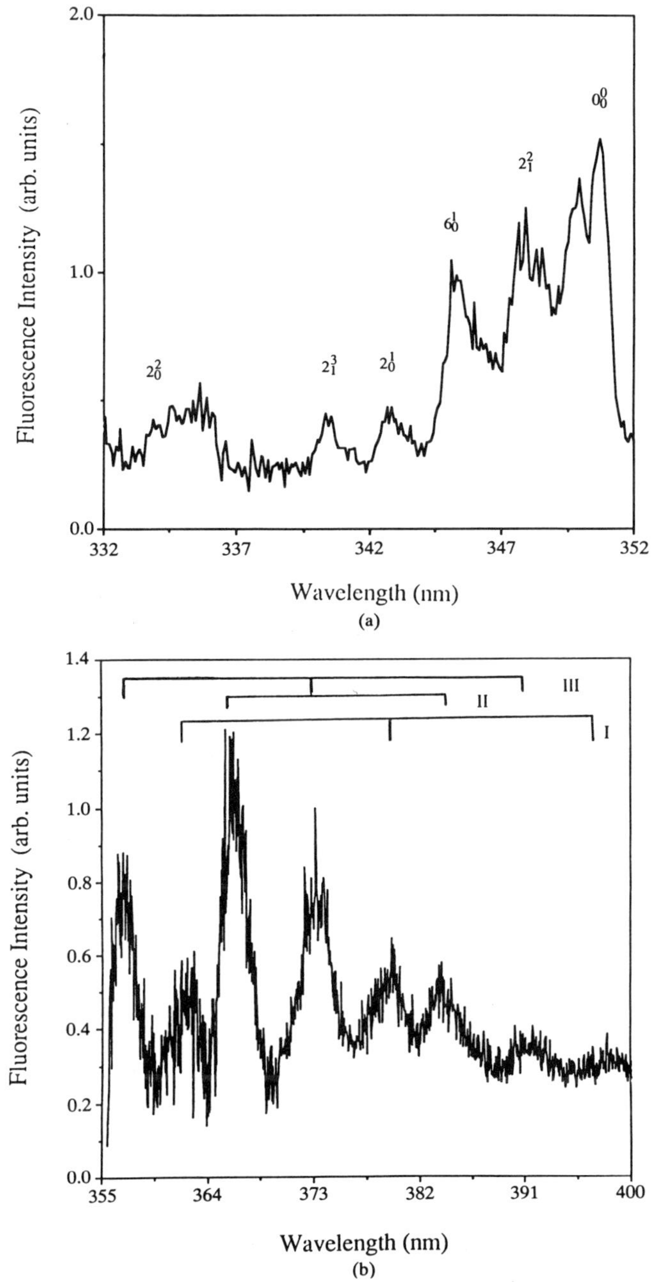

Figure 13. Laser-induced fluorescence spectra of the CF_3O radical: (*a*) laser excitation spectrum; (*b*) dispersed fluorescence spectrum.

Cycle III:

$$CF_3O_2 + O_3 \rightarrow CF_3O + 2O_2 \quad (77)$$

$$CF_3O + O_3 \rightarrow CF_3O_2 + O_2 \quad (78)$$

$$2O_3 \rightarrow 3O_2 \quad (79)$$

These cycles could be effective if the rates of reactions for (71), (72), (75), and (77) are fast. Or the impact of these proposed cycles could be small if the reaction rates for the aforementioned reaction are slow relative to the rates for reactions that remove the CF_3O radical. How these rates compete with processes that remove the CF_3O radical is a critical issue. Moreover, identification of the key removal processes for CF_3O radicals is an important concern.

The following processes are key removal reactions:

$$CF_3O + NO \rightarrow CF_2O + FNO \quad (80)$$

$$CF_3O + (CH_4, \text{alkanes}) \rightarrow CF_3OH + (CH_3, R') \quad (81)$$

$$CF_3O + H_2O \rightarrow CF_3OH + HO \quad (82)$$

First, we review recent studies of the removal reactions. These are summarized in Table 9. In the laser-induced fluorescence study of Li and Francisco [124], NO was added in the presence of the CF_3O radical and a decay of the CF_3O fluorescence signal was observed. A product analysis of the reaction mixture using Fourier transform-infrared spectroscopy (FT-IR) showed that the dominant reaction was to form CF_2O and FNO. This was later confirmed in a FT-IR product study of Chen et al. [142]. Howard and co-workers [106] used chemical ionization mass spectrometric detection in a flow tube reactor to measure a rate constant for the CF_3O + NO reaction of $2 \pm 1 \times 10^{-11}$ cm^3 s^{-1} at 297 K, a value that was later revised to 6×10^{-11} cm^3 s^{-1} [123]. The latter value is in good agreement with Sehested and Nielsen [107], who measured a rate constant of $5.2 \pm 2.7 \times 10^{-11}$ cm^3 s^{-1} at 295 K using pulse radiolysis/UV absorption. Very recently, Turnipseed et al. [122] have measured a rate constant of $3.3 \times 10^{-11} e^{160/T}$ cm^3 s^{-1}, showing a slight negative temperature dependence, whereas Fockenberg et al. [143] find the rate constant to be temperature independent from 220 to 300 K at 4.8×10^{-11} cm^3 s^{-1}. While there are some discrepancies between the individual measurements, there is good agreement that the CF_3O + NO reaction is fast and that it will form a sink for CF_3O.

Rate constants have recently been measured for the reaction of CF_3O with a variety of small hydrocarbons, which show that these reactions, too,

TABLE 9 CF_3O Removal Processes

Reaction	k ($cm^3 s^{-1}$)	Temp. Range (K)	Method[a]	Reference
Reactions with NO_x				
$CF_3O + NO \rightarrow FNO + CF_2O$	$5.2 \pm 2.7 \times 10^{-11}$	295	PR/UV	107
	$2 \pm 1 \times 10^{-11}$	297	CIMS	106
	$(3.3 \pm 0.7) \times 10^{-11} \times \exp[(160 \pm 45)/T]$	233–360	LIF	122
	$(4.8 \pm 1.0) \times 10^{-11}$	220–300	LIF	143
$CF_3O + NO_2 \rightarrow CF_3ONO_2$	$(1.5 \pm 0.2) \times 10^{-11}$	298	LIF	143
Reactions with hydrocarbons				
$CF_3O + CH_4 \rightarrow CF_3OH + CH_3$	2.2×10^{-14}	298	LIF	125
	$1.2 \pm 0.1 \times 10^{-14}$	298	RRM	146
	$(1.9 \pm 0.3) \times 10^{-12} \times \exp[-(1370 \pm 85)/T]$	240–360	LIF	147
	$3.3 \times 10^{-12} \exp[-(1430)/T]$		LIF	148
$CF_3O + C_2H_6 \rightarrow$ products	1.2×10^{-12}			146
	$\sim 10^{-14}$			144
	$(4.8 \pm 1.1) \times 10^{-12} \times \exp[-(400 \pm 70)/T]$	233–360	LIF	147
$CF_3O + C_3H_8 \rightarrow$ products	$4.7 \pm 0.7 \times 10^{-12}$			146
	$\sim 10^{-15}$			144
	$(5.1 \pm 1.1) \times 10^{-12} \times \exp[-(15 \pm 30/T]$	240–360	LIF	147
$CF_3O + C_2H_4 \rightarrow$ products	$3.0 \pm 0.1 \times 10^{-11}$			146
$CF_3O + C_6H_6 \rightarrow$ products	$3.6 \pm 0.1 \times 10^{-11}$			146
Reaction with H_2O				
$CF_3O + H_2O \rightarrow CF_3OH + HO$	$0.2–40 \times 10^{-17}$	296	FT-IR	145

[a]PR/UV, pulse radiolysis/UV absorption; CIMS, chemical ionization mass spectrometry; LIF, laser-induced fluorescence; RRM, relative rate method; FT-IR, FT-IR method.

can serve as CF_3O removal mechanisms. The simplest and atmospherically most abundant of these is methane. There is good agreement that the rate constant is approximately 2×10^{-14} $cm^3\,s^{-1}$ at 298 K and that the activation energy is near 1400 K (see Table 9). For larger hydrocarbons, the rate constant increases dramatically. Except for the unusually small values reported by Chen et al. [144], the rate constants for ethane and propane are, respectively, nearly 100 and 400 times larger than that of methane (see Table 9). Measurements of the rate constants for reactions between CF_3O and unsaturated hydrocarbons show that these are also fast reactions. In all cases the CF_3O acts as a hydroxyl radical, abstracting a hydrogen atom from the hydrocarbon to form CF_3OH. The fate of this species is currently unknown; however, smog chamber studies indicate that it decays slowly, most likely by a heterogeneous mechanism [145].

Trifluoromethanol is also produced by the reaction of CF_3O radicals with water:

$$CF_3O + H_2O \rightarrow CF_3OH + HO \tag{83}$$

Because there is an abundance of water vapor in both the troposphere and lower stratosphere where clouds are formed, it is likely that water may provide a facile way to remove CF_3O radicals from CF_3O_x catalytic cycles. Wallington et al. [145] estimated a rate coefficient to be in the range of 0.2 to 40×10^{-17} $cm^3\,s^{-1}$ at 296 K from FT-IR measurement following photolysis in a smog chamber. Comparing CF_3O reaction rates for CH_4 [146, 147], NO, and H_2O, shows that the reaction of CF_3O radicals with NO is a major chain-terminating reaction. However, because the concentration of CH_4 is greater than that of NO, reaction (81) also becomes a significant termination route for CF_3O radicals. The reaction rate of CF_3O radicals with larger hydrocarbons also leads to a similar major termination route [148].

The effectiveness of catalytic cycles in destroying stratospheric ozone depends on the efficiency of the chain propagation reactions in the cycles [i.e., reactions (71)–(79)] versus the rates of the chain termination reactions. In the proposed cycles the reactions of CF_3O and CF_3O_2 radicals with ozone are critical to assessing how propagation reactions compete with termination reactions. Rate constants for these reactions are summarized in Table 10.

The first measurement of CF_3 radicals with ozone was performed by Rossi et al. [149] using very low pressure photolysis. A rate coefficient of 9.3×10^{-12} cm^3 $molecule^{-1}$ s^{-1} was measured for the $CF_3 + O_3$ reaction.

TABLE 10 CF_3O_x Reactions with Ozone and Oxygen Atoms

Reaction	k ($cm^3 s^{-1}$)	Temp. Range (K)	Method[a]	Reference
$CF_3 + O_3$	9.3×10^{-12}		VLPϕ	149
$CF_3O + O_3$	$<1 \times 10^{-13}$	295	PR/UV	120
	$<3 \times 10^{-14}$	295	FT-IR	152
	$<5 \times 10^{-14}$	210–253	FP/TRUV	121
	$(2.5^{+0.7}_{-1.5}) \times 10^{-14}$	298	LIF	122
	$<4 \times 10^{-14}$	298	CIMS	123
	1×10^{-12}	298		151
$CF_3O_2 + O_3$	$<0.5 \times 10^{-14}$	295	PR/UV	120
	$<1 \times 10^{-14}$	210–253	FP/TRUV	121
	$<7 \times 10^{-15}$	298	LIF	122
	$<3 \times 10^{-15}$	298	CIMS	123
$CF_3O + O$	$\approx 2 \times 10^{-11}$			[b]
$CF_3O_2 + O$	$\approx 4 \times 10^{-11}$			[c]

[a] As in Table 9; also, VIPϕ, very low pressure photolysis; FP/TRUV, flash photolysis/time-resolved UV absorption.

[b] Based on the $CH_3O + O \rightarrow CH_3 + O_2$ reaction: (1) C. J. Cabos, H. Hippler, K. Cuather, A. R. Ravishankara, and J. Troe, *J. Phys. Chem.* **89**, 4332 (1985); (2) F. Ewig, D. Rhasa, R. Zellner, *Ber. Bungenges. Phys. Chem.* **91**, 708 (1987).

[c] Based on the $CH_3O_2 + O \rightarrow CH_3O + O_2$: R. Zellner, D. Hartmann, J. Karthauser, G. Weibring, *J. Chem. Soc. Faraday Trans. II* **84**, 549 (1988).

The reaction of CF_3O radicals with oxygen atoms has not been measured. However, an examination of the potential energy surface (see Figure 14) for the $CF_3O + O$ reaction suggests that the reaction should proceed without a barrier to form CF_3 radicals and molecular oxygen [15]:

$$CF_3O + O \rightarrow CF_3O_2^* \rightarrow CF_3 + O_2 \tag{84}$$

Although cycle II may be energetically plausible, the effectiveness of the cycle is diminished by the rapid removal of CF_3 radicals by the reaction with molecular oxygen.

Reaction 75 is important in the removal of odd-oxygen atoms. Li and Francisco [150] pointed out that the midstratosphere, where oxygen atom concentrations are greatest, the coupling of CF_3O_x and NO_x chemistry could form an effective catalytic cycle for the removal of oxygen atoms

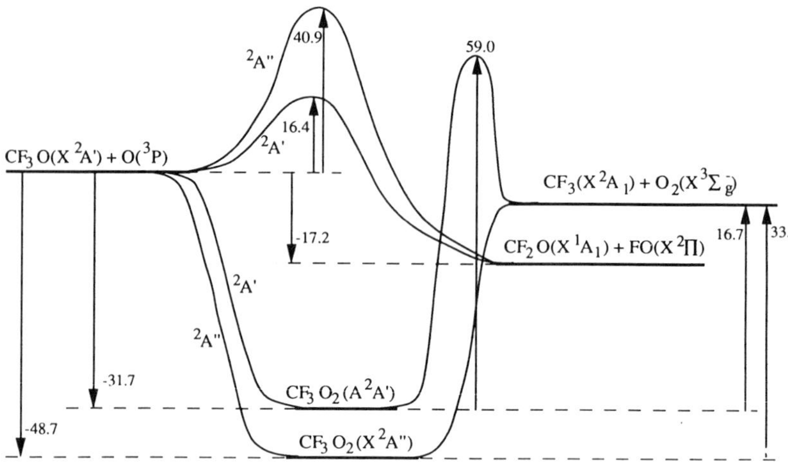

Figure 14. Potential energy surface for the $CF_3O + O$ reaction.

through the following coupling reactions:

Cycle IV:

$$CF_3 + O_2 \rightarrow CF_3O_2 \quad (85)$$

$$CF_3O_2 + NO \rightarrow CF_3O + NO_2 \quad (86)$$

$$CF_3O + O \rightarrow CF_3 + O_2 \quad (87)$$

$$NO_2 + O \rightarrow NO + O_2 \quad (88)$$

$$O + O \rightarrow O_2 \quad (89)$$

The rate constant for reaction (85) is 2.5×10^{-12} cm^3 s^{-1}, for reaction (86) it is 1.5×10^{-11} cm^3 s^{-1} [106], and for reaction (87) it is estimated as 2.0×10^{-11} cm^3 s^{-1}. These rate constants suggest that the CF_3O_x/NO_x coupled cycle for the removal of odd-oxygen atoms may be plausible; however, because NO concentrations are over an order of magnitude higher than O atom concentrations at the altitudes where peak ozone levels occur, CF_3O termination reactions will dominate this cycle.

The effectiveness of cycles I and III is critically dependent on the reaction of CF_3O radicals with O_3. Initially, Biggs et al. [151] reported a rate coefficient of 1×10^{-12} cm^3 s^{-1} for this reaction, which was of considerable concern for its ozone-depletion potential. Subsequently, Nielsen and Schested [120], Wallington et al. [152], and Maricq and Szente [121] simultaneously reported upper limits of $k < 1 \times 10^{-13}$, 3×10^{-14}, and 5×10^{-14} cm^3 s^{-1} for the $CF_3O + O_3$ reaction rate constant. The small rate constant has been verified by the recent measurements of Turnipseed et al. [122], $k = 2.5 \times 10^{-14}$ cm^3 s^{-1}, and Ravishankara et al. [123], $k < 4 \times 10^{-14}$ cm^3 s^{-1}. In light of these findings, it appears that the ozone-depletion effectiveness of cycles I and III is minimal.

From another perspective it is interesting to compare the results above with the analogous reaction of HO radicals with ozone. The rate coefficient for the latter reaction is 6.8×10^{-14} cm^3 s^{-1} [35]. Consequently, the measured rates for CF_3O and HO with ozone are comparable considering that both reactions involve an oxygen-atom transfer from ozone to RO or HO radical. An *ab initio* examination of the potential energy surface shows that these reactions proceed via a linear RO_4 intermediate [153]. Such an intermediate is formed by the attachment of the oxygen of the RO radical to the terminal ozone oxygen.

There is a number of measurements of the rate coefficient for the reaction of CF_3O_2 radicals with ozone. Nielsen and Sehested [120] measured an upper limit of the rate as 0.5×10^{-14} cm^3 s^{-1}, while Maricq and Szente [121] measured an upper limit of 1×10^{-14} cm^3 s^{-1}. More recently, Turnipseed et al. [122] find an upper limit of 7×10^{-15} cm^3 s^{-1} and Ravishankara [123] report an upper limit of 3×10^{-15} cm^3 s^{-1}. All these measurements show that the reaction of CF_3O_2 radicals with ozone is slow and that the impact of this reaction on ozone depletion, via cycle III, is small.

B. Fluoroformyl Radicals (FCO)

A major fate of CF_3O radicals is conversion to carbonyl difluoride (CF_2O). This results from several removal processes: (1) dissociation of CF_3O radicals:

$$CF_3O \rightarrow CF_2O + F \tag{90}$$

(2) the reaction of CF_3O radicals with NO, which is a major chain termination reaction:

$$CF_3O + NO \rightarrow CF_2O + FNO \tag{91}$$

or (3) the reaction of CF_3O radicals with hydrocarbons to yield trifluoromethanol (CF_3OH), which heterogeneously decomposes to form HF and CF_2O:

$$CF_3O + RH \rightarrow CF_3OH + R \tag{92}$$

$$CF_3OH \rightarrow CF_2O + HF \tag{93}$$

Carbonyl difluoride is not sufficiently unstable to dissociate thermally at a significant rate, nor are there strong forces to drive reactions with likely partners such as O_2, NO, or NO_2. The photochemistry of carbonyl difluoride is not completely understood. Workman and co-workers [154, 155] published the gas-phase absorption spectrum of CF_2O between 238 and 122 nm. Judge and Moule [156] made vibrational assignments for 150 bands and determined the origin of the first excited state to be at 39,252 cm^{-1}. More recently, Francisco et al. [157] examined the interpretation of the spectra measured by Judge and Moule and by Workman and co-workers using *ab initio* computations of the excited state of CF_2O. They find the assignment of Judge and Moule to be consistent with the computational results. Absorption cross sections have been measured by Molina and Molina [158] and more recently by Nolte and co-workers [159] with a diode array spectrometer.

The energetics of CF_2O is the subject of serious disagreement. The heat of formation for CF_2O reported in JANAF is -152.7 ± 0.4 kcal mol^{-1} [160]. However, recently *ab initio* calculations by Montgomery et al. [161] find the heat of formation for CF_2O to be -144.8 ± 1 kcal mol^{-1}, indicating that this quantity needs to be reexamined experimentally. An accurate value is required, for example, to determine the stability of CF_3OH toward decomposition via reaction (93). In shock-tube studies of the kinetics of the decomposition of CF_2O at high temperatures, Modica [162] found an activation barrier of 91.9 kcal mol^{-1}; Weston and Matala [163] determined a value of 93.3 kcal mol^{-1}. Both experiments attribute these activation barriers to the energy needed to break the carbon–fluorine bond to produce fluoroformyl radical (FCO):

$$CF_2O \rightarrow F + FCO \tag{94}$$

Molina and Molina [158] determined a quantum yield for dissociation of CF_2O to be 0.25 at 206 nm. Nolte and co-workers [159] observed a yield of 0.47 ± 0.03. They point out further that if the FCO radicals recombine via the reaction

$$FCO + FCO \rightarrow CO + CF_2O \tag{95}$$

the true quantum yield could be nearly unity. Because a fluoroformyl radical is a likely product of the photodissociation of CF_2O and of the photodissociation of CFClO and HFCO, its chemistry is important in understanding the impact of the degradation products of alternative halocarbons. In 1989, the atmospheric concentration of CF_2O was estimated at 1.4×10^{14} cm^{-3} s^{-1} at an altitude of 18 km, increasing by about 10% each year [164]. With the increasing use of HFC-134a, the concentration of CF_2O could increase beyond 10% each year, and consequently the significance of understanding FCO chemistry becomes more important. Fluoroformyl radicals have been shown to be moderately stable [165]. In the atmosphere, FCO will react with O_2 to give $FC(O)O_2$ radicals [82, 166, 167], which in turn react with NO to form FC(O)O radicals [168, 169].

$$FCO + O_2 \rightarrow FC(O)O_2 \tag{96}$$

$$FC(O)O_2 + NO \rightarrow FC(O)O + NO_2 \tag{97}$$

The atmospheric fate of FCO, FC(O)O, and $FC(O)O_2$ radicals is the subject of much current research. There has been speculation that $FC(O)O_x$ radicals [these include FCO, FC(O)O, and $FC(O)O_2$] can participate directly in catalytic cycles analogous to those for CF_3O [170]. Before the atmospheric chemical consequences of FCO radicals are discussed, we outline what is known and what has yet to be determined about these radicals.

JANAF thermochemical tables [160a] estimate the heat of formation of FCO radicals to be -41 ± 15 kcal mol^{-1}. MacNeil and Thynne [171] determined the F—CO bond energy from electron impact ionization study of CF_2O, arriving at a heat of formation of -42 kcal mol^{-1}. Henrici et al. [172] suggested a value of -34 ± 5 kcal mol^{-1} from shock tube experiments. Bowers and Chau [173] estimated a value of -42 ± 4 kcal mol^{-1}. Using isogyric and isodesmic reaction schemes and *ab initio* methods, Francisco and Zhao [174] estimated the heat of formation of FCO as -42.9 ± 0.8 kcal mol^{-1}, consistent with the measurements of Bower and Chau.

The spectroscopy of FCO is reasonably well understood. The first infrared spectrum of FCO formed in a CO matrix was obtained by Milligan and co-workers [175]. Three peaks due to FCO were assigned to the three fundamentals. Jacox [176] reexamined the spectra of FCO using isotopic substitution to refine the vibrational assignment of FCO. The CO stretch mode for FCO occurs at 1861 cm^{-1} (ν_1), the FCO bend is at 628 cm^{-1}, and the CF stretch occurs at 1026 cm^{-1}. Nagai and co-workers [177] used infrared diode laser spectroscopy to obtain a high-resolution spectra of the CO and CF stretching modes. There have been several studies of the ultraviolet spectra obtained in both the matrix and gas phase [175, 176, 178–180]. Fluoroformyl radical absorbs in the region 220 to 340 nm with a spacing of about 650 cm^{-1}. Three progressions are observed; two from the B ← X transition and one from the C ← X transition. Maricq and co-workers [180] reported a low-resolution gas-phase spectrum which was similar to that of Jacox. However, more recently, Maricq and co-workers [181] have obtained a high-resolution spectrum of FCO and found the origin of the B ← X transition to be at 24,000 to 27,000 cm^{-1} and that of the C ← X state at 28,500 ± 25 cm^{-1}. This refined spectrum (see Figure 15) is consistent with the *ab initio* prediction of these transitions by Francisco and co-workers [182].

The major atmospheric sources of FCO are expected to be the photolysis of CF_2O and FClCO; however, with the input of alternative halocarbons such as HCF-134a, an additional source of FCO is carbonyl monofluoride (HFCO). Another source is the reaction of fluorine atoms with atmospheric CO. As aforementioned, the photolysis of CF_2O produces FCO with quantum yields that range from 0.47 to 1. Reaction of HFCO with HO radicals, F, and Cl atoms are also sources of FCO radicals.

$$\text{HFCO} + \text{HO} \rightarrow \text{FCO} + \text{H}_2\text{O} \tag{98}$$

$$\text{HFCO} + (\text{F, Cl}) \rightarrow \text{FCO} + (\text{HF, HCl}) \tag{99}$$

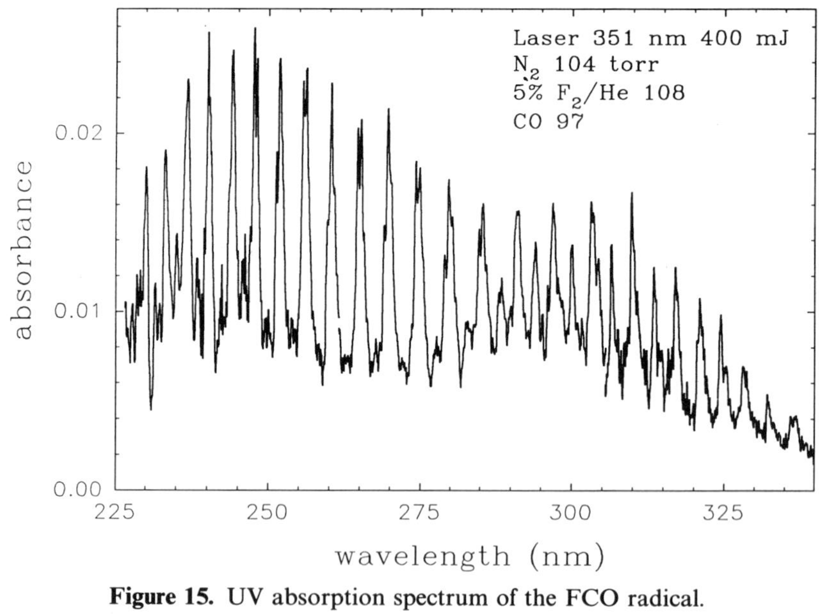

Figure 15. UV absorption spectrum of the FCO radical.

The reaction rate coefficient for the reaction of Cl atoms with HFCO measured at room temperature was reported as 2.0×10^{-15} cm^3 molecule^{-1} s^{-1} by Wallington and co-workers [99]. The reaction of HFCO with fluorine atoms is faster, with a measured rate coefficient of $2.3 \pm 0.2 \times 10^{-12}$ cm^3 s^{-1} [183]. More recently, a rate constant of $8.8 \pm 1.4 \times 10^{-13}$ cm^3 s^{-1} has been measured for the reaction of F atoms [184]. Using *ab initio* molecular orbital calculations, Francisco and Mina-Camilde [185] showed that the difference between the rate constants is a result of the difference in the activation barriers for the extraction of hydrogen by fluorine vs. chlorine atoms. The barrier for fluorine abstraction of hydrogen from HFCO is 1.6 kcal mol^{-1}, and that for chlorine abstraction of hydrogen is 3.2 kcal mol^{-1}. Francisco [186] has estimated, using *ab initio* methods, a rate constant of 7.0×10^{-16} cm^3 s^{-1} for the reaction of HO radicals with HFCO. This value is consistent with the upper limit of 4×10^{-15} cm^3 s^{-1} reported for this reaction by Wallington and Hurley [130]. The reaction is expected to be slow because the predicted activation barrier is 4.2 kcal mol^{-1}. There has also been interest in another reaction that produces fluoroformyl radicals, namely the reaction of difluoromethylene (CF_2) radicals with oxygen atoms:

$$CF_2 + O \rightarrow FCO + F \tag{100}$$

which has been studied by several groups [187–189].

There are several loss processes for FCO radicals:

$$FCO + O_2 \rightarrow FC(O)O_2 \tag{101}$$

$$FCO + O \rightarrow FO + CO \tag{102}$$

$$FCO + FCO \rightarrow CF_2O + CO \tag{103}$$

$$FCO + ClO \rightarrow FC(O)OCl \rightarrow FCl + CO_2 \tag{104}$$

$$FCO + NO \rightarrow FNO + CO \tag{105}$$

The most abundant reaction partner in the stratosphere is molecular oxygen [reaction (101)]. Maricq et al. [82] have measured a rate constant of $(8.0 \pm 0.8) \times 10^{-13}$ cm^3 s^{-1} for this reaction and found it to be temperature independent over the range 213 to 385 K. This suggests that the reaction is essentially barrierless, in agreement with *ab initio* calculations [167]. Another removal reaction is with atomic oxygen [reaction (102)]. This reaction has been studied experimentally by Ryan and Plumb [187], and computationally by Francisco and Ostafin [190]. The measured rate constant for the process is $(9.3 \pm 2.1) \times 10^{-11}$ $cm^3\,s^{-1}$. The reactions of FCO radicals with ClO radicals have not been studied experimentally. Francisco [191] examined the reaction computationally and found it to be exothermic with no activation barrier. Moreover, the exothermicity is sufficient to induce molecular dissociation via

$$FC(O)OCl \rightarrow FCl + CO_2 \tag{106}$$

Maricq et al. [180] have studied the self-reaction of FCO radicals. Although the self-reaction will be of minor importance for upper atmospheric chemistry, this reaction is critical in laboratory studies involving FCO radicals. For example, in the measurement of quantum yields from CF_2O, CFClO, and HFCO photodissociation, the self-reaction could lead to a gross underestimation of the quantum yields. The rate coefficient of $1.9 \pm 0.2 \times 10^{-11}$ $cm^3\,s^{-1}$ suggests that the self-reaction is quite rapid. In the self-reaction, the FCO radicals recombine to form oxylfluoride, which is produced with excess energy above the barrier for molecular dissociation to carbonyl fluoride and carbon monoxide [192].

C. Fluoroformyloxyl [FC(O)O] and Fluoroformylperoxy [$FC(O)O_2$] Radicals

The dominant removal process for FCO radicals is also the major formation reaction of $FC(O)O_2$ radicals, that is, reaction (101). It has been suggested that the interconversion reaction of $FC(O)O_2$ radicals with ozone could lead to the possible involvement of $FC(O)O_2$ and FC(O)O radicals in catalytic ozone destruction cycles.

Cycle V:

$$FC(O)O_2 + O_3 \rightarrow FC(O)O + 2O_2 \tag{107}$$

$$FC(O)O + O_3 \rightarrow FC(O)O_2 + O_2 \tag{108}$$

$$2O_3 \rightarrow 3O_2 \tag{109}$$

Francisco and co-workers [170] proposed that these radicals could couple to the NO_x cycle to create a catalytic cycle for the destruction and regeneration of odd oxygen in the upper atmosphere via

Cycle VI:

$$FCO + O_2 \rightarrow FC(O)O_2 \tag{110}$$

$$FC(O)O_2 + NO \rightarrow FC(O)O + NO_2 \tag{111}$$

$$FC(O)O + O \rightarrow FCO + O_2 \tag{112}$$

$$NO_2 + O \rightarrow NO + O_2 \tag{113}$$

$$O + O \rightarrow 2O_2 \tag{114}$$

Cycle VII:

$$FCO + O_2 \rightarrow FC(O)O_2 \tag{115}$$

$$FC(O)O_2 + NO \rightarrow FC(O)O + NO_2 \tag{116}$$

$$FC(O)O + NO \rightarrow FC(O)ONO \tag{117}$$

$$FC(O)ONO \rightarrow FCO + NO_2 \tag{118}$$

$$2\left\{\begin{array}{l} NO_2 \rightarrow NO + O \quad (119) \\ O + O_2 \rightarrow O_3 \quad (120) \end{array}\right\}$$

$$3O_2 \rightarrow 2O_3 \tag{121}$$

It has been only recently that the chemistry of FC(O)O and $FC(O)O_2$ radicals has attracted attention, because of the potential importance of these radicals in understanding the atmospheric chemical impact of HCFCs and HFCs. Before these recent studies, there had been no detection of these radicals. However, they were postulated to exist in reactions of fluorine and carbon monoxide by Schumacher and co-workers [193] and by Fox and Franz [194]. Both FC(O)O and $FC(O)O_2$ had their first characterization from computations. Francisco and Ostafin [168] studied the structure and vibrations of FC(O)O. They suggested that the ground state of FC(O)O is a $^2B_2$ state. They predicted two low-lying excited states; the first excited, $\tilde{A}^2A_2$, state was predicted to lie in the near-infrared region of about 3.4 μm. The second excited state, which is the $\tilde{B}^2A_1$ state, was predicted to lie near 725 nm. Maricq and co-workers [169] provided the first experimental characterization of the FC(O)O radical. The visible absorption spectrum of the $\tilde{B}^2A_1 \leftarrow \tilde{X}^2B_2$ transition of FC(O)O (see Figure 16) was analyzed with the aid of *ab initio* calculations. Six vibrational progressions were resolved with an origin near 760 nm. The spectrum was dominated by the OCO

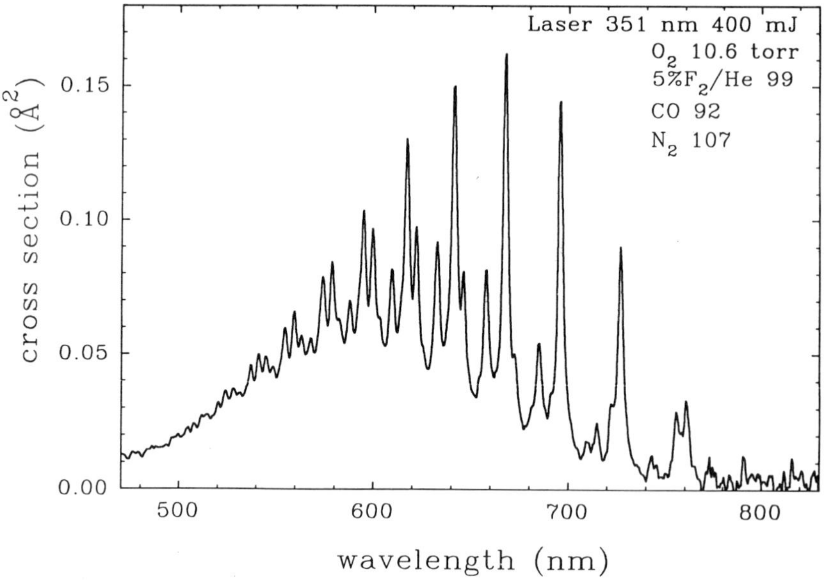

Figure 16. Visible absorption spectrum of the FC(O)O radical.

bending mode of the B state, which has a frequency of $610\,cm^{-1}$. The ground-state frequency of that mode was determined to be $520\,cm^{-1}$.

The first characterization of $FC(O)O_2$ was by the *ab initio* calculations of Francisco and Williams [195]. The excited state was predicted to occur in the near-infrared region at 3606 to $4703\,cm^{-1}$. The experimental characterization of the $FC(O)O_2$ radical was made by Maricq et al. [82]. Two broad and featureless bands were observed in the range 190 to 300 nm and absolute absorption cross sections were reported (see Figure 11). The lower-energy band was assigned to the $\tilde{X}^2A'' \rightarrow 2^2A''$ transition; while the higher-energy band was tentatively assigned to the $\tilde{X}^2A'' \rightarrow 3^2A''$ transition.

The estimates of the heats of formations for FC(O)O and $FC(O)O_2$ are -79.1 and $-69.1\,kcal\,mol^{-1}$ [196]. These are derived from *ab initio* calculations with PMP4/6-31G*//UMP2/6-31G* levels of theory. The probable uncertainty is within $\pm 5\,kcal\,mol^{-1}$. Antonik [166] estimated the experimental heat of formation for $FC(O)O_2$ as $-71 \pm 2\,kcal\,mol^{-1}$ and is in reasonable agreement with the *ab initio* estimate of $-69 \pm 1\,kcal\,mol^{-1}$.

The reaction that interconnects $FC(O)O_2$ radicals to FC(O)O radicals is

$$FC(O)O_2 + NO \rightarrow FC(O)O + NO_2 \qquad (122)$$

Rate constants for $FC(O)O_x$ reactions are listed in Table 11. The reaction above has been studied by Wallington and co-workers [83]. The rate coefficient at 296 ± 2 K for this reaction is $2.5 \pm 0.8 \times 10^{-11}$ cm^3 s^{-1}, which is consistent with the measured rate coefficient for other peroxy radical reactions with NO [9].

A preliminary upper limit of $k < 1.5 \times 10^{-15}$ cm^3 s^{-1} has recently been observed [87] for the reaction of $FC(O)O_2$ radicals with ozone:

$$FC(O)O_2 + O_3 \rightarrow FC(O)O + 2O_2 \qquad (123)$$

For the reaction of FC(O)O radicals with ozone,

$$FC(O)O + O_3 \rightarrow FC(O)O_2 + O_2 \qquad (124)$$

two upper limits in good agreement have been measured. Wallington and co-workers [83] found the rate constant to be $<6 \times 10^{-14}$ cm^3 s^{-1}, whereas the upper limit from Maricq et al. [197] is $k < 3 \times 10^{-15}$ cm^3 s^{-1} at 295 K. A comparison with the rate coefficient for the reaction of $CF_3O + O_3$ (see Table 10) suggests that these limits should be reliable. Although somewhat preliminary, these data suggest that cycle V should have minimal impact and hence $FC(O)O_x$-induced ozone destruction is probably unimportant.

Among the potential loss reactions is the reaction of FC(O)O radicals with methane:

$$FC(O)O + CH_4 \rightarrow FC(O)OH + CH_3 \qquad (125)$$

TABLE 11 FC(O)O and $FC(O)O_2$ Reactions

Reaction	k^a	Temp. Range (K)	Reference
$FC(O)O + O_3 \rightarrow FC(O)O_2 + O_2$	$\leqslant 6 \times 10^{-14}$		83
	$\leqslant 3 \times 10^{-15}$		197
$FC(O)O + NO \rightarrow FNO + CO_2$	$(1.3 \pm 0.7) \times 10^{-10}$	296 ± 2	83
	$(3.9 \pm 0.6) \times 10^{-11}$		198
	$(3.6 \pm 0.5) \times 10^{-11} (T/300)^{-0.1 \pm 0.5}$	230–323	197
$FC(O)O + CH_4 \rightarrow FC(O)OH + CH_3$	$\leqslant 2 \times 10^{-16}$		198
$FC(O)O_2 + HO \rightarrow FC(O)O + NO_2$	$2.5 \pm 0.8 \times 10^{-11}$	296 ± 2	83

^{a}Units are cm^3 s^{-1}.

Only one measurement has been made [198]. The reported upper limit of $\leqslant 2 \times 10^{-16}$ $cm^3\,s^{-1}$, excludes CH_4 as a major reagent for the loss of $FC(O)O_x$ and raises the question of whether a low reactivity of FC(O)O radicals toward hydrogen abstraction processes is general.

A key reaction that is critical to assessing the impact of potential catalytic activity of $FC(O)O_x$ radicals is the reaction of FC(O)O radicals with NO:

$$FC(O)O + NO \rightarrow \text{products} \tag{126}$$

There are several reaction pathways for this reaction:

$$FC(O)O + NO \rightarrow FCO + NO_2 \tag{127}$$

$$\rightarrow FC(O)ONO \tag{128}$$

$$\rightarrow FC(O)NO_2 \rightarrow FNO_2 + CO \tag{129}$$

$$\rightarrow FCO + NO_2 \tag{130}$$

Dibble and Francisco [199] have examined the details of the potential energy surface for the FC(O)O + NO reaction (see Figure 17). They find that the reaction of FC(O)O + NO should pass through an FC(O)ONO adduct. However, the exothermicity of the formation of the adduct directly from FC(O)O + NO is more than twice the activation energy for its dissociation to FNO + CO_2. The net reaction

$$FC(O)O + NO \rightarrow FNO + CO_2 \tag{131}$$

is exothermic by ~ 51 kcal mol^{-1}. Because the exothermicity is less than the N—F bond energy of FNO, FNO is expected to be a stable product of this reaction. There have been three experimental studies of the FC(O)O + NO reaction. Wallington and co-workers [83] measured the reaction rate by following the visible absorption change of FC(O)O radicals in the presence of NO. The observed rate coefficient for the reaction was found to be $1.3 \pm 0.7 \times 10^{-10}$ $cm^3\,s^{-1}$; however, this number is too high owing to corrections that were later shown to be unnecessary. The FNO yield was $78 \pm 4\%$. Zellner [198] measured a rate constant of $3.9 \pm 0.7 \times 10^{-11}$ $cm^3\,s^{-1}$. Maricq et al. [197] identified the product yield of this reaction to be essentially 100% FNO plus CO_2. No evidence of the FC(O)ONO adduct was found spectroscopically. These authors determined a rate constant of 3.6×10^{-11} $(T/300)^{-0.1}$ $cm^3\,s^{-1}$ over the temperature range of 233 to

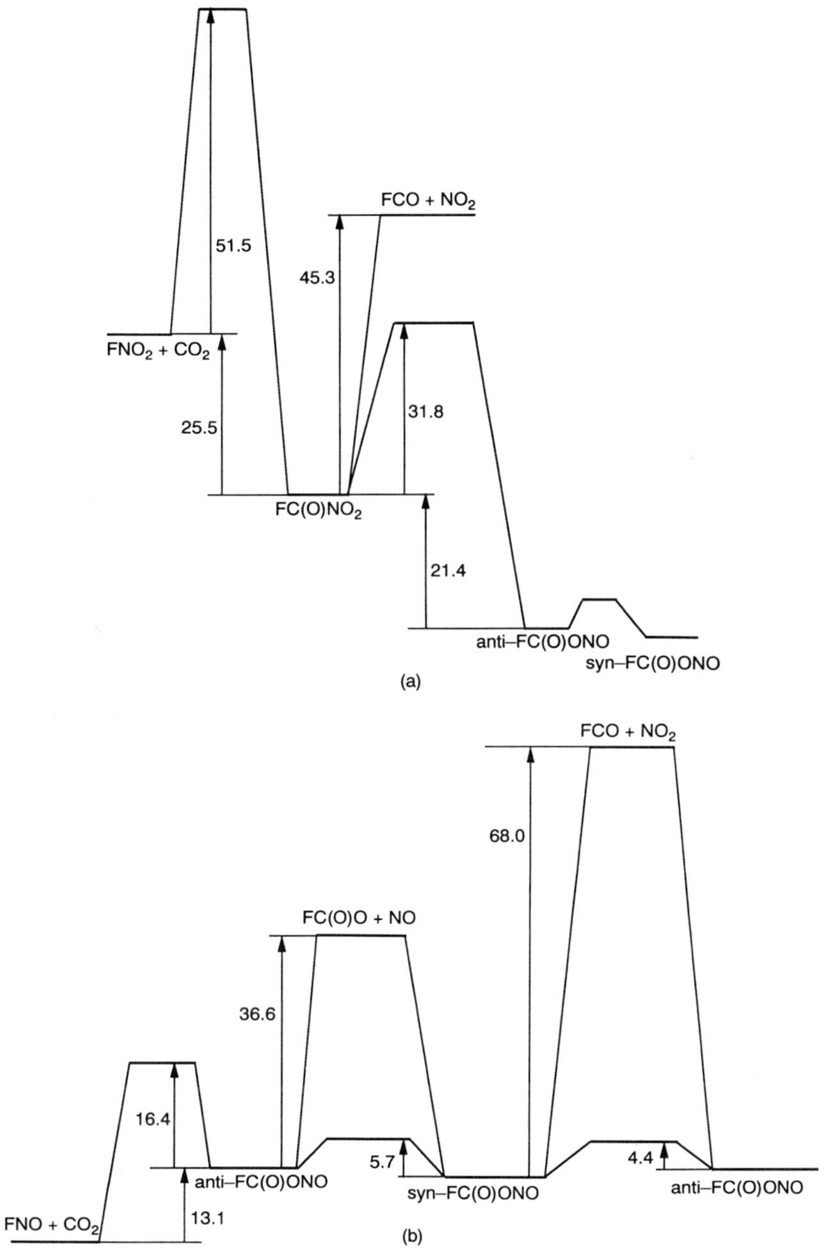

Figure 17. Potential energy surface for the formation and dissociation of (*a*) FC(O)NO_2 and (*b*) FC(O)ONO. Energies are in kcal mol^{-1} at the QCISD(T)/6-31G*//MP2/6-31g* + ΔZPE level of theory.

323 K [197]. This rate constant is nearly independent of temperature and is in good agreement with the result of Zellner [198]. These experimental and theoretical results suggest that cycles V and VII are unfeasible, because rather than facilitating the conversion of FC(O)O to FCO by means of oxidizing NO, the FC(O)O + NO reaction provides a terminal step in the photooxidation process. Moreover, a comparison of the rates for removal of FC(O)O radicals with NO, O_3, and CH_4 suggest that the dominant removal process for FC(O)O radicals is the FC(O)O + NO reaction (131), and hence that this terminates CX_3 photooxidation.

D. Chemistry of $CF_3C(O)O_x$ Radicals

In the degradation of hydrofluorocarbons and hydrochlorofluorocarbons of the type $CF_3CF_{3-x-y}Cl_xH_y$, the photooxidation process leads to the formation of $CF_3C(O)X$ species (where X = H, F, or Cl). In the case of $CF_3C(O)H$, which is an expected by-product of the photooxidation of CF_3CClH_2 (HCFC-133a) and CF_3CH_3 (HFC-143a), the reaction with HO radicals forms trifluoroacetyl radical (CF_3CO):

$$CF_3C(O)H + HO \rightarrow CF_3CO + H_2O \qquad (132)$$

In contrast, the photolysis of $CF_3C(O)Cl$ produces CF_3 radicals and carbon monoxide via [200,201]

$$CF_3C(O)Cl + h\nu \rightarrow CF_3 + CO + Cl \qquad (133)$$

The existence of CF_3CO radicals was first postulated by Tucker and Whittle [202]. These authors photolyzed hexafluoroacetone in the presence of Br_2 and Cl_2. To explain the observed products, a mechanism involving the production of CF_3CO radicals was proposed. Kerr and Wright [203] studied the gas phase decomposition of CF_3CO radicals over the range 338 to 417 K and showed that the activation energy of CF_3CO radicals is 19.8 kcal mol^{-1}, suggesting that the radicals are quite stable. This is contrary to the findings of Amphlett and Whittle [204], who found an activation energy of about 10 kcal mol^{-1}. An *ab initio* study [205] of the stability of CF_3CO radicals shows that the activation energy for the radicals is 11.4 kcal mol^{-1}; close to the results of Amphlett and Whittle [204].

There are only two spectroscopic studies of the CF_3CO radical. The first direct observation of the CF_3CO radical comes from infrared absorption-spectroscopy using a rapid-scan infrared spectrometer [206]. The radical was produced by flash photolysis of CF_3COI and four absorption bands of the radical were detected. They are the CO stretching mode at 1574 cm^{-1}, the CF_3 stretching (a symmetry) at 1338 cm^{-1}, the CF_3 stretching mode (e symmetry) at 1115 cm^{-1}, and the CC stretching mode at 1024 cm^{-1}. Li and Francisco [207] observed the laser-induced fluorescence spectrum of the CF_3CO radical following the IR multiphoton dissociation of $CF_3C(O)Cl$. The band origin of the $\tilde{A} \rightarrow \tilde{X}$ transition lies at 26,023 cm^{-1} (384.28 nm). The spectrum shows a progression involving the CO stretching mode. This mode in the excited state is 818 cm^{-1}. A dispersed fluorescence spectrum was obtained and the observed fundamental CO stretching mode in the ground state is 1525 cm^{-1}. This result is in reasonable agreement with the frequency of the CO mode observed by Kozuka and Isobe [85]. More recently, Maricq and co-workers [96] observed a weak UV absorption band attributed to CF_3CO, with an onset at 250 nm and increasing in intensity below 200 nm. This could be due to the $\tilde{B} \rightarrow \tilde{X}$ transition of CF_3CO.

The atmospheric formation of CF_3CO radicals is primarily via hydrogen abstraction from $CF_3C(O)H$ by OH radicals (reaction 132). Table 12 collects the rate constants measured for this reaction as well as those for some other pertinent reactions. Dobe et al. [208], using the discharge flow resonance fluorescence method, measured the rate coefficient to be $1.1 \pm 0.7 \times 10^{-12}$ $cm^3\,s^{-1}$. Scollard et al. [209] measured the rate with two independent techniques: by laser photolysis/resonance fluorescence and by the relative rate method. The methods, respectively, yield rate constants of 0.65 ± 0.05, and $0.55 \pm 0.12 \times 10^{-12}$ $cm^3\,s^{-1}$, nearly a factor 2 slower than the value of Dobe et al. The only temperature-dependent study for the reaction is the one by Dobe et al. [208]. They measured an activation energy for the $CF_3C(O)H + HO$ reaction of 1.4 kcal mol^{-1}. Francisco [210] finds from *ab initio* studies that the major pathway for the reaction is the hydrogen atom abstraction from the aldehyde group. In the transition state, the hydrogen is positioned between the attacking electrophilic HO radical and the CF_3CO radical. Francisco also found using PMP4/6-31G**// UMP2/6-31G* wavefunctions that the activation energy is 2.5 ± 2 kcal mol^{-1}. However, the reaction is expected to have a considerable tunneling contribution.

TABLE 12 $CF_3C(O)O_x$ Radical Reactions

Reaction	k^a	Temp. Range (K)	Method[b]	Reference
$CF_3C(O)H + HO \rightarrow CF_3CO + H_2O$	$1.1 \pm 0.7 \times 10^{-12}$		DF-RF	208
	$0.65 \pm 0.05 \times 10^{-12}$	298 ± 2	LP-RF	209
	0.55 ± 0.12	298 ± 2	RRM	209
$CF_3C(O)H + F \rightarrow CF_3CO + HF$	$2.7 \pm 0.1 \times 10^{-11}$	295 ± 2	RRM	211
	$8 \pm 2 \times 10^{-12}$	295	LPC	212
$CF_3C(O)H + Cl \rightarrow CF_3CO + HCl$	$1.8 \pm 0.4 \times 10^{-12}$	295 ± 2	RRM	211
	$2.7 \pm 0.1 \times 10^{-12}$	298 ± 2	RRM	209
$CF_3CO \rightarrow CF_3 + CO$	$k_\infty = 1.5 \times 10^{11} e^{-(4240 \pm 660)/T}$	223–353		96
	$k_0 = 3.9 \times 10^{-11} e^{-(2550 \pm 650)/T}[N_2]$			
$CF_3CO + O_2 \rightarrow CF_3C(O)O_2$				

[a]Units are $cm^3\ s^{-1}$, except for the CF_3CO decomposition, which has units of s^{-1}.
[b]DF-RF, discharge flow–resonance fluorescence; LP-RF, laser photolysis–resonance fluorescence; RRM, relative rate method; LPC, laser photolysis–chemiluminescence.

Hydrogen abstraction by halogen atoms such as fluorine and chlorine

$$CF_3C(O)H + (F, Cl) \rightarrow CF_3CO + (HF, HCl) \quad (134)$$

also yields the CF_3CO radical. The reaction of fluorine atoms with $CF_3C(O)H$ is approximately a factor of 10 faster than the reaction of F atoms with most alternative halocarbons [211]. In fact, Wallington suggested that to detect $CF_3C(O)H$ as an oxidation product of the alternative halocarbons, it is desirable to use F-atom-initiated oxidation. Using the relative rate method, Wallington [211] measured the rate coefficient to be $2.7 \pm 0.1 \times 10^{-11}$ $cm^3\,s^{-1}$. Li et al. [212] used laser photolysis of SF_6 as a source of fluorine atoms in the presence of $CF_3C(O)H$ and followed the reaction by monitoring the temporal behavior of the chemiluminescence from the HF reaction product. They measured a rate constant of $8 \pm 2\ 10^{-12}\,cm^3\,s^{-1}$.

The rate constant for chlorine attack on the hydrogen of $CF_3C(O)H$ is nearly an order of magnitude smaller than that of fluorine. Both Scollard [209] and Wallington [211] have measured the rate constant for abstraction by chlorine. Their results are in discrepancy by nearly 50% even though they use the same relative rate method. Nevertheless, a comparison between fluorine and chlorine shows that the rate for hydrogen abstraction is faster for fluorine than for chlorine. Dibble and Francisco [213], using *ab initio* calculations, found that this is attributed largely to differences in the activation energies for the two processes.

Maricq and co-workers [96] have studied the decomposition of CF_3CO radicals into $CF_3 + CO$ both by following the decay of the UV absorption by CF_3CO radicals and the appearance of CO by diode laser absorption. At 295 K the decomposition is in the falloff region above 40 torr N_2, and only approaches the high-pressure limit at well over 1 atm (see Table 12). Conversely, Li and Francisco [207] found that the reaction of CF_3 radicals with CO produces CF_3CO radicals in low yields:

$$CF_3 + CO \rightarrow CF_3CO \quad (135)$$

For the CF_3CO radicals produced in the atmosphere, their most probable fate is to react with molecular oxygen to yield $CF_3C(O)O_2$ radicals:

$$CF_3CO + O_2 \rightarrow CF_3C(O)O_2 \quad (136)$$

This reaction has been observed by Maricq et al. [96]. $CF_3C(O)O$ radicals are formed from $CF_3C(O)O_2$ radical disproportionation,

$$CF_3C(O)O_2 + CF_3C(O)O_2 \rightarrow 2CF_3C(O)O + O_2 \qquad (137)$$

and via reaction of $CF_3C(O)O_2$ radicals with NO,

$$CF_3C(O)O_2 + NO \rightarrow CF_3C(O)O + NO_2 \qquad (138)$$

However, the $CF_3C(O)O$ radicals are not long lived. These radicals are unstable with respect to dissociation (i.e., $\Delta H_{r,0} = 32.5\,\text{kcal mol}^{-1}$) [205]. Moreover, the barrier for their dissociation into CF_3 radicals and CO_2

$$CF_3C(O)O \rightarrow CF_3 + CO_2, \qquad (139)$$

has been estimated by Francisco [205] to be quite low, $4.9\,\text{kcal mol}^{-1}$.

A theoretical examination of reactions between $CF_3C(O)O_x$ radicals and major stratosphere species, such as O, O_2, O_3, and NO_3, has led to the conclusion that these radicals will not play a direct role in catalytic cycle processes for the removal of ozone. The net impact of the formation of $CF_3C(O)O_x$ radicals in the atmosphere is to yield CF_3 radicals and either CO or CO_2. These theoretical conclusions are confirmed by the experimental results of Maricq et al. [96]; however, more work is required to understand the details of $CF_3C(O)O_x$ chemistry. The CF_3 radical that is produced goes on to form CF_3O, but as discussed in Section IVa, at present this appears to have a negligible impact on stratospheric ozone depletion. Nevertheless, the chemistry of $CF_3C(O)O_x$ species provide an important connection to the CF_3CX_2H and CX_3H degradation chemistry.

V. SUMMARY

Chlorofluorocarbons have unique chemical and thermodynamic properties that make them very attractive for a wide variety of applications, ranging from refrigeration and air conditioning to foam blowing and medical sterilants. Unfortunately, as mounting evidence shows, their inertness allows CFCs to be transported to the stratosphere, where their photolysis releases chlorine atoms that participate in catalytic ozone destruction. Consequently, the manufacture and use of these compounds is being phased out.

Hydrochlorofluorocarbon and hydrofluorocarbon compounds are being introduced as first-generation substitutes. They have been presumed "ozone friendly" because they retain one or more hydrogen atoms, making them susceptible to tropospheric degradation by the attack of OH radicals. A great deal of research has been performed within the last five years to explore the atmospheric implications of the substitute compounds. The purpose has been to ensure, as best possible, that the CFC replacements do not themselves contribute to ozone depletion or contribute to some new environmental problem. The summary of this research has been the objective of this review chapter.

The research has shown that not all HFCs and HCFCs are equally environmentally acceptable from a gas-phase atmospheric chemistry standpoint. Most are only slowly removed, allowing the possibility that a fraction of what is released at the earth's surface reaches the stratosphere. For HCFCs, subsequent photolysis releases chlorine, making these compounds less attractive as CFC substitutes. Even if the parent HCFC undergoes tropospheric degradation, some of the chlorine-containing by-products can reach the stratosphere and release Cl atoms. It is up to the atmospheric modelers to ascertain the ozone depletion potentials implied by these chemical processes.

Because of the absence of chlorine, HFCs pose no direct ozone depletion threat. However, some of their degradation products (i.e., the alkoxy radicals) possess unusual properties. The prime example is CF_3O, which cannot undergo any of the usual chemical reactions that remove alkoxy radicals from the atmosphere. Another example is the FC(O)O radical. Because of their unusual character, possibilities were raised that these species could partake in catalytic ozone depletion cycles. If true, these processes would render the relevant HFCs unsuitable. As demonstrated by a number of research groups, the chain propagation reactions are slow and the chain termination reactions are fast; thus this is not the case.

The gas-phase chemistry reviewed in this chapter shows how the pattern of substitution of chlorine and fluorine atoms on the parent hydrocarbon affects its atmospheric degradation. Although numerous reaction details need to be investigated and some discrepancies resolved, a core of reliable data is available from which one can deduce which HFC or HCFC may have deleterious atmospheric consequences and which potential substitute appears environmentally acceptable. However, the work is not finished. This work has dealt only with the gas-phase chemistry. Research is still needed into the heterogeneous chemistry of HFC and HCFC degradation products and into the biological effects that they may have. It is then up to the atmospheric modelers to combine this scientific information with emission scenarios and meterology to ascertain the feasibility of a particular CFC substitute.

ACKNOWLEDGMENTS

We would like to thank J. J. Szente for help in generating a number of the tables and figures that appear in this work. We have benefited from helpful discussions with T. J. Wallington and W. F. Schneider at Ford and S. P. Sander and R. R. Friedl at JPL on a variety of HCFC atmospheric chemistry issues. We are grateful for the preprints and private communications received from C. J. Howard, A. R. Ravishankara, R. Zellner, O. J. Nielsen, and G. D. Hayman, which have helped us to incorporate the more recent results into the present review.

REFERENCES

1. J. P. Jessen, *Proc. NATO Advanced Study Institute on Atmospheric Ozone: Its Variation and Human Influences*, U.S. Dept. Trans. Report FAA-EE-80-20 (1980).
2. F. S. Rowland and M. J. Molina, *Rev. Geophys. Space Phys.* **13**, 1 (1975).
3. (a) R. A. Rasmussen and M. A. K. Khali, *Chemosphere* **12**, 371 (1983). (b) P. J. Crutzen and L. T. Gidel, *J. Geophys. Res.* **88**, 6641 (1983).
4. J. D. Farman, B. G. Gardiner, and J. D. Shanklin, *Nature* **315**, 207 (1985).
5. S. Solomon, *Nature* **347**, 6291 (1990).
6. *World Meteorological Organization Global Ozone Research and Monitoring Project*, Report 20; *Scientific Assessment of Stratospheric Ozone*, Vol. 1, 1989.
7. R. Atkinson, *Atmos. Environ.* **24a**, 1 (1990).
8. R. Atkinson, *Chem. Rev.* **86**, 69 (1986).
9. R. Atkinson, D. L. Baulch, R. A. Cox, R. F. Hampson, Jr., J. A. Kerr, and J. Troe, *J. Phys. Chem. Ref. Data*, **21**, 1125 (1992).
10. A. F. Wagner, I. R. Slagle, D. Sarynski, and D. Gutman, *J. Phys. Chem.* **94**, 1853 (1990).
11. T. J. Wallington, P. Dagaut, and M. J. Kurylo, *Chem. Rev.* **92**, 667 (1992).
12. P. D. Lightfoot, R. A. Cox, J. N. Crowley, M. Destriau, G. D. Hayman, M. E. Jenkin, G. K. Moortgat, and F. Zabel, *Atmos. Environ.* **26A**, 1805 (1992).
13. R. P. Wayne, *Chemistry of Atmospheres*, 2nd ed., Clarendon Press, Oxford, 1991.
14. M. M. Maricq, J. Shi, J. J. Szente, L. Rimai, and E. W. Kaiser, *J. Phys. Chem.* **97**, 9686 (1993).
15. J. Shi, T. J. Wallington, and E. W. Kaiser, *J. Phys. Chem.* **97**, 6184 (1993).

16. J. Sehested and T. J. Wallington, *Environ. Sci. Technol.* **27**, 146 (1993).
17. V. L. Orkin and V. G. Khamaganov, *J. Atmos. Chem.* **16**(2), 157 (1993).
18. T. Gierczak, R. Talukdar, G. L. Vaghjiani, E. R. Lovejoy, and A. R. Ravishankara, *J. Geophys. Res. Atmos.* **96**(D3), 5001 (1991).
19. A. C. Brown, C. E. Canosa-Mas, A. D. Parr, and R. P. Wayne, *Atmos. Environ.* **24A**, 2499 (1990).
20. R. Liu, R. E. Huie, and M. J. Kurylo, *J. Phys. Chem.* **94**, 3247 (1990).
21. M. A. A. Clyne and P. M. Holt, *J. Chem. Soc. Faraday Trans. 2* **75**, 582 (1979).
22. R. T. Watson, A. R. Ravishankara, G. Machado, S. Wagner, and D. D. Davis, *Int. J. Chem. Kinet.* **11**, 187 (1979).
23. C. J. Howard and K. M. Evenson, *J. Chem. Phys.* **64**, 4303 (1976).
24. R. Talukdar, A. Mellouki, T. Gierczak, J. B. Burkholder, S. A. McKeen, and A. R. Ravishankara, *J. Phys. Chem.* **95**, 5815 (1991).
25. J. P. Martin and G. Paraskevopoulos, *Can. J. Chem.* **61**, 861 (1983).
26. K. M. Jeong, K. J. Hsu, J. B. Jeffries, and F. Kaufman, *J. Phys. Chem.* **88**, 1222 (1984).
27. V. Handwerk and R. Zellner, *Ber. Bunsenges. Phys. Chem.* **82**, 1161 (1978).
28. W. B. DeMore, *Geophys. Res. Lett.* **20**, 1359 (1993).
29. Z. Zhang, R. E. Huie, and M. J. Kurylo, *J. Phys. Chem.* **96**, 1533 (1992).
30. G. Paraskevopoulos, D. L. Singleton, and R. S. Irwin, *J. Phys. Chem.* **85**, 561 (1981).
31. R. T. Watson, G. Machado, B. Conaway, S. Wagner, and D. D. Davis, *J. Phys. Chem.* **81**, 256 (1977).
32. W. S. Nip, D. L. Singleton, R. Overend, and G. Paraskevopoulos, *J. Phys. Chem.* **83**, 2440 (1979).
33. K. M. Jeong and F. Kaufman, *J. Phys. Chem.* **86**, 1808 (1982).
34. C. J. Howard and K. M. Evenson, *J. Chem. Phys.* **64**, 197 (1976).
35. W. B. DeMore, S. P. Sander, D. M. Golden, R. F. Hampton, M. J. Kurylo, C. J. Howard, A. R. Ravishankara, C. E. Kolb, and M. J. Molina, *Chemical Kinetics and Photochemical Data for Use in Stratospheric Modeling*, JPL Publication 92-20 (1992).
36. P. H. Wine and D. H. Semmes, *J. Phys. Chem.* **87**, 3572 (1983).
37. E. C. Tuazon, R. Atkinson, and S. B. Corchnoy, *Int. J. Chem. Kinet.* **24**, 639 (1992).
38. T. J. Wallington and M. D. Hurley, *Chem. Phys. Lett.* **189**, 437 (1992).
39. O. J. Nielsen, T. Ellermann, J. Sehested, and T. J. Wallington, *J. Phys. Chem.* **96**, 10875 (1992).
40. E. Tschuikow-Roux, T. Yano, and J. Niedzielski, *J. Chem. Phys.* **82**, 65 (1985).
41. J. P. Sawerysyn, A. Talhaoui, B. Meriaux, and P. Devolder, *Chem. Phys. Lett.* **198**, 197 (1992).

42. J. L. Jourdain, G. Poulet, G. Barassin, B. Lebras, and J. Combourieu, *Pollut. Atmos.* **75**, 256 (1977).
43. T. J. Wallington, M. D. Hurley, J. C. Ball, T. Ellermann, O. J. Nielsen, and J. Sehested, *J. Phys. Chem.* **98**, 5435 (1994).
44. T. Yano and E. Tschuikow-Roux, *J. Photochem.* **32**, 25 (1986).
45. S. Glavas and J. Heicklen, *J. Photochem.* **31**, 21 (1985).
46. E. Tschuikow-Roux, F. Faraji, S. Paddison, J. Niedzielski, and K. Miyokawa, *J. Phys. Chem.* **92**, 1488 (1988).
47. T. J. Wallington, J. C. Ball, O. J. Nielsen, and E. Bartkiewicz, *J. Phys. Chem.* **96**, 1241 (1992).
48. K. Y. Choo, *J. Kor, Chem. Soc.* **23**, 349 (1979).
49. R. G. Manning and M. J. Kurylo, *J. Phys. Chem.* **81**, 291 (1977).
50. T. J. Wallington, M. D. Hurley, J. Shi, M. M. Maricq, J. Sehested, O. J. Nielsen, and T. Ellermann, *Int. J. Chem. Kinet.* **25**, 651 (1993).
51. O. J. Nielsen, unpublished data.
52. T. J. Wallington and O. J. Nielsen, *Chem. Phys. Lett.* **187**, 33 (1991).
53. O. J. Nielsen, T. Ellermann, J. Sehested, E. Bartkiewicz, T. J. Wallington, and M. D. Hurley, *Int. J. Chem. Kinet.* **24**, 1009 (1992).
54. M. M. Maricq and J. J. Szente, *J. Phys. Chem.* **96**, 4925 (1992).
55. M. A. A. Clyne and A. Hodgson, *J. Chem. Soc. Faraday Trans. 2*, 443 (1985).
56. O. J. Nielsen, T. Ellermann, E. Bartkiewicz, T. J. Wallington, and M. D. Hurley, *Chem. Phys. Lett.* **192**, 82 (1992).
57. A. S. Manocha, D. W. Setser, and M. A. Wichramaaratchi, *Chem. Phys.* **76**, 129 (1983).
58. M. A. Wickramaaratchi, D. W. Setser, H. Hildebrandt, B. Korbitzer, and H. Heydmann, *Chem. Phys.* **94**, 109 (1985).
59. M. A. A. Clyne, D. J. McKenny, and R. F. Walker, *Can. J. Chem.* **51**, 3596 (1973).
60. R. Warren, T. Gierczak, and A. R. Ravishankara, *Chem. Phys. Lett.* **183**, 403 (1991).
61. R. G. Green and R. P. Wayne, *J. Photochem.* **6**, 371 (1977).
62. J. A. Davidson, H. I. Schiff, T. J. Brown, and C. J. Howard, *J. Chem. Phys.* **69**, 4277 (1978).
63. I. S. Fletcher and D. Husain, *J. Phys. Chem.* **80**, 1837 (1976).
64. J. N. Pitts, Jr., H. L. Sandoval, and R. Atkinson, *Chem. Phys. Lett.* **29**, 31 (1974).
65. J. J. Orlando, J. B. Burkholder, S. A. McKeen, and A. R. Ravishankara, *J. Geophys. Res.* **96**, 5013 (1991).
66. D. Gillotay and P. C. Simon, *J. Atmos. Chem.* **13**, 289 (1991).
67. R. G. Green and R. P. Wayne, *J. Photochem.* **6**, 375 (1976).
68. C. Hubrich and F. S. Stuhl, *J. Photochem.* **12**, 93 (1980).
69. G. Herzberg, *Electronic Spectra of Polyatomic Molecules*, Van Nostrand, New York, 1966, p. 619.

70. J. Troe, *J. Phys. Chem.* **83**, 114 (1979).
71. K. R. Ryan and J. C. Plumb, *Int. J. Chem. Kinet.* **16**, 591 (1984).
72. R. Danis, F. Carlap, M. T. Rayez, and R. Lesclaux, *J. Phys. Chem.* **95**, 7300 (1991).
73. J. J. Russel, J. A. Seetala, D. Gutman, F. Danis, F. Carlap, P. D. Lightfoot, R. Lesclaux, C. F. Melius, and S. M. Senkan, *J. Phys. Chem.* **94**, 3277 (1990).
74. R. Cooper, J. B. Cumming, S. Gordon, and W. A. Mulac, *Radiat. Phys. Chem.* **16**, 169 (1980).
75. F. Carlap and R. Lesclaux, *Chem. Phys. Lett.* **102**, 54 (1983).
76. K. R. Ryan and I. C. Plumb, *J. Phys. Chem.* **86**, 4678 (1982).
77. F. Carlap, R. Lesclaux, and A. M. Dognon, *Chem. Phys. Lett.* **129**, 433 (1986).
78. W. F. Schneider and T. J. Wallington, *J. Phys. Chem.* **97**, 12783 (1993).
79. M. M. Maricq, J. J. Szente, and E. W. Kaiser, *Chem. Phys. Lett.* **197**, 149 (1992).
80. E. W. Kaiser, T. J. Wallington, and J. Andino, *Chem. Phys. Lett.* **168**, 309 (1990).
81. P. Neta, R. E. Huie, and A. B. Ross, *J. Phys. Chem. Ref. Data* **19**, 413 (1990).
82. M. M. Maricq, J. J. Szente, G. A. Khitrov, and J. S. Francisco, *J. Chem. Phys.* **98**, 9522 (1993).
83. T. J. Wallington, T. Ellermann, O. J. Nielsen, and J. Sehested, *J. Phys. Chem.* **98**, 2346 (1994).
84. J. A. Jafri and D. H. Phillips, *J. Am. Chem. Soc.* **112**, 2586 (1990).
85. G. Herzberg, *Spectra of Diatomic Molecules*, Van Nostrand, New York, 1950, pp. 388–394.
86. J. G. Winans and E. C. G. Stueckelberg, *Proc. Natl. Acad. Am.* **14**, 867 (1928).
87. M. M. Maricq and J. J. Szente, unpublished data.
88. T. J. Wallington, T. Ellermann, and O. J. Nielsen, submitted to *Res. Chem. Intermed.* **20**, 265 (1994).
89. A. A. Jemi-Alade, P. D. Lightfoot, and R. Lesclaux, *Chem. Phys. Lett.* **179**, 119 (1991).
90. G. D. Hayman, M. E. Jenkin, T. P. Murrells, and S. J. Shalliker, *STEP-Halocside/AFEAS Workshop*, Dublin, 1991.
91. J. Sehested, T. Ellermann, O. J. Nielsen, T. J. Wallington, and M. D. Hurley, *Int. J. Chem. Kinet.* **25**, 701 (1993).
92. M. M. Maricq and J. J. Szente, *J. Phys. Chem.* **96**, 10862 (1992).
93. T. J. Wallington and O. J. Nielsen, *Int. J. Chem. Kinet.* **23**, 785 (1991).
94. P. Dagaut, T. J. Wallington, M. J. Kurylo, *Chem. Phys. Lett.* **146**, 589 (1988).
95. P. Dagaut, T. J. Wallington, and M. J. Kurylo, *Int. J. Chem. Kinet.* **20**, 815 (1988).
96. M. M. Maricq, J. J. Szente, G. A. Khitrov, T. S. Dibble, and J. S. Francisco, *J. Phys. Chem.*, submitted.

97. M. M. Maricq, J. J. Szente, E. W. Kaiser, and J. Shi, *J. Phys. Chem.* **98**, 2083 (1994).
98. T. J. Wallington, J. M. Andino, and S. M. Japar, *Chem. Phys. Lett.* **165**, 189 (1990).
99. T. J. Wallington, M. D. Hurley, J. C. Ball, and E. W. Kaiser, *Environ. Sci. Technol.* **26**, 1318 (1992).
100. E. Sanhueza and J. Heicklen, *J. Phys. Chem.* **79**, 7 (1975).
101. H. Niki, P. D. Maker, C. M. Savage, and L. P. Breitenbach, *Int. J. Chem. Kinet.* **12**, 1001 (1980).
102. E. W. Kaiser and T. J. Wallington, *J. Phys. Chem.* **98**, 5679 (1994).
103. J. Sehested, O. J. Nielsen, and T. J. Wallington, *Chem. Phys. Lett.* **213**, 457 (1993).
104. A. M. Dognon, F. Carlap, and R. Lesclaux, *J. Chim. Phys. Phys. Chim. Biol.* **82**, 349 (1985).
105. R. Lesclaux and F. Carlap, *Int. J. Chem. Kinet.* **16**, 1117 (1984).
106. T. J. Bevilacqua, D. R. Hanson, and C. J. Howard, *J. Phys. Chem.* **97**, 3750 (1993).
107. J. Sehested and O. J. Nielsen, *Chem. Phys. Lett.* **206**, 369 (1993).
108. J. Peeters, J. Vertommen, and I. Langhans, *Ber. Bunsenges. Phys. Chem.* **96**, 431 (1992).
109. I. C. Plumb and K. R. Ryan, *Chem. Phys. Lett.* **92**, 236 (1982).
110. R. Atkinson, W. P. L. Carter, and A. M. Winer, *J. Phys. Chem.* **87**, 2012 (1983).
111. I. R. Slagle and D. Gutman, *J. Am. Chem. Soc.* **107**, 5342 (1985).
112. S. W. Benson, *Thermochemical Kinetics*, Wiley, New York, 1976.
113. F. Carlap, R. Lesclaux, M.-T. Rayez, J.-C. Rayez, and W. Frost, *J. Chem. Soc. Faraday Trans. 2* **84**, 569 (1988).
114. F. Wu and R. W. Carr, *Int. J. Chem. Kinet.* **23**, 701 (1991).
115. D. Köppenkastrop and F. Zabel, *Int. J. Chem. Kinet.* **23**, 1 (1991).
116. F. Kirchner, F. Zabel, and K. H. Becker, *STEP-Halocside/AFEAS Workshop*, Dublin, 1991.
117. N. Haahr, J. Hjorth, and G. Ottobrini, *STEP-Halocside/AFEAS Workshop*, Dublin, 1991.
118. M. M. Maricq, J. J. Szente, M. O. Hurley, and T. J. Wallington, *J. Phys. Chem.* **98**, 8962 (1994).
119. G. D. Hayman, *AFEAS Workshop on Atmospheric Degradation of HCFCs and HFCs*, Boulder, Colo., Nov. 1993.
120. O. J. Nielsen and J. Sehested, *Chem. Phys. Lett.* **213**, 433 (1993).
121. M. M. Maricq and J. J. Szente, *Chem. Phys. Lett.* **213**, 449 (1993).
122. A. A. Turnipseed, S. B. Barone, and A. R. Ravishankara, *J. Phys. Chem.* **98**, 4594 (1994).
123. A. R. Ravishankara, A. A. Turnipseed, N. R. Jensen, S. Barone, M. Mills, C. J. Howard, and S. Solomon, *Science*, 1993, **263**, 71 (1994).
124. Z. Li and J. S. Francisco, *Chem. Phys. Lett.* **186**, 336 (1991).
125. H. Saathoff and R. Zellner, *Chem. Phys. Lett.* **206**, 349 (1993).

126. J. S. Francisco, Z. Li, A. Bradley, and A. E. W. Knight, *Chem. Phys. Lett.* **214**, 77 (1993).
127. E. O. Edney, B. W. Gay, Jr., and D. J. Driscoll, *J. Atmos. Chem.* **12**, 105 (1991).
128. E. O. Edney and D. J. Driscoll, *Int. J. Chem. Kinet.* **24**, 1067 (1992).
129. E. C. Tuazon and R. Atkinson, *J. Atmos. Chem.* **16**, 301 (1993).
130. T. J. Wallington and M. D. Hurley, *Environ. Sci. Technol.* **27**, 1448 (1993).
131. Z. Li and J. S. Francisco, *J. Am. Chem. Soc.* **111**, 5660 (1989).
132. J. S. Francisco and I. H. Williams, *Proceedings of the 16th Annual Meeting of the National Organization of Black Chemists and Chemical Engineers*, Chicago, Vol. 16, 1989, pp. 59–72.
133. *World Meterological Organization, Global Ozone Research and Monitoring Project Rep. 20*, U.N. Environmental Program, World Meterological Organization, Geneva, 1990.
134. J. Czarnowski and J. H. Schumacher, *Int. J. Chem. Kinet.* **13**, 639 (1981).
135. R. C. Kennedy and J. B. Levy, *J. Phys. Chem.* **76**, 3480 (1972).
136. J. S. Francisco, M. Finders, and J. I. Steinfeld, *Int. J. Chem. Kinet.* **13**, 627 (1981).
137. J. S. Francisco and I. H. Williams, *Chem. Phys. Lett.* **110**, 240 (1984).
138. L. Andrews, M. Hawkins, and R. Witherall, *Inorg. Chem.* **24**, 4234 (1985).
139. K. C. Clemitshaw and J. R. Sodeau, *J. Phys. Chem.* **93**, 3551 (1989).
140. J. S. Francisco, Z. Li, and I. H. Williams, *Chem. Phys. Lett.* **140**, 531 (1987).
141. J. S. Francisco, Z. Li, and I. H. Williams, *Chem. Phys. Lett.* **186**, 343 (1991).
142. J. Chen, T. Zhu, and H. Niki, *J. Phys. Chem.* **96**, 6115 (1992).
143. C. Fockenberg, H. Saathoff, and R. Zellner, to be published.
144. J. Chen, T. Zhu, H. Niki, and G. J. Mains, *Geophys. Res. Lett.* **19**, 2215 (1992).
145. T. J. Wallington, M. D. Hurley, W. F. Schneider, J. Sehested, and O. J. Nielsen, *J. Phys. Chem.* **97**, 7606 (1993).
146. C. Kelley, J. Treacy, H. W. Sidebottom, and O. J. Nielsen, *Chem. Phys. Lett.* **207**, 498 (1993).
147. S. B. Barone, A. A. Turnipseed, and A. R. Ravishankara, *J. Phys. Chem.* **98**, 4602 (1994).
148. G. Bednarek, J. P. Kohlman, H. Saathoff, and R. Zellner, *Z. Phys. Chem.*, submitted.
149. M. J. Rossi, J. R. Barker, and D. M. Golden, *J. Chem. Phys.* **71**, 3722 (1979).
150. Z. Li and J. S. Francisco, *Chem. Phys. Lett.* **209**, 369 (1993).
151. P. Biggs, C. E. Canosa-Mas, D. E. Shallenoss, R. P. Wayne, C. Kelley, and H. W. Sidebottom, abstract presented at *Alternative Fluorocarbon Environment Acceptability Study Workshop*, Dublin, Mar. 1993.
152. T. J. Wallington, M. D. Hurley, and W. F. Schneider, *Chem. Phys. Lett.* **213**, 442 (1993).
153. M. Dupuis, G. Fitzgerald, B. Hammond, W. A. Lester, and H. F. Schaefer III, *J. Chem. Phys.* **84**, 2691 (1986).
154. G. L. Workman and C. W. Mathews, *J. Chem. Phys.* **53**, 857 (1970).

155. G. L. Workman and A. B. F. Duncan, *J. Chem. Phys.* **52**, 3204 (1970).
156. R. H. Judge and D. C. Moule, *J. Chem. Phys.* **78**, 4806 (1983).
157. J. S. Francisco, Z. Li, M. Head, and I. H. Williams, *Chem. Phys. Lett.* **214**, 591 (1993).
158. L. T. Molina and M. J. Molina, *182nd ACS National Meeting*, New York, 1982.
159. A. Nolte, H. Heydtmann, R. Meller, W. Schneider, and G. R. Moortgat, *Geophys. Res. Lett.* **19**, 281 (1992).
160. (a) D. R. Stall and H. Prophet, Eds., *JANAF Thermochemical Tables*, 2nd ed., NSRDS-NB537, Washington, D.C., 1972. (b) M. W. Chase, Jr., C. A. Davies, J. R. Downey, D. J. Frurip, R. A. McDonald, and A. N. Syverud, *J. Phys. Chem. Ref. Data* **14** (Suppl.) (1985).
161. J. Montgomery, H. H. Michels, and J. S. Francisco, *Chem. Phys. Lett.* **220**, 391 (1994).
162. A. P. Modica, *J. Phys. Chem.* **74**, 1194 (1970).
163. D. M. Weston and R. A. Matula, *Combust. Sci. Technol.* **4**, 37 (1971).
164. C. P. Rinsland, A. Goldman, F. J. Marcray, R. D. Blathwerwick, J. J. Kosters, D. G. Murcray, N. D. Sze, and S. T. Massie, *J. Geophys. Res.* **95**, 16477 (1990).
165. J. S. Francisco, A. N. Goldstein, and I. H. Williams, *J. Chem. Phys.* **89**, 3044 (1988).
166. S. Antonik, *Bull. Soc. Chim. Fr.*, I-83 (1982).
167. J. S. Francisco and Y. Zhao, *Chem. Phys. Lett.* **153**, 296 (1988).
168. J. S. Francisco and A. Ostafin, *Mol. Phys.* **68**, 255 (1989).
169. M. M. Maricq, J. J. Szente, Z. Li, and J. S. Francisco, *J. Chem. Phys.* **98**, 784 (1993).
170. J. S. Francisco, A. N. Goldstein, Z. Li, Y. Zhao, and I. H. Williams, *J. Phys. Chem.* **94**, 4791 (1990).
171. K. A. G. MacNeil and J. C. J. Thynne, *Int. J. Mass. Spectrosc. Ion Phys.* **3**, 35 (1969).
172. H. Henrici, M. C. Lin, and S. H. Bauer, *J. Chem. Phys.* **52**, 5834 (1970).
173. M. T. Bowers and M. Chau, *J. Phys. Chem.* **80**, 1739 (1976).
174. J. S. Francisco and Y. Zhao, *J. Chem. Phys.* **93**, 9203 (1990).
175. D. E. Milligan, M. E. Jacox, A. M. Bass, J. J. Comeford, and D. E. Mann, *J. Chem. Phys.* **42**, 3187 (1965).
176. M. E. Jacox, *J. Mol. Spectroscosc.* **80**, 257 (1980).
177. K. Nagai, C. Yamada, Y. Endo, and E. Hirotia, *J. Mol. Spectrosc.* **90**, 249 (1981).
178. K. K. W. Wang and W. E. Jones, *J. Photochem.* **1**, 147 (1972).
179. S. Toby and F. S. Toby, *J. Phys. Chem.* **85**, 4071 (1981).
180. M. M. Maricq, J. J. Szente, G. A. Khitrov, and J. S. Francisco, *Chem. Phys. Lett.* **199**, 71 (1992).
181. M. M. Maricq, J. J. Szente, Yi Su, and J. S. Francisco, *J. Chem. Phys.* **100**, 8673 (1994).

182. J. S. Francisco, A. N. Goldsten, M. A. Robb, and I. H. Williams, *Chem. Phys. Lett.* **191**, 13 (1992).
183. J. S. Francisco and Y. Zhao, *J. Chem. Phys.* **93**, 276 (1990).
184. P. Behr, K. Goldbach, and H. Heydtmann, *Int. J. Chem. Kinet.* **25**, 957 (1993).
185. J. S. Francisco and N. Mina-Camilde, *Can. J. Chem.* **71**, 135 (1993).
186. J. S. Francisco, *J. Chem. Phys.* **96**, 7597 (1992).
187. K. R. Ryan and I. C. Plumb, *Plasma Chem. Plasma Process* **4**, 271 (1984).
188. G. Hancock and D. E. Heard, *Chem. Phys. Lett.* **158**, 167 (1989).
189. C. P. Tsai and D. L. McFadden, *J. Phys. Chem.* **94**, 3298 (1990).
190. J. S. Francisco and A. Ostafin, *J. Phys. Chem.* **94**, 6337 (1990).
191. J. S. Francisco, *Chem. Phys. Lett.* **163**, 375 (1989).
192. Y. Zhao and J. S. Francisco, *Chem. Phys. Lett.* **199**, 65 (1992).
193. A. J. Arivia, P. J. Aymsaino, and H. J. Schumacher, *An. Asoc. Quim. Arg.* **50**, 135 (1962).
194. W. B. Fox and G. Franz, *Inorg. Chem.* **5**, 946 (1966).
195. J. S. Francisco and I. H. Williams, *J. Phys. Chem.* **92**, 5347 (1988).
196. Z. Li, T. S. Dibble, and J. S. Francisco, *Adv. Phys. Chem.*, in press.
197. M. M. Maricq, J. J. Szente, T. S. Dibble, and J. S. Francisco, *J. Phys. Chem.* **98**, 12294 (1994).
198. V. Mors, A. Hoffman, G. A. Arguello, and R. Zellner, in press.
199. T. S. Dibble and J. S. Francisco, *J. Phys. Chem.*, **98**, 5010 (1994).
200. R. Meller, D. Boglu, and G. K. Moortgat, STEP-HALOSIDE/AFEAS Workshop, Dublin, 130 (1993).
201. M. M. Maricq and J. J. Szente, *J. Phys. Chem.*, in press (1995).
202. B. G. Tucker and E. Whittle, *Trans. Faraday. Soc.* **63**, 80 (1967).
203. J. A. Kerr and J. P. Wright, *J. Chem. Soc. Faraday Trans. I* **81**, 1471 (1985).
204. J. C. Amphlett and E. Whittle, *Trans. Faraday Soc.* **63**, 80 (1967).
205. J. S. Francisco, *Chem. Phys. Lett.* **191**, 7 (1992).
206. M. Kozuka and T. Isobe, *Bull. Chem. Soc. Jpn.* **49**, 1766 (1976).
207. Z. Li and J. S. Francisco, *Mol. Phys.* **79**, 1127 (1993).
208. S. Dobe, L. A. Kachatryan, and T. Berces, *Ber. Bungenges. Phys. Chem.* **93**, 847 (1989).
209. D. J. Scollard, J. J. Treacy, H. W. Sidebottom, C. Balestra-Garcia, G. Laverdet, G. LeBras, H. MacLeod, and S. Teton, *J. Phys. Chem.* **97**, 4683 (1993).
210. J. S. Francisco, *J. Chem. Soc. Faraday. Trans.* **88**, 1943 (1992).
211. T. J. Wallington and M. D. Hurley, *Int. J. Chem. Kinet.* **25**, 819 (1993).
212. Z. Li, T. S. Dibble, and J. S. Francisco, *Chem. Phys. Lett.*, in press.
213. T. S. Dibble and J. S. Francisco, *Chem. Phys. Lett.* **215**, 409 (1993).

PHOTOCHEMISTRY AND PHOTOELECTROCHEMISTRY OF QUANTIZED MATTER: PROPERTIES OF SEMICONDUCTOR NANOPARTICLES IN SOLUTION AND THIN-FILM ELECTRODES

Horst Weller and Alexander Eychmüller
Institut für Physikalische Chemie, Universität Hamburg,
20146 Hamburg, Germany

CONTENTS

Advances in Photochemistry, Volume 20, Edited by Douglas C. Neckers, David H. Volman, and Günther von Bünau
ISBN 0-471-11469-3

I. INTRODUCTION: SOMETHING IN BETWEEN

Light-induced processes are described quite differently in molecular photochemistry and solid-state photophysics. In photochemistry one is used to an atomistic picture in which the arrangement of the atoms in the structure of a single molecule determines the electronic levels and thus the photochemical behavior. In contrast, the electronic levels of a solid are determined by the infinite periodicity of the atomic sequence in the crystal lattice. This leads to a basic concept according to which the solid can be treated as a dielectric continuum. Atomistic irregularities in the crystalline structure, such as lattice defects or impurities, are treated as perturbations of the spatially independent states in the energy bands.

These concepts work very well in the case of small molecules on one side and large solids on the other. Both approaches, however, reach their limits for nanometer-sized particles (i.e., for conglomerates of a few dozen up to a few thousands of atoms). In this very size regime, a gradual transition from solid-state to molecular structures occurs, and consequently, the chemical and physical properties of the particles strongly depend on their size. In the case of semiconductors and metals, the deBroglie wavelength of free charge carriers becomes comparable or even smaller than the diameter of the particles, leading to the *size quantization effect*. The energy of the optical transition increases gradually with decreasing particle size, which gives rise to beautiful color changes of the particles. For semiconductor particles, which are the subject of this chapter, this condition is fulfilled in a range of diameters typically between 2 and 50 nm, whereas for metal particles quantization effects usually occur in the subnanometer range.

During the last decade, the basic principles of nanochemistry and nanophysics have been explored in many fields of science, mostly independent of each other. A wide variety of nanometer-sized particles has been prepared up to the gram scale via colloid chemistry, either homogeneously dispersed in solution or embedded in solid matrices and glasses, in micelles, vesicles, and Langmuir–Blodgett films. Synthetic inorganic cluster chemistry has made remarkable progress in preparing relatively "large" and well-defined metal and semiconductor clusters such as $[Au_{55}(PPh_3)_{126}]^{6+}$ [1], $Cu_{70}Se_{35}(PEt_3)_{22}$ [2], and $Cd_{32}S_{14}(SC_6H_5)_{36} \cdot DMF_4$ [3]. Gas-phase cluster physics succeeded in the preparation of large clusters of various materials in supersonic beams, and sophisticated techniques allow detailed investigations of their electronic and structural properties; the isolation of C_{60} from carbon vapor even started the new branch of fullerene chemistry. Modern thin-film techniques such as molecular beam epitaxy and chemical vapor deposition in combination with advanced lithography enabled solid-state physicists to prepare smaller and smaller nanostructured solids, such as multiple quantum wells, quantum wires, and quantum dots. Presently, these different scientific disciplines come close together, and nanochemistry and nanophysics is a blooming field in materials science. Due to the highly interdisciplinary character and the different scientific roots of this research, many names for the same matter are still found in literature: *quantum-sized particles*, *Q-particles*, *quantum crystallites*, *nanocrystallites*, *clusters*, *nanoclusters*, and *quantum dots*.

A series of excellent review articles on Q-particles has appeared during the last years. Henglein reports in his article [4] on preparative aspects of oxide and sulfide semiconductor particles as well as on metal particles. Topics such as photoanodic and photocathodic dissolution of particles, water splitting; absorption and fluorescence, interfacial photo and redox reactions, as well as nonlinear optical effects and some theoretical aspects are addressed. In another review [5], the same author discusses in more detail characteristics of small metal particles in addition to the question of "magic agglomeration numbers" of semiconductor particles, some aspects of surface modification, "sandwich" particles, and photoelectron emission. A very recent review by Henglein describes the physicochemical properties of small metal particles, including microelectrode reactions, chemisorption, composite metal particles, and the atom-to-metal transition [6]. Wang and Herron report in their review [7] on the synthesis of small semiconductor crystallites (with first indications on "cluster" species) and in detail on the theoretical basis of the quantum size effect, including dielectric confinement and the role the surface of the particles plays for the energetics and dynamics. In addition to synthetic aspects, Brus [8] summarizes the knowledge about linear and nonlinear optical properties, including the work on superradiance in CuCl nanocrystals [9, 10], the electrooptical Stark effect, and third-order optical nonlinearities in semiconductor crystallites. Recent reviews [11, 12] focus on general features of size quantization, different approaches to reach monodispersity in the synthesis, fluorescence properties, layers formed from particles, excess charge carriers on the particles, and photoionization and summarize the various theoretical treatments published so far.

In this chapter we report on properties of nanometer-sized semiconductor particles in solution and in thin films and thereby concentrate on the photochemical, photophysical, and photoelectrochemical behavior of these particles. We shall, very briefly, describe the energetic levels in semiconductors and the size quantization effect. The bottleneck in small-particle research is the preparation of well-defined samples. As many preparative aspects are already reviewed in several actual assays, we present here only the preparative highlights of the last two years. In Section IV we describe the fluorescence properties of the particles. We report on different models for the description of the very complex fluorescence mechanism and we show how fluorescence can be utilized as a tool to learn about surface chemistry. Moreover, we present complex nanostructures consisting of either linked particles or multiple shells of different nanosized materials. The other large paragraph describes experiments with particles that are deposited on conductive substrates. We show how the combination of photoelectrochemistry and optical spectroscopy provides important information on the electronic levels as well as on charge transport properties in quantized particle films. We report on efficient charge separation processes in nanostructured films and discuss the results with respect to possible applications as new materials for optoelectronics and photovoltaics.

II. ENERGY LEVELS OF SEMICONDUCTORS AND SIZE QUANTIZATION EFFECT

The semiconductor properties observed in many inorganic substances are not properties of individual molecules or atoms but are a result of the arrangement of the constitutive elements in an ordered crystal lattice. The overlapping of the atomic orbitals in the crystal leads to the formation of a continuum of electronic energy levels, the energy bands. A typical energy scheme of a semiconductor is shown in Figure 1. The valence band in an undoped semiconductor is completely filled with electrons at 0 K, whereas the conduction band is empty. The energy difference between the valence and conduction band is called the *bandgap*. An electron can be excited from the valence band into the conduction band by the absorption of a quantum of light. A positive charge (a hole) remains in the valence band. The electron

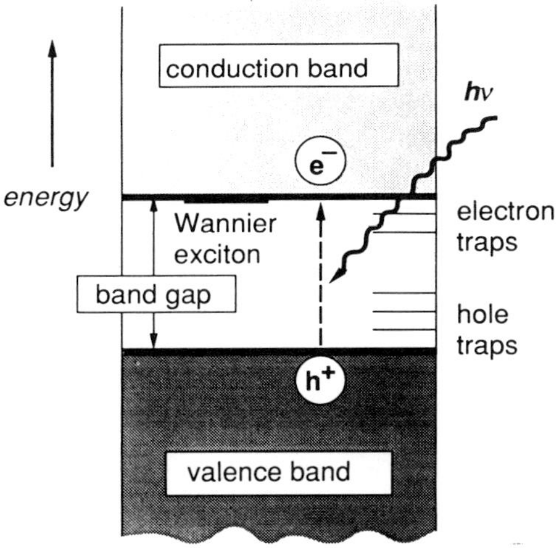

Figure 1. Energy diagram of a bulk semiconductor.

and hole can move independent of one another in the crystal, resulting in electrical conductivity. The electron and hole also experience coulombic forces and can form a Wannier exciton, a state that is very similar to a hydrogen atom. In molecular terminology an exciton would be called the *first excited state*. However, from solid-state physics one knows that because of the interaction of the electron and hole with the core levels of the atoms in the crystal, the masses of the electrons and holes must be replaced by what are called *effective masses* (m^*). This is the ratio of the masses of the electrons and holes in a semiconductor to the mass of the electron in vacuum. As the effective masses are usually small and, additionally, because the crystal with its electrons screens the charge carriers in the exciton (expressed by the high-frequency dielectric constant), the binding energy of the exciton is small and its radius large. For example, the binding energy of the excitons in CdS is only 0.05 eV and its radius is approximately 2.5 nm (for comparison, the binding energy of a hydrogen atom is 13.51 eV and its radius is 0.053 nm). The small binding energy results in dissociation of the excitons into free charge carriers at room temperature. Usually, excitons are observed via their sharp absorption and emission lines in optical spectra only at temperatures close to absolute zero.

Beside bands and excitonic energy levels, semiconductors contain traps for electrons and holes. These are interstitial atoms, impurities, dislocations, and defects. Traps are frequently located on the surface as a result of the disturbed stoichiometry. The trapped charge carriers can recombine, either radiatively or nonradiatively, or undergo transfer to adsorbed electron donors or acceptors. They can also recombine from the bands and from the excitonic state. The traps in nanometer-sized particles are found mainly on the surface as a result of the large surface-to-bulk ratio.

In the case of nanometer-sized particles the unusual situation exists that the particle size is the same as or actually smaller than the size of the excitons in macrocrystalline material. The electron–hole pair can "fit" into such a particle only when the charge carriers assume a state of higher kinetic energy. In solid-state terminology this results in the bands splitting into discrete quantized levels and the bandgap (or, better, HOMO-LUMO or $1s$–$1s$ transition energy) increasing with decreasing particle size. This size quantization effect can amount to several electron volts in nanometer-sized semiconductor Q-particles. Whether one can speak of bands or excitons at all in this situation is a purely semantic question. On the one hand, the electron and hole, because of their increased kinetic energy, move practically freely within the particle, a situation similar to that in the energy bands. On the other hand, they experience a coulombic attraction, which is why this state is frequently called *1s–1s excitonic* (we shall use this terminology in this chapter). Approaching a size range of 1 to 2 nm, in which each particle consists of a few dozen molecules, it is better to talk about molecular orbitals. From the difficulties experienced in the terminology, one can see that the properties of such particles lie between those of the solid state and those of their molecules.

III. NOVEL PREPARATIVE ASPECTS

As mentioned above, the synthesis of highly monodisperse and well-characterized samples is the prerequisite for advances in small-particle research. Besides optimization of reaction parameters, several size fractionation techniques from polymer and biochemistry, including exclusion chromatography and gel electrophoreses, have been applied successfully to colloidal Q-particles. As already reviewed in Ref. 11, these techniques allow the preparation of samples with standard size deviations of only a few percent. The disadvantage is, however, the relative small amount of sample obtained by the fractionation. A remarkable advance was made in 1993 by Chemseddine and Weller [13] and Murray et al. [14], who described utilization of the size-dependent solubility for a size-selective precipitation of Q-particles. This technique yielded samples of nearly single-sized CdS, CdSe, and CdTe nanoclusters on a gram scale. Following the pioneering work of Dance and co-workers [15], Herron et al. [3] succeeded in the preparation and crystallization of pure $Cd_{32}S_{14}(SC_6H_5)_{36} \cdot DMF_4$.

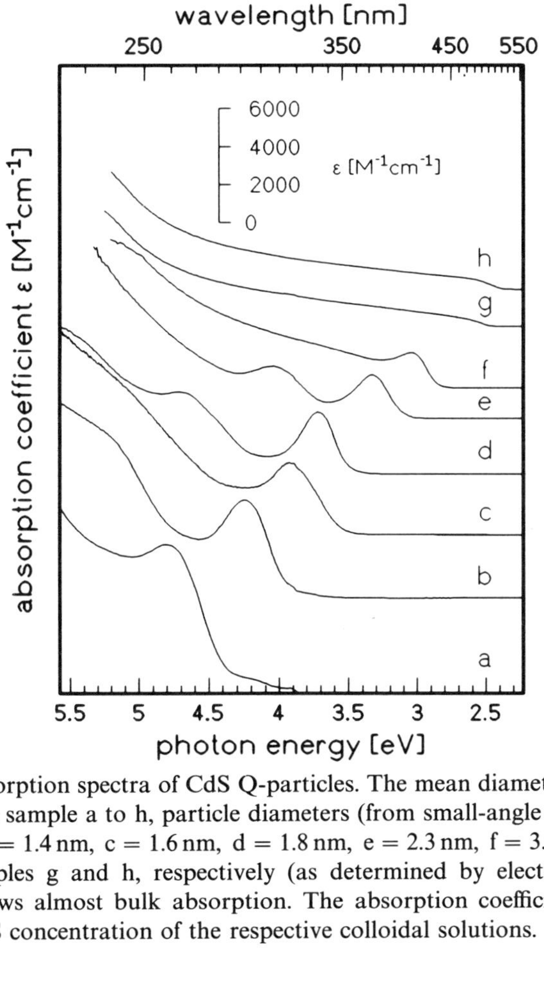

Figure 2. Absorption spectra of CdS Q-particles. The mean diameter of the samples increases from sample a to h, particle diameters (from small-angle x-ray scattering): a = 1.3 nm, b = 1.4 nm, c = 1.6 nm, d = 1.8 nm, e = 2.3 nm, f = 3.8 nm, and 6 and 9 nm for samples g and h, respectively (as determined by electron microscopy). Sample h shows almost bulk absorption. The absorption coefficients refer to the analytical CdS concentration of the respective colloidal solutions.

Samples obtained via these chemical methods exhibit well-structured absorption spectra which reflect mostly two or more excitonic transitions of the particles. This is readily visible in Figure 2, where the absorption spectra of different-sized CdS nanoclusters are shown [16]. Samples a to f contain clusters that are capped with thioglycerine ligands, whereas sodium polyphosphate was used as the stabilizer for samples g and h. The respective particle sizes obtained from small-angle x-ray scattering and electron microscopy are indicated in the caption of Figure 2. As a consequence of the size quantization effect, the 1*s*–1*s* excitonic transition is shifted toward higher photon energies with decreasing particle size. Moreover, it is seen that the integral absorption coefficients of the $1s$–$1s$ transition, which are a measure of the respective oscillator strength, increase greatly with decreasing particle size. This behavior is a direct consequence of the spatial restriction of electron and hole in the diminishing volume of the nanoparticles, as has been predicted for several years [17, 18]. Rajh et al. [19] also reported on the size-dependent oscillator strength of Q-CdTe particles; the size distribution of their samples was, however, quite broad.

Another remarkable synthetic effort has been made by the preparation of colloidal Q-particles of the technically more relevant IV-IV materials (i.e., silicon and germanium) [20–25]. Silicon nanoparticles, especially, are currently drawing a lot of attention, since it was found by Canham [26] that nanostructured silicon formed under anodic etching of silicon wafers (called "porous silicon") exhibits bright red fluorescence. Due to the indirect nature of the band transition, bulk silicon shows, by contrast, almost no fluorescence and thus cannot be utilized for optoelectronic devices.

The best preparation techniques for silicon nanoparticles presently involve silicon aerosol formation via controlled combustion, microwave discharge, or photochemical plasma decomposition of silane. These samples exhibit luminescence in all colors from blue to red, depending on the size and the surface treatment of the particles. The fluorescence mechanism in these samples is still discussed controversally. Although a large number of investigations have been published, it could not yet be fully clarified whether the luminescence is due to size quantization effects, surface states, or the formation of low-molecular-weight siloxene species. A detailed discussion of the fluorescence in silicon nanostructures would exceed the scope of this chapter; interested readers are referred to a very recent review by Brus [27].

IV. FLUORESCENCE OF Q-PARTICLES

A. Fluorescence Mechanisms

Soon after the discovery that fluorescence occurs in size-quantized semiconductor particles, the question of the underlying mechanism has been addressed. Without the claim of completeness some articles dealing with this topic are listed in Refs. 28 to 35. On the basis of this work, which, in part, has already been reviewed, three studies appeared in the early 1990s whose results are outlined here in some detail.

McLendon and co-workers carried out experiments on the luminescence decay of size-quantized CdS and Cd_3As_2 particles with picosecond time resolution [36]. CdS particles of various sizes ranging from 2.6 to

10 nm were examined at room temperature. The luminescence decays were recorded as a function of the emission wavelength, both with standard hexametaphosphate-stabilized CdS colloids in water and KOH-surface-treated clusters. Temperature-dependent fluorescence decays were monitored between 4.8 and 225 K with 3.2-nm CdS clusters prepared in 2-propanol. Irrespective of size, solvent, surface treatment, and temperature, the fluorescence decay profiles were found to be multiexponential. The authors of Ref. 36 divided the decays into a fast and a slow component. The first was found to depend only weakly on size, emission energy, and temperature, the latter being strongly dependent on these parameters. Only in the case of surface-modified CdS particles was the slow component virtually independent on the emission wavelength at room temperature. A large paragraph in Ref. 36 is devoted to kinetic analysis of the fluorescence decay curves. The fast component is fit to a single exponential; the slow component is described well with distributed kinetics either using the Williams–Watts function or the Thomas–Hopfield model. As a result of this analysis a fluorescence mechanism is presented which accounts for the wealth of experimental data. After the formation of an exciton in the particles, it quickly decays into a trapped electron–hole pair. The electron is assumed to be trapped in an energetically narrow distribution of trap states close to the conduction band (or LUMO), while the hole traps are assumed to have a wide distribution of energies within the bandgap of the particles. From this energetic situation an electron may either thermally repopulate the LUMO and recombine radiatively with a hole, giving rise to the fast decay component, or it can tunnel to a hole in a donor–acceptor sense, accounting for the slow decay, which is fitted well to distributed kinetics. The thermal repopulation mechanisms applies equally well to strongly size-quantized Cd_3As_2.

The second paper on the fluorescence mechanism in Q-sized CdS colloids to be discussed here appeared in 1991 [37]. In this work use was made of two preparations, one that yields particles emitting exclusively excitonic fluorescence (7.0 nm diameter, standard deviation $<10\%$) [38] and one giving somewhat smaller particles which show both excitonic and red-shifted trapped fluorescence. This is different from the particles investigated by McLendon et al., in which a clear distinction between excitonic and trapped fluorescence could not be made. Both types of samples of Ref. 37 exhibit high fluorescence quantum yields at room temperature, increasing to close to unity at low temperatures. Thus in these samples nonradiative processes are not the dominating deactivation channels at lower temperatures. The integrated fluorescence intensities of the first sample are plotted in Figure 3 as a function of temperature. From this plot it is deduced that the fluorescence in small semiconductor particles does not follow a simple

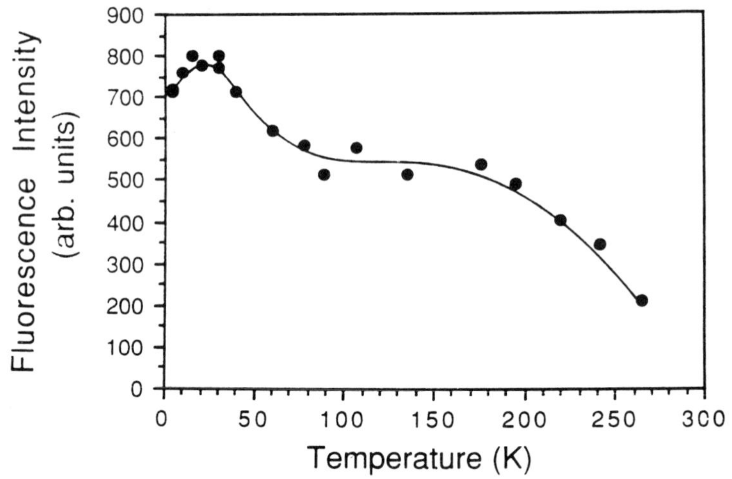

Figure 3. Intensity of the excitonic emission of CdS Q-particles as a function of temperature.

mechanism, as this temperature dependence exhibits several distinctly different regions and not a single linear dependence. The situation becomes even more puzzling by inspecting Figure 4, in which some fluorescence decay curves at the temperatures indicated are plotted. As in the work described above, the fluorescence decays are found to be multiexponential at all temperatures. It is realized that the overall fluorescence decay is faster when the fluorescence intensity is higher and is slower when the fluorescence intensity is lower. This is completely different from what is generally observed in molecular photochemistry. A model accounting for these findings is depicted in Figure 5. Excitonic fluorescence arises after detrapping of electrons from shallow surface traps, giving rise to delayed fluorescence (which has been named "repopulation of the LUMO" by McLendon et al.). Since the fluorescence quantum yields are high, detrapping must be an effective process, meaning that the trap depths for electrons cannot be large. The multiexponential fluorescence decay results from the dispersion in trap energy levels, which leads to a spectrum of rate constants for electron release into the conduction band. It was proposed that the trap population depends on the temperature, as shown schematically in Figure 5 for two temperatures. In the following, we distinguish between shallow and deeper traps (knowing that deeper traps are also relatively shallow). Nonradiative processes ($k_{\text{n.r.}}$) take place primarily out of deeper traps. At 270 K most of these deeper traps are populated. The fluorescence decay rate is determined by detrapping (k_d) and $k_{\text{n.r.}}$. With decreasing temperature two

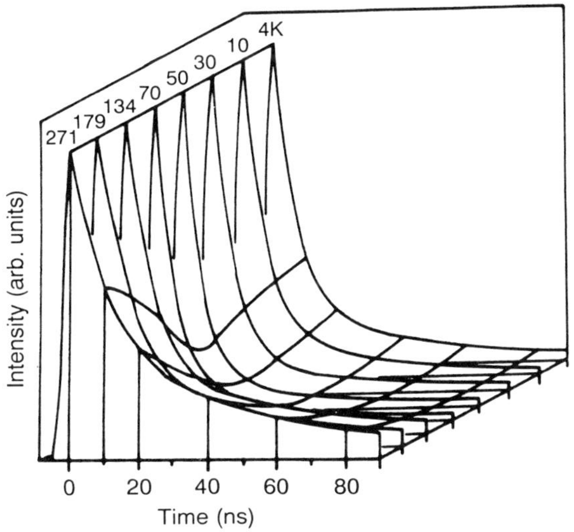

Figure 4. Fluorescence decay profiles of the excitonic emission ($\lambda_{obs} = 500\,nm$) of CdS Q-particles.

effects influence the fluorescence decay in opposite ways: detrapping out of a given trap becomes slower, as it is a thermally activated process. On the other hand, the trap population shifts closer to the conduction band [i.e., most of the electrons are then trapped in very shallow traps (4 K in Figure 5)]. Between 271 and 50 K the fluorescence decay becomes faster. In this temperature range the shift of the trap population overcompensates the decrease in the rate of the detrapping processes. Below 50 K, the decrease in the rate of detrapping becomes the dominant factor, and the decay rate becomes slower again.

The fluorescence intensity increases between 270 and 30 K. This is understood by the diminishing population of deeper traps (which are the origin for nonradiative processes) with decreasing temperature. Below 30 K the fluorescence intensity goes down slightly. At these lowest temperatures only extremely shallow trapped electrons can be detrapped again. A small fraction of electrons that might be present in deeper traps can only recombine in a nonradiative manner.

Using simple assumptions the depths of the traps populated are estimated to be between 285 meV at room temperature and only 4 meV at 4 K. Trap depths of only a few meV imply that the chemical environment around such shallow traps cannot be very different from that in the "bulk" of the particle. Tentatively, this situation is assigned to an electron that is delocalized at the surface of the particle. Thus trapping into shallow traps is

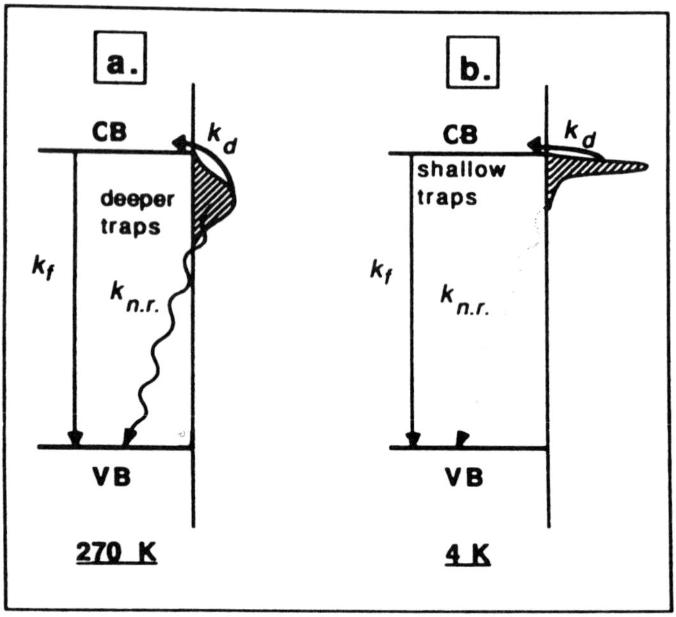

Figure 5. Fluorescence mechanism in size-quantized CdS particles.

understood as an increase of the probability of presence of an electron at the surface of the particle.

In the smaller particles investigated, radiative recombination from trapped charge carriers is observed besides the excitonic fluorescence. The model involving trapping and detrapping of electrons can also be applied for this sample since the temperature dependence of the fluorescence intensity and decay rate of the excitonic fluorescence is found to be very similar to the phenomena described above for particles that show excitonic fluorescence exclusively. For positive holes an additional trapping process has to be involved. Trapping of holes leads to the formation of localized surface S^- species. The crucial parameter for the radiative recombination rate of trapped charge carriers is the distance between them. If a positive hole is trapped very close to the electron, and immediate recombination is expected in the moment of trapping, whereas a distant charge carrier pair should recombine relatively slowly. Thus the multiexponential decay of the trapped fluorescence that is observed is understood in terms of different distances between the charge carriers.

The third study on the fluorescence mechanism to be inspected in some detail was published in 1992 by Brus and co-workers [39]. They prepared 3.2-nm CdSe particles with a standard deviation below 8%. By separating the nucleation step from the growth step, particles with good crystalline quality and monodispersity were received. They exhibit a very pronounced 1*s*–1*s* optical transition at about 520 nm and a symmetric and narrowband edge emission (fluorescence quantum yield >0.1 at 15 K) close to the onset of absorption. A small amount of emission further in the red spectral region is also observed. Excitation at the low-energy edge of the absorption spectrum, thus exciting selectively only the largest crystallites, yields a beautifully structured fluorescence spectrum exhibiting a Franck–Condon LO phonon emission progression with the bulk spacing of $200\,cm^{-1}$. Time-resolved emission spectra reveal that the observed shift of $75\,cm^{-1}$ between the excitation wavelength and the peak of the LO zero phonon emission occurs on a time scale of hundreds of picoseconds. Fluorescence decay curves show two distinctly different components, a fast temperature-insensitive decay (approx. 100 ps) and a long multiexponential microsecond component whose average lifetime decreases by a factor of about 10 from 10 K to 50 K. The quantum yield of the band edge emission is found to decrease by a factor of 3 from 10 K to 50 K, whereas the red deep trap luminescence increases in this temperature range. Since the fluorescence excitation spectrum of the trap emission is blue shifted from that of the edge emission and the deep trap emission appears without a resolvable (<500 ps) rise time, the authors of Ref. 39 conclude that the two emissions originate in separate groups of particles. As the temperature is increased from 15 K to 75 K, the edge emission is found to shift to higher energies, which is explained by the thermal population of emitting states. Thus the authors of Ref. 39 also conclude that a fast thermal equilibrium between electronic states is active in small group II–VI semiconductor particles, which governs the emission properties. The multiexponential character of the long emission decay is attributed to a distribution of rates among the particles. The data received from the time-, wavelength-, and temperature-dependent measurements, together with experiments on the polarization-resolved luminescence, leads the authors of Ref. 39 to suggest a strong resonant mixing between molecular orbitals delocalized inside the particles and intrinsic surface Se lone-pair states. This model is analogous to the two models described above if the roles of electron and hole are reversed.

To answer the question of whether the electrons or the holes are shallowly trapped at the surface of the small semiconductor particles stationary and time-resolved quenching experiments of the excitonic and the trapped fluorescence of differently sized CdS colloids were carried out [40]. The idea of these measurements was to utilize the size of the particles for shifting the electron traps in a controlled manner through the acceptor levels of the quenchers, and thereby, to scan the trap distribution. It was found that nitromethane is only capable of quenching the excitonic fluorescence nearly completely, whereas a fraction of the trapped fluorescence, the amount of which depends on the particle size, always remains. Methylviologen, with a slightly more positive acceptor potential, is able to quench both types of fluorescence nearly completely in all samples investigated. The experiments prove the existence of shallow electron traps in Q-CdS. With the assumption of deep electron traps (and consequently, arising from the spectral data, shallow hole traps) the quenching behavior would not be completely explicable. This interpretation, however, does not exclude the possibility of a state mixing of the valence band with chalcogenide surface states as proposed for Q-CdSe [39]. In a complete fluorescence model such mixed states might be involved to explain the size-dependent Stokes shift of the excitonic fluorescence.

A number of other groups also addressed the question of the fluorescence mechanism in group II–VI semiconductor small particles [41–44]. The main findings of a thermal repopulation mechanism for one charge carrier and a localization for the other are not altered substantially irrespective of the particle preparation and environment, such as reversed micelles and glasses.

B. Composite Particles

In this section some optical properties of composite semiconductor particles are reviewed. We shall restrict ourselves to aspects of surface modification of given particles and the formation of core–shell structures. We are leaving out of consideration the field of sandwich colloids (i.e., two different semiconductor particles attached to each other) since their properties have been reviewed in detail before [5, 11, 12].

The first example in the literature was presented in 1984 by Henglein and co-workers [29]. Small amounts of Cd^{2+} ions added to a solution of ZnS colloids were found sufficient to quench the ZnS bandgap fluorescence (λ_{em} = 420 nm) completely. At higher concentrations of added Cd^{2+}, a new fluorescence band appears centered around 500 nm. "We explain the phenomenon by the formation of small spots (at low Cd^{2+} concentrations) and later of a thin layer (larger Cd^{2+} concentrations) of 1:1 co-colloid on the ZnS particles. When the recombination occurs at these spots, the new fluorescence is emitted. There seems to be efficient channeling of the electron and/or the positive hole to these spots, which could be understood in terms of the less negative potential of the conduction band (and the less positive potential of the valence band) in co-colloids as compared to pure ZnS." This explanation was a signpost for the investigations that followed.

White et al. [45] reported in 1985 on experiments involving the same two sulfide semiconductors, CdS and ZnS. Although the focus of this work was on the photocatalytic hydrogen evolution as a function of shape and composition rather than on the optical properties of the particles, it should be mentioned here since it contains the first x-ray photoelectron spectroscopy (XPS) measurements on spherically layered semiconductor particles. By successive precipitation of cadmium and zinc salts with H_2S onto SiO_2 supports, both composites have been prepared, ZnS on a CdS core and CdS on a ZnS core. The surface composition was analyzed by XPS taking the Cd(3*d*)/Zn(2*p*) peak area ratio as the measure of the surface atomic ratio in the composite particles. The results are such that, indeed, spherically layered particles are formed in the preparative procedure.

Two years later it was again the Henglein group that carried out experiments on the system CdS–Ag_2S [46]. In this study it was found that strong fluorescence in the red and near-infrared spectral region is generated by the addition of Ag^+ ions to a CdS colloidal solution. The various preparative approaches with different stabilizers, varying concentrations of stabilizers, Ag^+, and additional metal ions, together with the results of experiments on the separate colloids, yielded the conclusion that an interfacial region containing anion vacancies and incorporated silver centers is present in the composite particles. This region is held responsible for delivering electrons into the conduction band of CdS upon illumination of the composites, even with photon energies less than the bandgap energy of CdS.

Fendler et al. [47] carried out experiments on vesicle-stabilized mixed crystals of $Zn_xCd_{1-x}S$ and on CdS particles coated with ZnS. Besides absorption and fluorescence spectroscopy, x-ray diffractrometry was used for structural characterization of the various pure and mixed sulfide particles. In this article, the authors discuss thoroughly the fine interplay between kinetics and thermodynamics governing the rather complicated reaction scheme, finally yielding either separate particles of CdS and ZnS, mixed crystals $Zn_xCd_{1-x}S$, or ZnS-coated particles of CdS. The latter may appear either as ZnS islands on the CdS particles or as a closed shell of ZnS surrounding a CdS core. In their experiments on capped particles, the authors of Ref. 47 do not attempt to decide which of the two morphologies occurs.

One report has appeared dealing with the silver halide colloidal solution of AgI and the system AgI–Ag_2S [48]. Besides the photochemical properties such as absorption, fluorescence, and photocorrosion, some experiments are outlined on the gradual conversion of AgI into Ag_2S upon the addition of H_2S to a polymer-stabilized solution of AgI. AgI colloids exhibiting a sharp first electronic transition at about 411 nm are fluorescing close to the onset of absorption at 425 nm. Subsequent addition of H_2S to this solution quenches this fluorescence and yields up to three new fluorescence bands centered at wavelengths between 850 and 1100 nm, depending on the amount of added H_2S. Together with the loss of structured absorption around 400 nm and an extension of absorbance to beyond 700 nm, this is interpreted as the gradual conversion of AgI into Ag_2S. At intermediate stages of this reaction, composite AgI–Ag_2S particles occur which fluoresce at energies corresponding to the size of the Ag_2S deposits. The fluorescence is explained in terms of electron transfer from AgI to the hole captured at the AgI–Ag_2S interface.

In 1990 an article from the AT&T group headed by Brus appeared [49] in which the ZnS/CdSe system is investigated. The authors succeeded in the preparation of composite semiconductor crystallites in which CdSe was grown on a ZnS seed, and vice versa, both in inverse micelle solutions. The CdSe seeds exhibit a relatively broad excitonic absorption peak around 500 nm. The absorption spectrum is not changed significantly upon the growth of ZnS on the CdSe seeds, which is taken as an indication that alloying or surface etching of CdSe by ZnS does not occur. On the other hand, luminescence spectra indicate that, indeed, ZnS is grown on the CdSe cores. The initial CdSe seed emits a very weak and broad fluorescence with maxima at 580 nm and 680 nm. After capping by ZnS, the quantum yield of emission increases significantly and the fluorescence spectrum shifts blue with a narrow fluorescence band at 530 nm, close to the excitonic absorption of the CdSe seeds. This was interpreted as the filling of deep surface traps by growth of the ZnS layer on CdSe. The reverse preparation is somewhat more difficult since the addition of Cd^{2+} ions to a ZnS sol, intended to grow into a CdSe layer by the addition of Se(trimethylsilyl)$_2$, may lead to the partial replacement of Zn^{2+} by Cd^{2+}, thus forming CdS since the solubility product of CdS is smaller than that of ZnS. However, by careful analysis of x-ray fluorescence, Auger spectroscopy, transmission electron microscopy and x-ray powder diffraction, the authors conclude that the preparation yields tiny (approximately 1.5 nm) ZnS seeds capped with CdSe. The $1s$–$1s$ electronic transition of these composites appears to be a little broader than that of pure CdSe nanocrystals of the same size and comparable size distribution. The fluorescence of the composites is much weaker than that of the pure CdSe sample, which is attributed to the introduction of deep interior traps by the inclusion of ZnS where electron and hole nonradiatively recombine.

A very important step toward a better understanding of the electronic properties of layered semiconductor particles is presented in the same article. In the framework of the effective mass approximation, the authors modeled the optical transition of CdSe crystallites with small inclusions of

ZnS. The most remarkable result of this analysis is that depending on the size of the ZnS inclusion, the energy of the 1*s*–1*s* transition of the composite is shifted either to higher energy than that of CdSe (larger inclusions) or to lower energy (0.5 to 1 nm inclusion). In other words, shift of the 1*s*–1*s* transition energy is not monotonic with increasing size of the inclusion but passes an energetic minimum lying farther in the red than the transition energy of the CdSe crystallite without an inclusion.

An analogous synthetic approach has been used by Fox et al. [50] for the preparation of CdSe/ZnSe core–shell particles. X-ray photoelectron and Auger and absorption spectroscopies were applied to demonstrate the assumed structure of both composites, CdSe on ZnSe and ZnSe on CdSe. It is found that "the shell material dominates the optical properties of the particles, even when the narrower bandgap material was used for the core." This apparent contradiction to the above-mentioned statement of Brus et al. on the composite ZnS on CdSe is probably due to different sizes and different core-to-shell ratios in the two studies.

In 1992 and 1993, yet another combination of group II–VI semiconductor materials, HgS and CdS, was investigated [51, 52]. Two kinds of composite particles have been synthesized: (1) small islands of HgS in the surface of CdS particles, and (2) very small HgS particles surrounded by a closed shell of CdS. The first preparation is facilitated by the large difference in the solubility products of HgS and CdS, which amounts to more than 20 orders of magnitude. This difference allows the substitution reaction $CdS_m + nHg^{2+} \rightarrow CdS_{m-n}HgS_n + nCd^{2+}$ to take place readily. Probably, the second preparation is also favored by the choice of the two materials since both semiconductors appear in the colloidal state in the cubic crystal structure with almost identical lattice constants, avoiding a lattice mismatch between the core and the shell. On the other hand, this good matching excludes a direct proof of the assumed core–shell structure by, for example, x-ray or electron diffraction. Thus one is left to transmission electron microscopy as an aid to detect particle growth and optical spectroscopy, which turns out to be a powerful combination in deriving structural information.

Figure 6 shows a set of absorption spectra of colloidal CdS with different amounts of added $HgCl_2$. The absorption edge of colloidal size-quantized CdS is clearly seen at about 480 nm. With increasing amounts of added $HgCl_2$, the onset of absorption shifts to longer wavelengths reaching the red and finally, the infrared spectral region. This absorption behavior is attributed to the formation of size-quantized HgS. The more $HgCl_2$ is added, the larger the HgS particles are and the smaller the bandgap becomes. The absorption spectra of the attached composite CdS–HgS particles are not simply superpositions of the two size-quantized semiconductors, as is easily

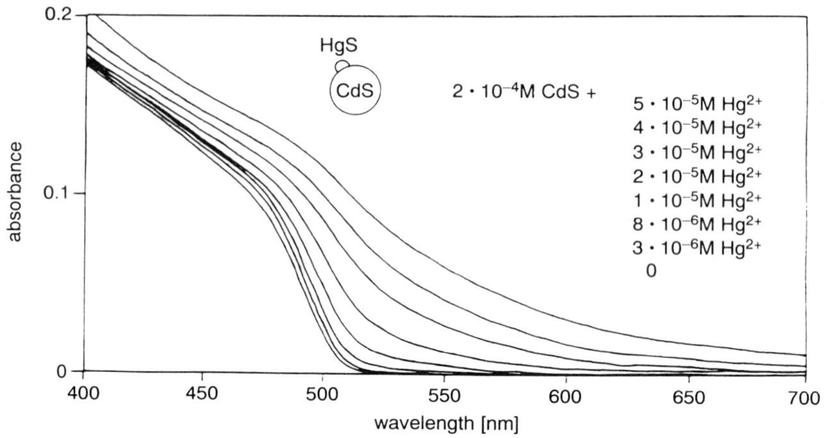

Figure 6. Absorption spectra of CdS particles with differently sized HgS islands incorporated into the surface.

seen in the uppermost spectrum in Figure 6. Although only 25% of the CdS is converted to HgS, there is only a hint of the CdS absorption edge in this spectrum. Thus it was concluded that the electronic structure of the CdS particles is strongly influenced by surface-bound HgS. As the amount of $HgCl_2$ used in the experiments was much less than that required to cover each CdS particle with a complete monolayer, the formation of a core–shell structure can be excluded.

The formation of an isotropic mixed crystal $Cd_xHg_{1-x}S$ can be excluded as a consequence of the fluorescence lifetime measurements together with the fluorescence excitation spectra outlined below. They show clearly that the particles exhibit local differences in composition that justify the use of a pictogram like the one shown in Figure 6. However, in the very early stages of the growth of HgS on CdS, one faces a situation that may be described as a locally mixed crystal. In such nanoheterogeneous systems, this raises the interesting question of how many HgS molecules have to be linked among themselves in an environment of CdS to justify the term "HgS island on CdS particle." Figure 7 shows a set of fluorescence spectra of colloidal CdS solutions with different amounts of added $HgCl_2$ as indicated. The intense green excitonic fluorescence of CdS at about 500 nm is effectively quenched and luminescence bands develop at longer wavelengths with maxima ranging between 550 and 770 nm, depending on the amount of $HgCl_2$ added. This luminescence is attributed to the fluorescence of size-quantized HgS particles formed on the surface of the CdS particles. The fluorescence intensity of the particles increases slowly after mixing the CdS

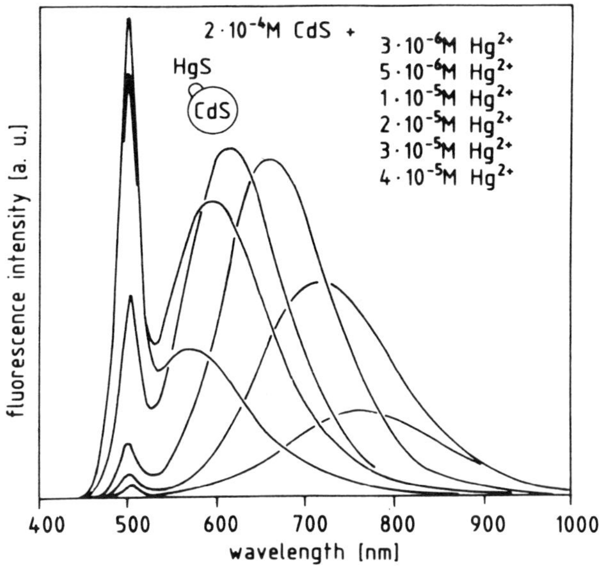

Figure 7. Fluorescence spectra of CdS particles with different-sized HgS islands incorporated into the surface.

and $HgCl_2$. After 2 days at room temperature, there is no further change for several months or more. The induction period of 2 days may indicate a slow annealing of defects on the surface of the HgS particles, which are assumed to act as centers for radiationless recombination. Heating accelerates the annealing process.

Fluorescence excitation spectra were recorded with a series of CdS colloids containing different amounts of HgS. An observation wavelength of 650 nm was chosen to obtain information about the origin of the long-wavelength luminescence. A comparison of these spectra with the respective set of absorption spectra shows that the main characteristics are identical (i.e., the CdS absorption edge at 480 nm, the long-wavelength absorption of HgS, and its shift to longer wavelengths with increasing particle size). From these experiments the formation of a significant fraction of separately formed HgS particles can be excluded. In Figure 8 a set of fluorescence decay curves recorded at 500 nm is plotted. The uppermost decay curve is recorded with a sample of pure CdS colloid, the other two with different amounts of added $HgCl_2$, as indicated in the figure. It is seen that the decay *kinetics* remain unchanged compared to the excitonic fluorescence of pure CdS irrespective of the quantity of added $HgCl_2$, whereas the *initial height* (or intensity) decreases with increasing HgS content.

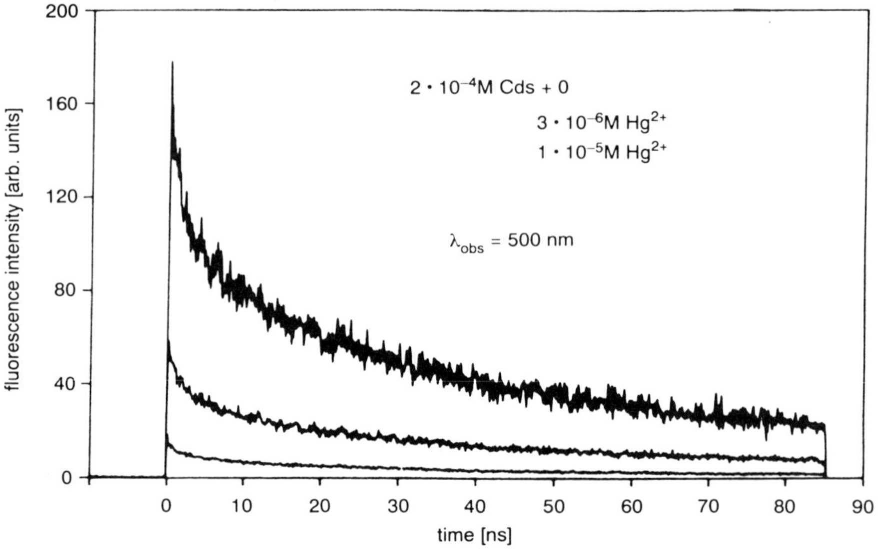

Figure 8. Fluorescence decay profiles of pure CdS particles (*top*) and after Hg^{2+} treatment.

Three energy diagrams suitable for explaining the experimental spectroscopic findings are depicted in Figure 9. Scheme 1 represents pure CdS colloids, which behave as described above. That is, after light absorption the photogenerated electrons are quickly trapped in shallow traps on the surface of the particles, from where they might be detrapped into the conduction band, giving rise to the delayed excitonic fluorescence of CdS nanocrystals. This mechanism does not change significantly when the HgS formed is small enough to exhibit an energetic situation like the one shown in scheme 2. Neither the electrons from the conduction band nor the holes from the valence band of CdS are transferred to the HgS moiety, and the fluorescence of the CdS remains unchanged. As the HgS particles grow by the addition of $HgCl_2$, more and more co-colloids reach the situation labeled 3 in Figure 9. In these structures hole transfer from the CdS to the HgS valence band is assumed to be much faster than the fluorescence in the CdS moiety. As a result, these particles do not show any CdS fluorescence at all. Considering the time resolution of about 50 ps, the hole transfer must be faster than $2 \times 10^{10}\,s^{-1}$, which is not unreasonable, bearing in mind the similarity of the bulk band structures of the CdS and HgS valence bands. After hole capture from the CdS, the electron is eventually transferred to the HgS part, where radiative electron–hole recombination can occur. Thus, in a certain

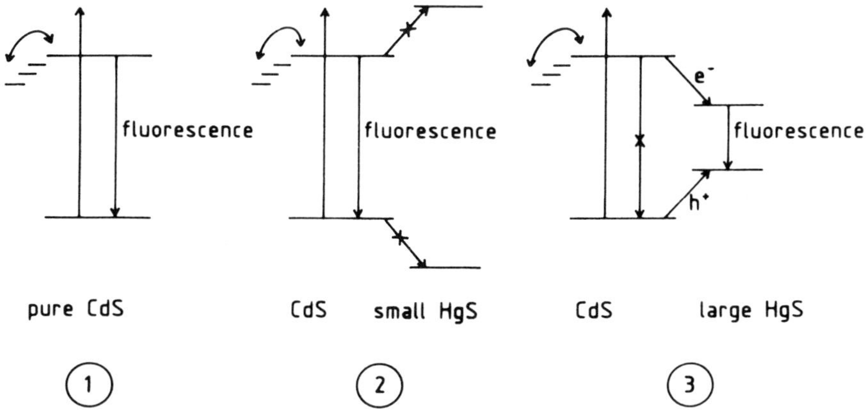

Figure 9. Energy diagrams illustrating the fluorescence properties of CdS particles with different-sized HgS islands on the surface.

sense, the solutions prepared are inhomogeneous. Even if the assumption holds that all the CdS particles carry the same number of HgS *molecules* (an assumption that does not have to be valid since the mixing process may itself introduce inhomogenities), the sizes of the HgS *particles* may vary from co-colloid to co-colloid. Some may carry a larger number of very small HgS particles, leading to an energetic situation such as the one shown as scheme 2 in Figure 9. Others carry a smaller number of larger HgS particles and so have a more type 3 character.

The second type of preparation yields particles with 1.8-nm HgS cores surrounded by approximately two monolayers of CdS, giving total diameters of about 3.0 nm, as has been shown by transmission electron microscopy. The formation of core–shell particles is nicely reflected in the absorption and fluorescence spectra. The unstructured absorption spectrum of pure HgS can be seen in Figure 10, spectrum A. The addition of $2 \times 10^{-4}\,M$ $Cd(ClO_4)_2$ yields spectrum B. The absorption decreases slightly over the entire spectral range of absorption, suggesting that adsorbed Cd^{2+} ions influence the electronic properties of the HgS particles. The effect of stepwise addition of portions of $2.2 \times 10^{-5}\,M$ H_2S to this solution can be seen in Figure 10, spectra C to H. The absorbance increases in the entire spectral range from spectra B to E, whereas an absorption edge emerges around 480 nm in spectra F to H and no additional increase in the absorbance occurs at longer wavelengths. In TEM micrographs of sample H, pure CdS particles are seen with diameters of about 6 nm in addition to the composite particles. Thus it was concluded that up to an addition of

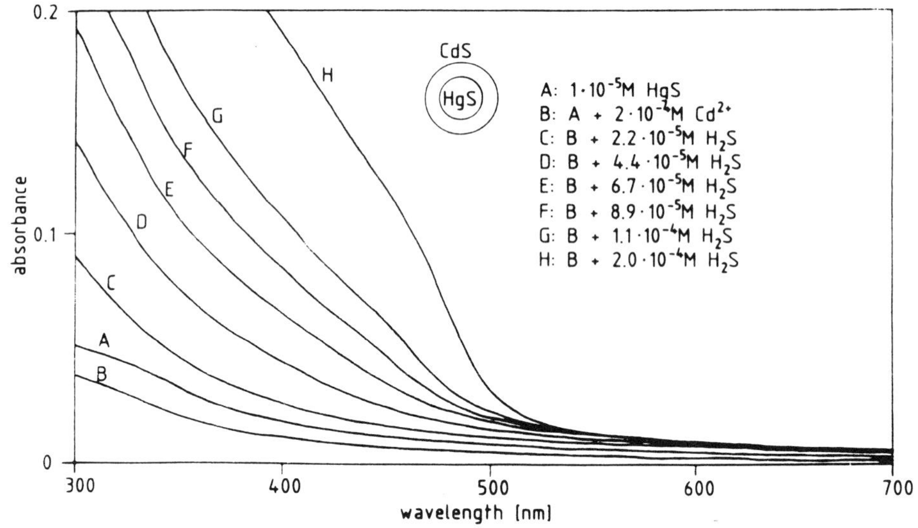

Figure 10. Absorption spectra of HgS particles surrounded with CdS.

about 7×10^{-5} *M* CdS to the original HgS sol, mainly coated particles, and at higher concentrations, additional separate CdS particles, are formed. The absorption spectra of the coated particles (up to sample E in Figure 10) do not reflect a superposition of HgS and CdS spectra. It appears that the absorbance of the HgS core is enhanced as a result of coating with CdS. Figure 11 shows the fluorescence spectra ($\lambda_{exc.} = 360$ nm) corresponding to the absorption spectra shown in Figure 10. Solutions A and B (pure HgS and HgS with Cd^{2+} ions) do not fluoresce at all. A strong fluorescence band with a maximum at 950 nm occurs after the first addition of H_2S (the slight dent in the HgS fluorescence band is due to absorption of the solvent water). Increasing amounts of H_2S do not change its spectral position. The intensity, however, increases until approximately 7×10^{-5} *M* H_2S is added, which is nearly the same amount of additive after which the absorption at $\lambda > 500$ nm also no longer increases on addition of more H_2S. Obviously, fluorescence has been generated in the HgS particles by covering them with CdS, whereas the separately formed CdS particles do not contribute to the fluorescence.

A fluorescence "activation" of this kind was also observed in the systems $Cd(OH)_2$ on CdS [38] and ZnS on CdSe [49], as described above. Following the same arguments the fluorescence activation is interpreted as the removal of traps for nonradiative recombinations, mainly by a coor-

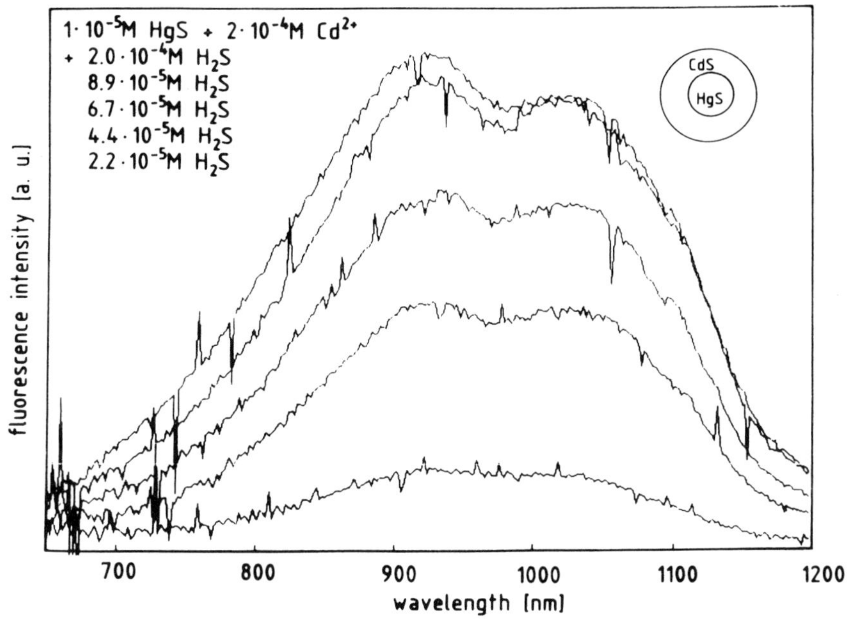

Figure 11. Fluorescence spectra of HgS particles surrounded with CdS.

dinative saturation of surface anions. Due to the shape and the spectral position, one could ascribe the observed fluorescence to the recombination of trapped charge carriers at the interface of the HgS core and the CdS shell. The fluorescence could just as well arise from band-to-band recombination in the HgS core. In this case the width of the fluorescence band would reflect the size distribution of the HgS cores. Evidence for the formation of a core–shell structure instead of a mixed solid phase is also supported by the fluorescence properties. Namely, if a $Hg_xCd_{1-x}S$ phase would be formed during the growth of CdS on HgS, the molar ratio of Hg and Cd should decrease, and as a consequence, the fluorescence band should shift spectrally. On the contrary, it was found that the spectral position of the fluorescence remains fixed with increasing amounts of deposited CdS and that only the intensity of the luminescence increases concomitantly. As outlined above, the latter was understood as a gradual removal of nonradiative recombination centers during the formation of a compact CdS layer on the surface of the HgS core.

The last example of this class of composite particles published so far is the system CdS/PbS [53]. A polymer-stabilized CdS colloid of 6 to 8 nm diameter is prepared to which different amounts of Pb^{2+} ions are given.

Since the reaction $Pb^{2+} + CdS \rightarrow PbS + Cd^{2+} + \Delta G$ is exothermic, the lead ions replace the cadmium ions in the CdS particle surface. The thickness of the PbS shell around the remaining CdS cores could be controlled by the concentration of the added Pb^{2+}. A careful transmission electron microscopy and electron diffraction study revealed that, indeed, coated and not separate particles have been formed during the synthesis. Indications are given for highly strained interfacial regions between the CdS cores and the PbS shells, which are attributed to the large lattice mismatch of the two materials. Interestingly, in this study hexagonal CdS has been prepared, whereas in most others a cubic crystal structure was obtained. The absorption of the CdS/PbS composites extends farther into the red spectral region the larger the shell-to-core ratio. This effect has been studied theoretically by the same authors [54] based on the concept mentioned earlier used by Brus et al. [49] for calculating the $1s$–1s electronic transition in coated semiconductor particles.

C. Quantum Dot Quantum Wells

The latest development in the subfield of surface-modified semiconductor nanocrystals is the synthesis of three-layered colloidal particles [55–58]. The novel structures consist of a size-quantized semiconductor particle acting as the core spherically covered by several monolayers of another semiconductor material, which by themselves are surrounded by several monolayers of, again, the core material acting as the outermost shell. These particles are called quantum dot quantum wells (QDQWs) or, metaphorically, nano-onions. The more scientific naming is motivated by the analogy to "real" quantum wells, which are semiconductor structures with alternating layers of two semiconductor materials exhibiting quantum confinement in one dimension in at least one of the materials.

The first example of QDQWs described in the literature is the system CdS/HgS/CdS. The synthesis, which is outlined in detail in Ref. 56, starts with a standard polyphosphate-stabilized CdS colloid in aqueous solution. To these particles with an average diameter of 5.3 nm and a fairly narrow distribution of sizes, $Hg(ClO_4)_2$ is added at neutral pH. The mercury salt dissociates readily in this medium, allowing the mercury ions to react with the CdS colloids. Since the solubility product of HgS is more than 20 orders of magnitude smaller than that of CdS, Cd^{2+} ions are readily replaced by the Hg^{2+} ions. As is demonstrated by polarography and mass spectrometry, this substitution reaction proceeds until the surface Cd^{2+} is replaced by Hg^{2+}, thus forming a surface layer of HgS on the CdS core, which then passivates the particle toward further substitution. Addition of H_2S to this solution leads to homogeneous precipitation onto the CdS/HgS composites of the Cd^{2+} ions released in the form of CdS, thus yielding a QDQW of CdS/HgS/CdS.

From this point in the preparation, one of two outcomes may be chosen: either thickening of the HgS layer or an increase in the outermost CdS layer thickness. The first alternative is achieved by repeated addition of mercury ions and H_2S to the former solution, the second by successive addition of Cd^{2+} and H_2S. Both branches have been followed, giving rise to a family of new materials having a 4.7-nm CdS core, one to three layers of HgS as the well and one to five monolayers of CdS acting as the outermost shell. The course of the preparation has been followed by optical spectroscopy, polarography, mass spectrometry, and electron microscopy.

In Figure 12 a gallery of TEM micrographs is shown which demonstrates the particle growth by successive repetition of the surface substitution and sulfidization steps. The size histograms of the same samples shown in Figure 13 clearly indicate that during the preparation the distribution of sizes remains constant (the standard deviation amounts to approximately 18%); that is, homogeneous growth is achieved by the synthetic approach

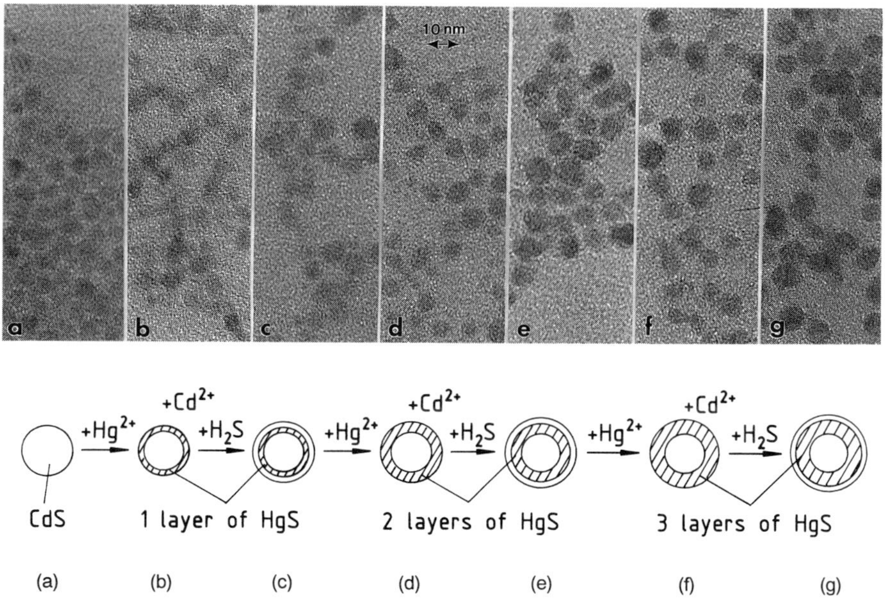

Figure 12. TEM micrographs and schematic drawings illustrating the growth of a QDQW CdS/HgS/CdS.

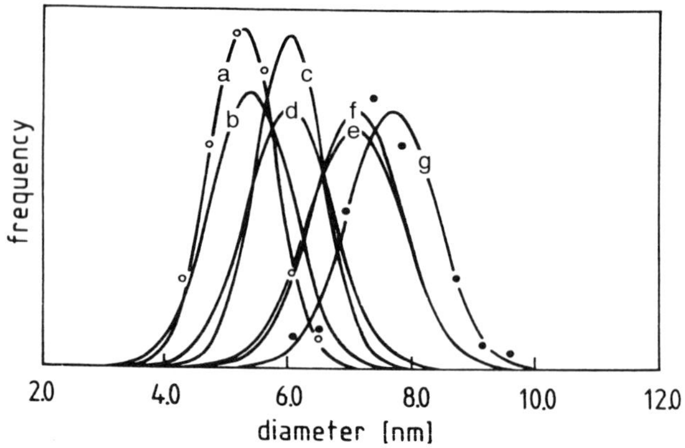

Figure 13. Size histograms corresponding to the micrographs shown in Figure 12.

described. As expected, the histograms show that the mean diameter does not change during the substitution reaction (a → b, c → d, e → f). The diameters increase by the expected amount by the formation of an additional layer of CdS on the particles (b → c, d → e, f → g).

In Figure 14 a set of absorption spectra of some representative QDQW species is depicted. The absorption spectrum of the starting CdS colloid is labeled a. The effect of addition of Hg^{2+} to this colloidal solution is reflected in spectrum b. The absorption extends to 600 nm and the remarkably pronounced shoulder indicating the energy of the 1*s*–1*s* electronic transition is shifted by 30 nm, to 500 nm. The absorption spectrum after the addition of H_2S to the colloids is seen in spectrum c, Figure 14. The absorption increases over the entire spectrum and extends to 700 nm, while the structure in the absorption spectrum observed before is lost gradually. After the addition of Cd^{2+} and slow sulfidization by H_2S spectrum d, Figure 14 evolves. Repetition of this step yields spectrum e. Again, an increase of absorption over the entire spectrum is observed together with a slight red shift at the onset of absorption. These last particles prepared have a total diameter of 8.0 nm, from which 4.7 nm is allotted to the CdS core, 0.6 nm to the HgS well, and 2.7 nm to the outermost CdS shell.

The colloidal solutions labeled a, b, and c in Figure 14 do not fluoresce at all in the range 450 to 1800 nm. Excitation of solution d (7.2-nm particles) at 450 nm yields the weak fluorescence seen in Figure 15, spectrum d. This fluorescence is enhanced dramatically by a factor of 20 by the addition of 1×10^{-3} *M* Cd^{2+} ions and raising the pH to pH 11.5 (Figure 15, spectrum da). This procedure does not change the absorption spectrum of the crystallites. As earlier, the chemistry of this step is interpreted as the

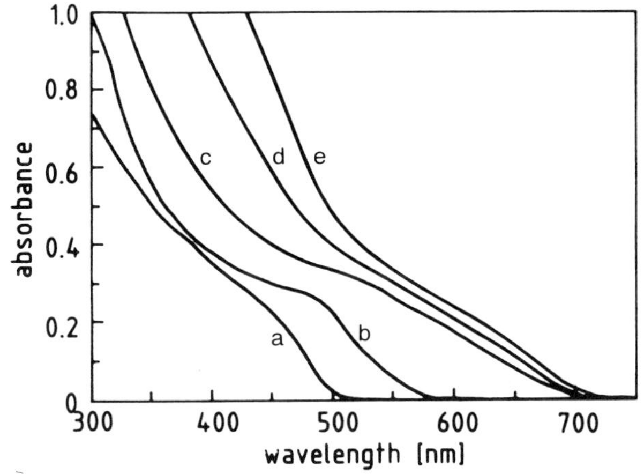

Figure 14. Absorption spectra of pure CdS (spectrum a), CdS capped with a monolayer of HgS (b), and (b) surrounded with (c) one, (d) two, and (e) three monolayers of CdS.

formation of a layer of $Cd(OH)_2$ on the surface of the particles [38]. In earlier publications the enhancement of the fluorescence intensity has also been called *fluorescence activation*, which is why the respective spectra are labeled with an additional "a".

Principally, the same phenomenon is observed with the somewhat larger particles (8.0 nm, solution e). In this case, the "nonactivated" colloid shows a detectable fluorescence (spectrum e, Figure 15). This fluorescence is enhanced by a factor of 5 by the addition of $1 \times 10^{-3}\,M$ Cd^{2+} and the raising of the pH to 11.5 (spectrum ea, Figure 15). The intensities of the fluorescence spectra shown in Figure 15 are comparable with each other since they have been corrected for the absorbance of the respective solutions at the excitation wavelength. In all cases a single fluorescence band is observed which is almost symmetrical and fairly narrow. Its maximum is located close to the onset of absorption.

The observed relatively sharp fluorescence band with its maximum at approximately 710 nm cannot be assigned to fluorescence from CdS, which in the colloidal phase either shows fluorescence close to the onset of absorption (i.e., at 500 nm or at shorter wavelengths, depending on the size of the particles) or a broad fluorescence band between 500 and 800 nm which is attributed to the radiative recombination of trapped charge carriers. Furthermore, the fluorescence of the QDQWs is also excitable with light of less energy than the CdS bulk bandgap (i.e., with $\lambda > 500$ nm). On

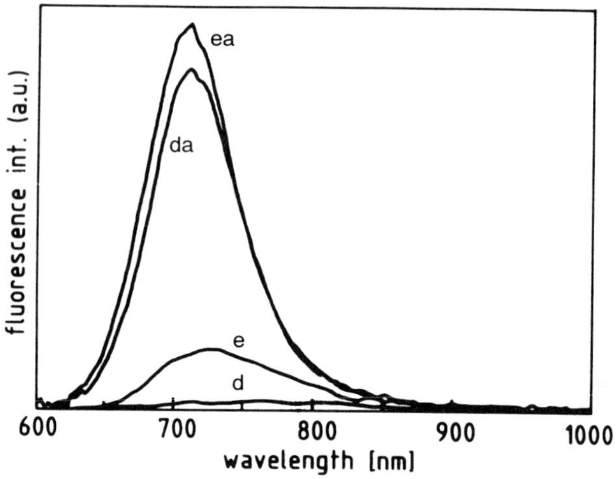

Figure 15. Fluorescence spectra of QDQWs corresponding to Figure 14 before (d and e) and after "fluorescence activation" (da and ea).

the other hand it cannot be assigned to fluorescence from either HgS alone (which is nonfluorescing in the colloidal state under these conditions [52]) nor from the CdS/HgS core–shell particle since solution b is nonfluorescing. The third layer is a prerequisite for the occurrence of the fluorescence at 710 nm. The interpretation is as follows: the photogenerated electrons and holes have a maximum overlap of their wavefunctions in the region of the HgS well, where they may either recombine via fluorescence or nonradiatively. As shown earlier, it is the surface of the colloidal particles that is the main channel for radiationless transitions. Thus it is not surprising that the fluorescence intensity increases with increasing layer thickness of the outer CdS layer since the centers for nonradiative transitions on the surface of the HgS sphere are blocked. Even more effective for fluorescence enhancement is the addition of a layer of $Cd(OH)_2$ on the surface of the particles, which acts exactly as described before, as an insulating layer saturating centers of radiationless transitions and pushing charge carrier density toward the well region.

Obviously, an effective scavenger of excitation energy has been introduced by incorporation of the HgS layer. This layer by itself may act as an efficient transformer of the excitation energy into fluorescence, the energy of which may be tuned from 700 nm to 950 nm, depending on the thickness of the HgS layer. The latter is shown nicely in Figure 16. At the top of the figure the absorption spectra of QDQWs with the 4.7-nm CdS core covered by one, two, and three monolayers of HgS are shown. Irrespective of the

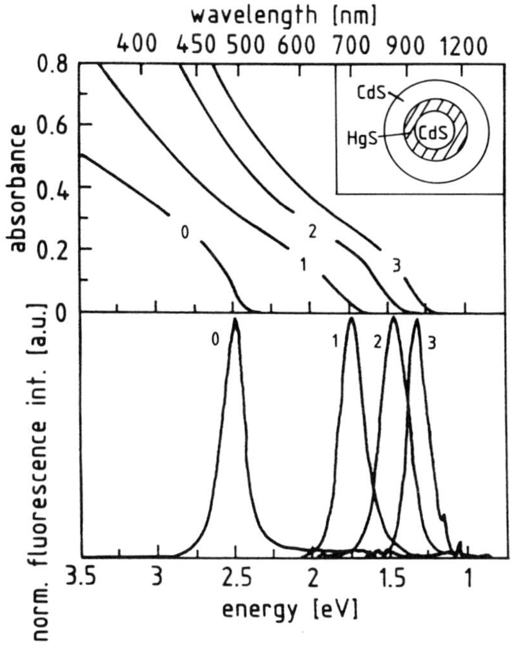

Figure 16. Absorption (*top*) and fluorescence (*bottom*) spectra of QDQWs CdS/HgS/CdS with one, two, and three monolayers of HgS between the CdS segments.

number of HgS layers, all particles are surrounded by (almost) one monolayer of CdS and, additionally, $Cd(OH)_2$ for fluorescence activation. The corresponding fluorescence spectra are shown in the lower part of Figure 16. The fluorescence intensities have been normalized to the CdS bandgap fluorescence intensity. Thus it is seen that all spectra consist of a single, relatively narrow band which occurs close to the onset of absorption. Related to this, the thicker the HgS well, the more the fluorescence maximum is shifted to lower energies.

The room-temperature fluorescence quantum yield decreases from about 10% for pure CdS to 5% for particles with one monolayer of HgS, to 4% and 3% for particles with two and three monolayers of HgS between the CdS segments, respectively. A feature similar to this has been observed by Zajicek et al. [59] for the photoluminescence from ultrathin ZnSe/CdSe quantum wells. By these authors the reduction of the photoluminescence intensity is attributed to the generation of misfit dislocations when the thickness of the CdSe layers exceeds a critical thickness of 4 ± 1 monolayers. In the current status of the investigations it is not yet clear whether this interpretation also holds for the QDQWs. So far, the fact is simply accepted that with increasing HgS layer thickness, radiationless transitions either in the interface between CdS and HgS or within the HgS comes more and more to dominate the relaxation process.

Temperature-dependent measurements have been carried out with sample c (including the outer CdS shell and "fluorescence activation"). With decreasing temperature the fluorescence intensity increases, reaching a maximum at about 80 K. At this temperature the fluorescence quantum yield is between 20 and 30%, depending on the specific sample. Below about 50 K the fluorescence intensity starts to decrease again, terminating at 4 K with a quantum yield of between 10 and 20%. Over the entire temperature range the shape of the fluorescence band is not affected except that in some samples a slight shoulder on the red wing of the fluorescence band emerges at low temperatures. Together with this a slight red shift of the fluorescence maximum is detectable.

Some photobleaching experiments have been carried out with QDQWs [58]. They reveal that the photobleaching spectrally follows the newly evolving 1*s*–1*s* electronic transition of the composite particles, thus pointing also in the direction that the electronic structure of QDQWs is not simply a superposition of the electronic properties of the segments of the particles. This is also concluded from theoretical modeling of the particles in the framework of the effective mass approximation [57]. This calculation fits the 1*s*–1*s* optical transition of the QDQWs of various compositions very well. A further description of the model goes beyond the scope of this chapter.

Future work on these unique particles, which may act as prototypic structures for investigations of semiconductor boundaries, inner surfaces, and charge carrier separation and localization, will include the systematic study of temperature-dependent charge carrier dynamics as a function of the composition of the particles as well as experiments with materials other than CdS and HgS. It is believed that the preparative approach should be of more generality, possibly yielding particles consisting of more than one well and/or more than two materials.

V. SPECTROELECTROCHEMISTRY OF Q-PARTICLES

A. Particles in Solution

As shown in the previous sections, optical spectroscopy is a convenient method to obtain data about the relative energetic positions of ground and excited states of molecules, clusters, and solids. To fix the electronic

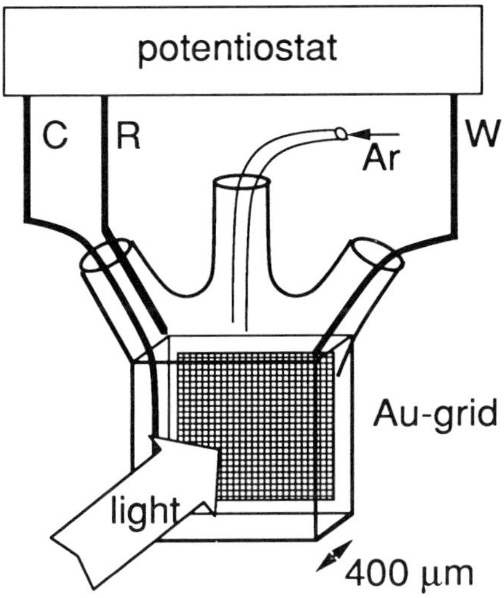

Figure 17. Optical transparent thin-layer electrode (OTTLE). W, working electrode; R, Ag/AgCl reference electrode; C, platinum counterelectrode.

levels at an absolute scale, one needs, however, measurements versus a well-known reference system. Such experiments are the typical domain of electrochemistry. Especially advantageous in the investigation of molecular solutions are measurements using the *optically transparent thin-layer electrode* (OTTLE, Figure 17), which was recently used also for colloidal particles [60]. It combines potentiostatic control and optical spectroscopy in solution. It consists of a very thin optical quartz cell (here 400 μm), which is equipped with a fine porous gold sieve (e.g., 300 mesh) as the working electrode, a platinum counterelectrode, and a reference electrode (e.g., Ag/AgCl). The cell is filled with solution and purged with argon. After applying a definite potential and allowing the solution to equilibrate with the working electrode (which takes usually a few minutes), absorption spectra can be measured through the gold grid (which absorbs only a moderate fraction of the incident light). Figure 18 shows the spectra that were obtained for colloidal alcoholic ZnO solutions [61]. At potentials more positive than −0.8 V, the spectra are identical to the one of the pure solution. At more negative potentials, however, a strong optical bleaching between 310 and 340 nm and an increase in the absorbance at shorter wavelength are seen. Such behavior is known from earlier photochemical and radiolytical measurements [62, 63], in which excess electrons were brought onto ZnO particles. The critical

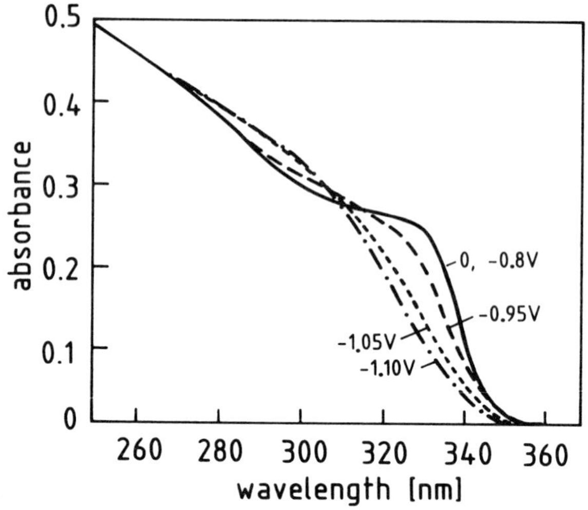

Figure 18. Absorption spectra of a colloidal ZnO solution in ethanol at various potentials vs. Ag/AgCl. The spectra were measured through the OTTLE.

potential at which optical bleaching starts can therefore be interpreted as the electron affinity of the particles. Thus, by measuring the critical potentials of colloidal solutions with differently sized particles, the size-dependent electron affinity of Q-sized particles is obtained in a very easy manner. The results of such measurements are seen as dots in Figure 19. The solid line represents the results from a quantum mechanical particle in the box calculation (with $m^* = 0.28$, barrier height: 3 eV, conduction band edge of bulk ZnO: -650 mV). The agreement between theory and experiment is obviously good, and it was therefore concluded that the electron injection occurs either directly into the quantized conduction band level or in shallow trap states which are energetically strongly coupled to the quantized levels themselves.

The optical bleaching by stored electrons is the basis for the occurrence of strong optical nonlinearity observed in Q-particles [64]. The physical reason for this optical bleaching is still not discussed conclusively in literature. The most obvious explanation comes from a state filling model. The stored electrons occupy the lowest electronic levels in the conduction band and, consequently, the optical transition has to occur to higher electronic levels (i.e., at shorter wavelength). This effect is known in solid-state semiconductor physics as the *Burstein shift* [65]. Other theoretical models describe the optical bleaching as a consequence of the polarization of the exciton in the electric field of the stored electron, which is then

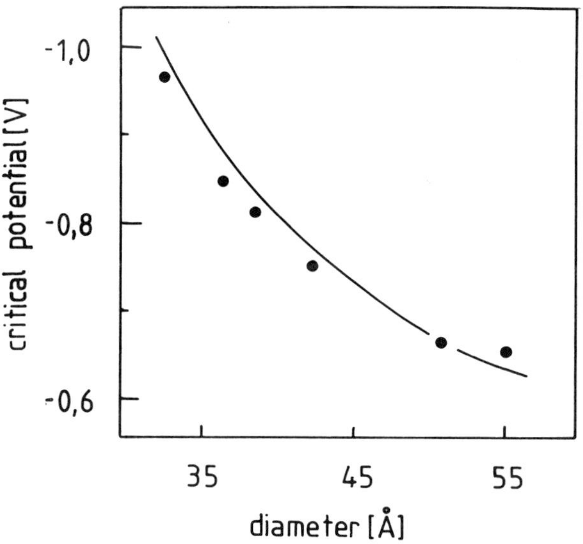

Figure 19. Critical potential (electron affinity) of colloidal ZnO Q-particles as a function of the particle diameter.

considered as a point charge at the surface [64, 66–68]. The main difference between the two models lies in a description of the stored electron itself. The state filling requires free electrons and the polarization model, trapped ones. The fluorescence measurements on CdS, which show clearly the evidence of electron trap states and fast trapping processes support the polarization model rather than the state filling. The spectral changes of the Q-particles should be slightly different for both mechanisms, and from a detailed analysis of the spectra in Figure 18 it was proposed that the first electron can be stored in a trap state, whereas quantized levels are occupied in doubly reduced ZnO Q-particles in solution.

Again, the conclusion that might be drawn from these findings is that future work on Q-particles should concentrate on the investigation and finally, on the chemical control of surface states.

B. Particles in Thin Layers

In 1984, Baranski and Fawcett found that the properties of electrochemically deposited thin films of CdS varied strongly depending on the deposition conditions [69]. Hodes and co-workers [70, 71] later detected that these films consist of aggregated nanoparticles the size of which (a few nanometers) determined the bandgap of the films via size quantization effects. With a total thickness of a few micrometers it is readily calculated that the films consist of some hundreds of particles lying on top of each other. Photocurrent quantum yields of up to 10% (i.e., 0.1 electron per incident photon) were measured for such films under monochromatic illumination in a spectral region where the incident light is absorbed almost completely.

At first sight it appears astonishing that reasonable photocurrents can be measured in such layers, since electrons and holes must be restricted to the volume of the individual particles within the film in order to observe size quantization. Hence electron transport through the entire thickness of the film cannot occur via free diffusion in the energy bands as in the case of bulk semiconductors, but must involve hopping processes from one particle to the other. Keeping in mind that electrons and holes get trapped more or less immediately after light excitation, it is most likely that these hopping processes involve surface traps. We will come back to this point later.

Chemical and electrochemical deposition of nanoparticles is a very easy method to use to obtain films. The disadvantage of this technique, however, lies in the relatively broad size distribution of the particles obtained. As shown above, colloid chemistry offers several methods to obtain well-defined particles with narrow size distribution. O'Regan and Grätzel [72], as well as Anderson and Spanhel [73], developed the preparation of thin films consisting of TiO_2 and ZnO nanoparticles by spin or dip coating the concentrated colloidal solutions on optical transparent and electrically conducting indium/tin oxide-coated glass (ITO). Final heating of these electrodes results in nanoporous films in which the colloidal particles were tightly aggregated to a three-dimensional framework.

1. Potential-Dependent Optical Properties of Nanoporous Zinc Oxide Electrodes. Figure 20 shows the experimental setup of a photoelectrochemical cell containing a nanoporous particle film electrode (W: working electrode). The electrochemical cell is also equipped with a platinum counterelectrode (C) and a Ag/AgCl reference electrode (R), and a redox couple electrolyte. As the individual particles of which the films consist are much smaller than the wavelength of visible light, the films are completely transparent. The whole cell can be mounted in an optical spectrometer. In this way it is easy to monitor spectral changes of the particle film when the potential at the back contact is varied. Such spectroelectrochemical measurements with colloidal TiO_2 films have been described by Grätzel and co-workers. The well-known blue color of Ti^{3+} centers was observed upon stepping the potential at the back contact into the cathodic region. The reduction and the flatband potential derived from these measurements were identical with the bulk values of TiO_2 [74].

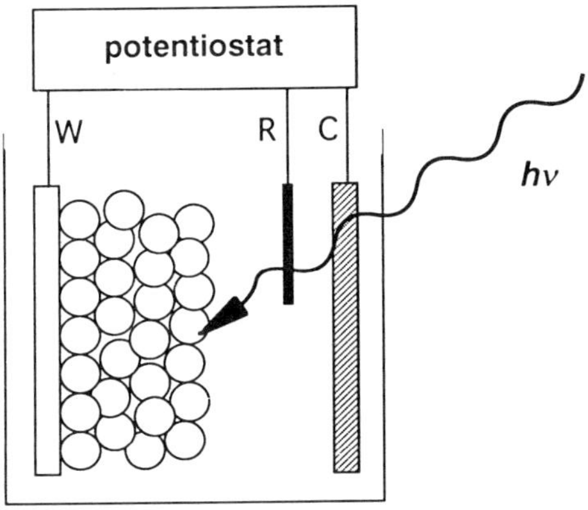

Figure 20. Nanoparticle (nanoporous) film electrode mounted in an electrochemical cell, W, R, C: working, reference, and counter electrodes.

Let us now consider the case of a colloidal semiconductor nanoparticle film in which the individual particles are in the size quantization regime and which display fluorescence under light excitation. The question we deal with is how the applied potential influences the optical properties of the film. We distinguish three different potential regions which are shown schematically in Figure 21. In the anodic region (*a*) the Fermi level at the back contact lies within the bandgap of the particles. Under these conditions, the photogenerated holes oxidize the electrolyte and the electrons are drawn to the back contact. Consequently, a photocurrent is observed in the outer circuit (i.e., via the potentiostat) and the radiative recombination of electrons and holes must be strongly suppressed. When the Fermi level is raised to the (quantized) conduction band levels of the particles (Figure 21*b*; a situation comparable to flatband conditions in bulk electrodes), no potential gradient is left that drives the electrons to the back contact. Thus recombination processes within the individual particles of the film can become predominate, giving rise to an increased luminescence efficiency. In the more cathodic region (Figure 21*c*), electrons will be injected into the film. For ZnO this should lead to optical bleaching of the particles, as shown above for colloidal samples.

Photocurrent and optical absorbance of a nanoporous ZnO film are shown in the upper part of Figure 22 as a function of the applied potential. As expected, the photocurrent drops with increasing negative potential to almost zero at the critical potential of $U = -540\,\text{mV}$, whereas the optical absorbance is constant up to this potential and decreases at even more negative values. The potential dependence of the trapped fluorescence is shown in the lower part of Figure 22 (curve "560 nm"). The strong decrease

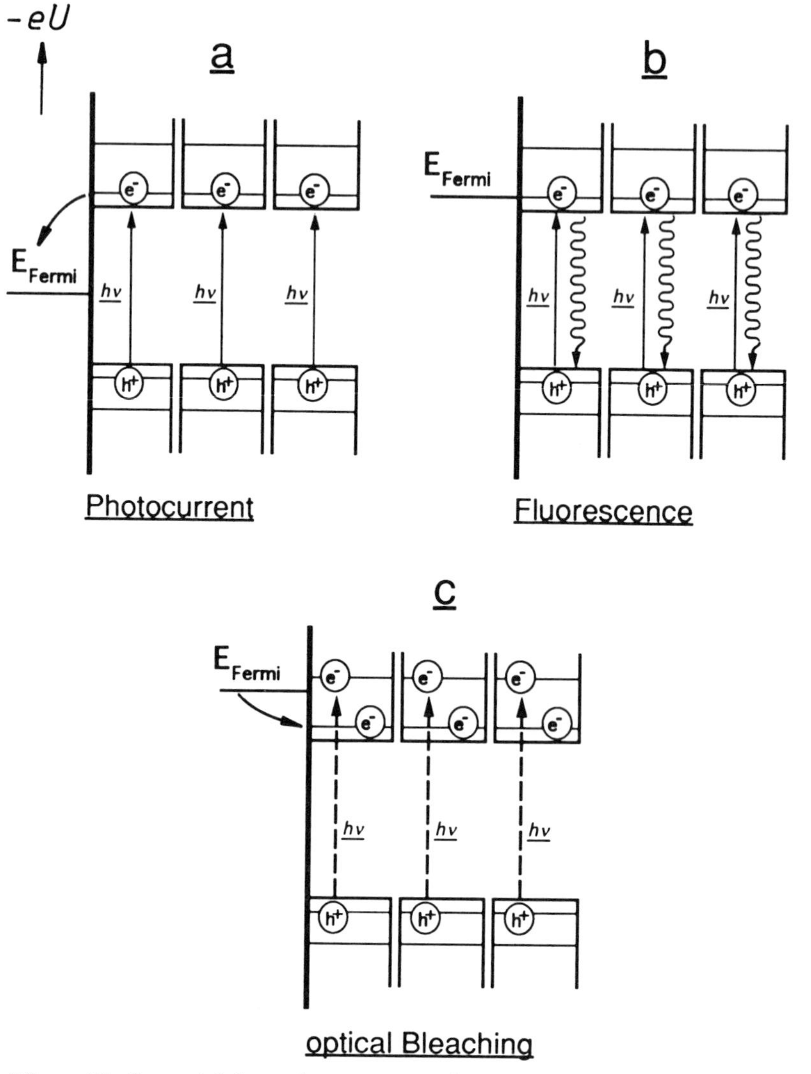

Figure 21. Potential-dependent processes in a nanoparticle film electrode.

in the anodic region, where charge separation and consequently photocurrent compete with radiative recombination, is clearly visible. At potentials negative from −540 mV, trapped fluorescence also declines, whereas excitonic fluorescence (curve "370 nm") appears at −540 mV and increases with increasing negative potential. It was shown that the intensity of the excitonic fluorescence is proportional to the charge deposited on the

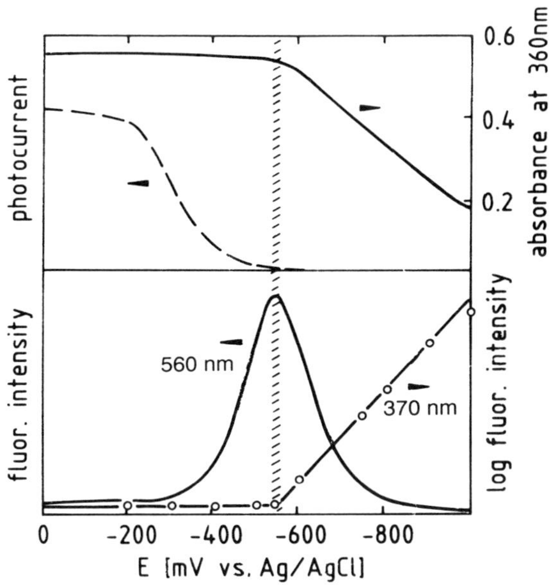

Figure 22. Photocurrent, optical absorption, trapped fluorescence (560 nm), and excitonic fluorescence of a nanoporous zinc oxide electrode as a function of the applied potential vs. Ag/AgCl, pH 8 buffer.

particles [75] and that the decay of the excitonic fluorescence occurs already in the subnanosecond region [76]. It was concluded that the stored electrons occupy the traps (and possibly conduction band levels) and thus the photogenerated holes quickly recombine with free electrons prior to deep-hole trapping.

The well-pronounced potential dependence of the optical absorbance and fluorescence are typical features of nanoporous ZnO films, bulk ZnO shows only weak potential dependencies [77]. As seen from Figure 22, the critical potential of nanoporous films (i.e., flatband potential, electron affinity) can be obtained easily from spectroscopic data. It was found that the critical potential of ZnO films consisting of different-sized Q-particles rise to more negative values with decreasing particle size. Thus the size quantization effect changes not only the optical properties of individual nanoparticles in a drastic manner but also the electrical properties of macroscopic nanocrystalline electrodes. Keeping in mind the enormous progress made in the synthesis of highly monodisperse nanoparticles in gram scale, one is tempted to dream about possible applications of quantized materials via colloid chemistry.

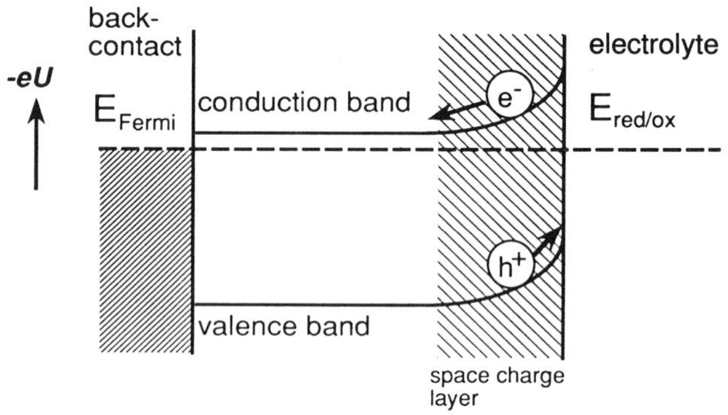

Figure 23. Energy diagram of an *n*-doped bulk semiconductor electrolyte junction.

2. Photochemically Induced Electron Transport. Let us now consider some basic concepts of photoelectrochemistry of the (bulk) semiconductor–electrolyte interface and their application to porous nanocrystalline electrodes. Figure 23 shows a typical example of the energetic levels of an *n*-doped semiconductor in thermal equilibrium with the electrolyte (i.e., the Fermi level of the semiconductor and the electrochemical potential of the electrolyte are equal). The energetic position of the band edges at the interface is determined by the chemical composition of the semiconductor and by the surface potential, which is built up due to the specific adsorption of ions (for oxide semiconductors mainly OH^- or H^+). On the other hand, the energy difference between the band edges and the Fermi level in the interior of the crystal is given, predominantly, by the doping densities. Thus a potential gradient is established, leading to a space-charge layer close to the surface of the crystal. At normal doping densities this space-charge layer extends typically over a distance of 0.1 to 1 μm.

The light penetration depth of bulk semiconductors is usually comparable to the length of the space-charge layer and thus photogenerated electron–hole pairs are generated to a large extent within the space-charge layer and separated by the strong electric field. The electrons are drawn to the back contact and the holes move to the surface. Band bending in the space-charge layer is thus responsible for the highly efficient and irreversible charge separation observed in many photoelectrochemical cells.

In the case of porous nanocrystalline electrodes the situation becomes complicated:

1. The concept of “doping” becomes questionable for nanoparticles since typical bulk doping densities correspond to a situation where only one doping atom is present in every thousand or more particles.

2. Since the electrolyte penetrates the pores of the electrode, like water a sponge, the surface potential should be constant over the entire thickness of the film.
3. The distance from the particle surface to the interior of the interconnected particles never exceeds a few nanometers (i.e., a distance that is too small to build up a space-charge layer).

Trap states, which "disturb" the processes of charge separation and charge carrier transport in bulk semiconductors, become essential in nanoporous electrodes. As a consequence of the enormous surface-to-volume ratio, the density of traps is very high. These traps can be detected by cyclovoltametry, as shown for nanoporous ZnO films [75]. Two main features are recognized in the cyclic voltamogram of Figure 24. First, the peaks are almost symmetric, indicating that the electron injection is a reversible process. Second, a considerable current flow is also observed at potentials positive from the critical potential of −540 mV. It was checked carefully that this current is due to trap filling in ZnO. To measure the total charge stored in the film at a given potential more accurately, a potential jump technique was applied, as outlined in Ref. 75. With the knowledge of the analytical amount of ZnO and the mean particle diameter, the number of electrons per ZnO particle in the film could be obtained as a function of the applied potential (Figure 25). It is seen that approximately six electrons per particle can be stored at −1 V. At more negative potentials, zinc reduction occurred. Below the critical potential of −540 mV, an average of only 0.2 electron is stored on each particle. One should, however, keep in

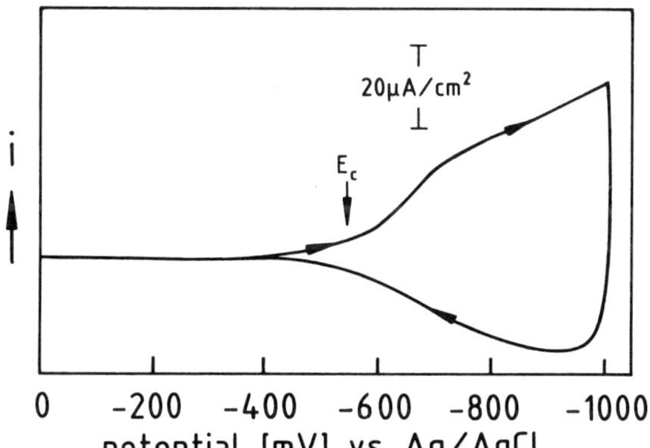

Figure 24. Cyclovoltamogram of a nanoporous zinc oxide electrode. Film thickness, 1.3 μm; scan speed, 10 mV s^{-1}, pH 8 buffer.

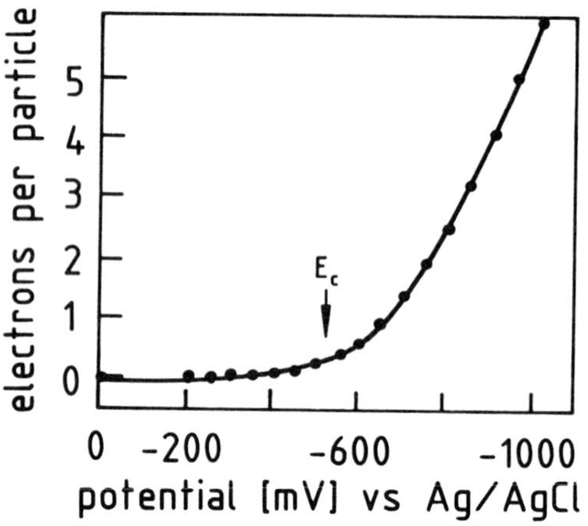

Figure 25. Number of stored electrons per particle in a nanoporous zinc oxide film as a function of the applied potential. Mean particle diameter, 5.9 nm; pH 8 buffer.

mind that these values are affected with a relative large error (small fluctuations in the size of the particles result in a large error in the determination of the number of particles).

The maximum number of electrons that can be stored on a nanoparticle before metal formation occurs depends on the size and the chemical composition of the particle. In the case of 3-nm CdS colloids, for instance, it was shown by pulse radiolysis that cadmium formation occurs when two electrons are present on a particle [64].

Let us now return to the question of how the electrons can be transferred so efficiently through the electrode. Willig and co-workers postulated a trap filling model for the electron transport in highly polycrystalline TiO_2 [78]. Accordingly, the photogenerated electrons first fill up the large number of traps within the electrode, and the kinetics of the photocurrent is then determined primarily by the release rate of the electrons out of the traps. The authors were able to fit time-resolved photocurrent rise curves obtained from chopped illumination experiments. Kamat and co-workers gave a qualitatively similar explanation for the shape of time-resolved photovoltage measurements after laser illumination of nanoporous ZnO electrodes [79].

The role of trap states during the electron transport is very nicely reflected in potentiostatically controlled and time-resolved photocurrent measurements in Ref. 75. Figure 26 shows such photocurrent profiles and, in the inset, the experimental setup of this measurement. The nanoporous ZnO electrode was mounted in a photoelectrochemical cell and the photocurrent was measured after excitation with a 308-nm flash of an excimer

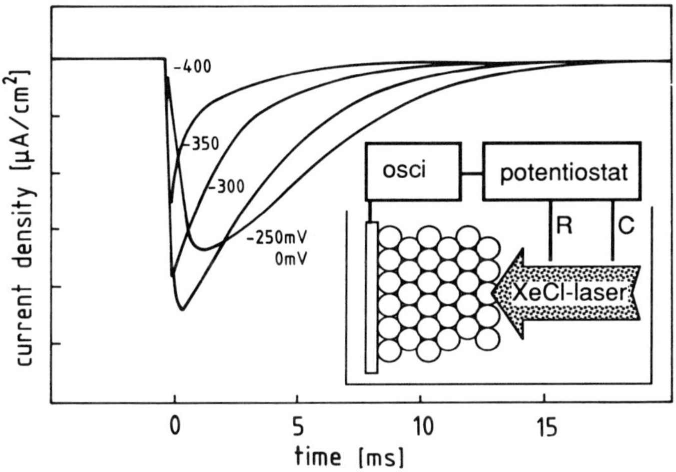

Figure 26. Time-resolved photocurrent profiles of a nanoporous zinc oxide electrode after excitation with a 308-nm excimer laser flash at various applied potentials. The experimental setup is shown schematically in the inset of the figure. R, C, reference and counter electrodes; pH 8 buffer; film thickness, 3 μm.

laser. Ninety percent of the incident light was absorbed within the first 300 nm of the film, whereas the total film thickness was chosen to be 3 μm. Thus it was confirmed that the electron–hole pairs were generated close to the film surface, and the photocurrent kinetics was determined by the transport of electrons through the entire thickness of the film. The laser intensity was kept low to guarantee the potentiostatic control of the Fermi level within the film, also under light excitation. Upon increasing the potential from 0 V vs. Ag/AgCl toward the critical potential of −540 mV, it is seen that up to −250 mV no changes are visible in photocurrent profiles, but upon further increase the rise as well as decay rate of the photocurrent increases.

The integrated photocurrents decline and become almost zero at the critical potential, which is consistent with the stationary measurements from Figure 22. The picture that arises from the experimental data is shown schematically in Figure 27. In the upper part of the figure, the Fermi level at the back contact lies under the energetic position of the trap levels. After light excitation, the positive hole is trapped in deep surface states (where it is electrostatically shielded) and finally, reacts with the electrolyte. The electron is drawn to the more positive back contact; however, it falls almost immediately into trap states. As long as the traps are empty, the electrons

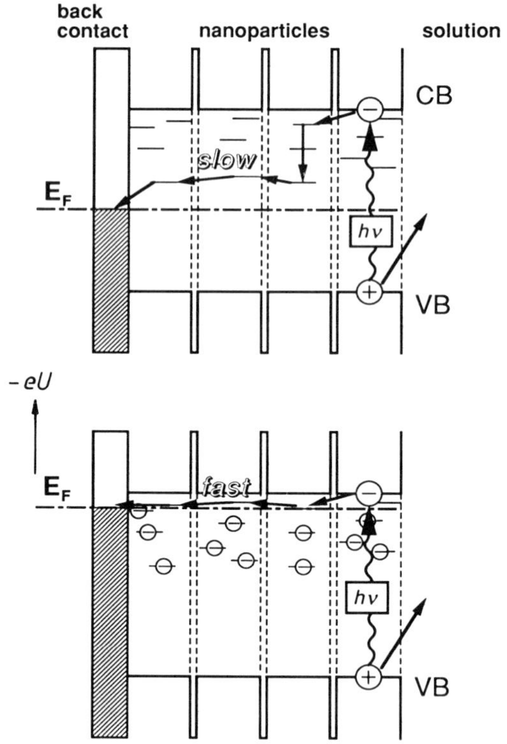

Figure 27. Electron transport through a nanoporous film electrode after flash excitation. The potential jumps symbolize the grain boundaries between the individual particles. Under strong positive polarization (*upper part*) the transport occurs via deep traps, whereas only shallow traps are involved when the Fermi level at the back contact rises close to the conduction band of the nanoparticle.

are finally captured in deep traps via a trapping cascade. To reach the back contact, the electron has to hop from trap to trap (or from one particle to the other). This may occur either directly by tunneling or is thermally assisted via the quantized conduction band level. Independent of which mechanism is predominant, the release rate out of a trap is slower the deeper the trap (and thereby the more localized the electron). Thus the rise and decay of the photocurrent after flash excitation must be slow under strong anodic polarization and almost constant, independent of how much the Fermi level lies energetically below the deepest traps. This is what was found experimentally (curve 0 and −250 mV in Figure 26). It should be pointed out that bulk semiconductor electrodes usually behave in the opposite manner. That is, the band bending of an *n*-doped material increases with increasing anodic polarization, and therefore the electron transport becomes faster.

In the lower part of Figure 27, the Fermi level at the back contact rises closer to the conduction band of the nanoparticles. As seen in the cyclovoltamogram, electron traps can be populated reversibly via the back contact. Thus all traps below the Fermi level adjusted at the back contact are filled with electrons. Consequently, the photogenerated electrons move to the back contact via more shallow states, meaning that the photocurrent response after flash excitation becomes faster. The efficiency of charge separation more and more decreases since the recombination rate increases upon filling the traps with excess electrons (cf. fluorescence and photocurrent in Figure 22).

As shown in Figure 26, electron transport through a thin nanocrystalline semiconductor electrode occurs in the micro- or millisecond range and is thus a relatively slow process (in bulk semiconductors of comparable film thickness the electron transport would, typically, occur in the nano- to picosecond range). Under conditions of maximum photocurrent (i.e., comparable to the situation in the upper part of Figure 27) and using "normal" stationary photochemical illumination sources such as xenon lamps or sunlight, the rate of charge carrier generation becomes comparable with the rate of charge flow through the film. Consequently, a significant steady-state concentration of majority carriers (i.e., electrons) is built up leading to a charging of the electrode. This charging will be especially strong in the region of the electrode, where most of the incident light is absorbed. Thus a charge gradient will be established over the light penetration depth and we will end up with a situation comparable to band bending in *n*-doped bulk semiconductors (Figure 28). However, we should keep in mind that the electrochemical background of these phenomena is quite different. Whereas in the case of bulk semiconductors, band bending is caused by an electrostatical potential drop in the thermodynamical equilibrium of semiconductor and electrolyte, the band bending-like situation in nanoporous electrodes arises under steady-state illumination and is due to a gradient in the electrochemical potential of the electrode. It is somehow comparable to the quasi-Fermi-level concept for bulk electrodes. The consequence of the latter, however, is that the quasi-Fermi-level within the depletion layer becomes more positive for an *n*-doped semiconductor under illumination leading to a diminishing band bending and thus to a reduced photocurrent. The main difference in the case of the nanoporous electrodes is that photogenerated holes are rapidly scavenged by the electrolyte (which is present almost everywhere "in" the electrode), and thus the photogenerated

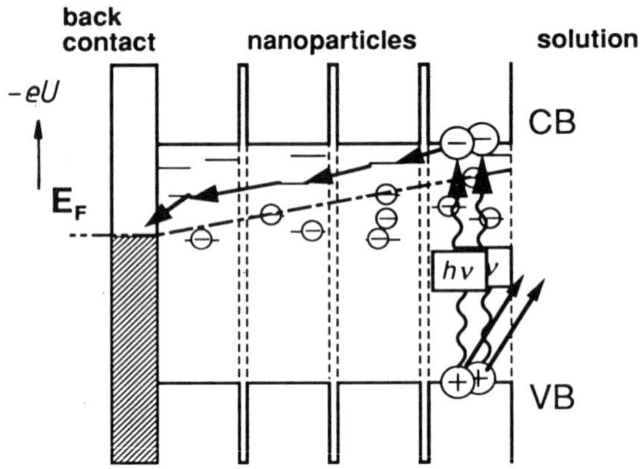

Figure 28. Electron transport through a nanoporous electrode under steady-state illumination conditions. A potential gradient is built up by trapped electrons.

electrons can build up an electrochemical gradient that *favors* electron transport through the layer. Hence these electrodes photocatalyze their own transport properties. In this sense, one might speak of a "smart material."

3. Sensitization of Nanoporous Electrodes. Nanoporous TiO_2 electrodes are currently being investigated for applications in solar energy conversion. The basic concept of these photoelectrochemical solar cells is the well-known sensitization of wide-bandgap semiconductors by organic dyes [80–82], which is shown in the pictogram of Figure 29. After absorption of visible light the dye molecules transfer electrons into the semiconductor substrate. The photoxidized dye, D_{ox} (the hole), is reduced by the electrolyte, the oxidized form of which is then reduced at the counter electrode by the electrons which flow through the outer current circuit. The main advantage of this type of a regenerative cell derives from the fact that light absorption and charge transport are spatially separated and recombination processes in the charge transport region are strongly suppressed. For more than two decades, practical applications in the field of solar energy conversion, however, could never arise. Efficient sensitization occurs only as long as much less than a mono layer of dye is brought onto the electrode surface. Using bulk semiconductor electrodes, the amount of adsorbed dye is, consequently, so small that only a negligible fraction of the incident light can be absorbed. Grätzel and co-workers [83] were the first to overcome this problem by using highly porous TiO_2 electrodes, the internal surface area of which was so large that less than a monolayer of dye was sufficient for almost complete light absorption. Although several groups worldwide

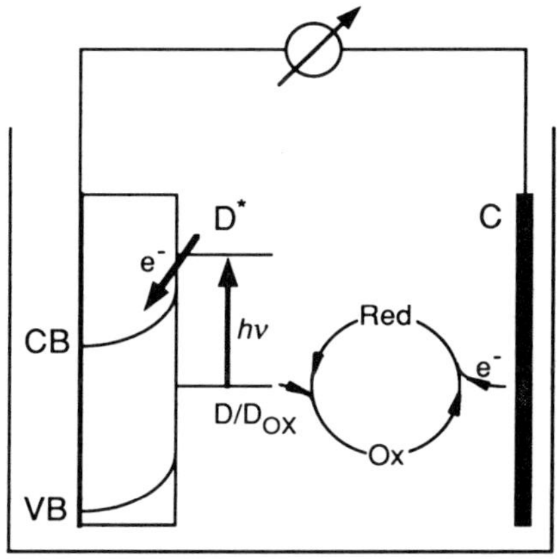

Figure 29. Schematic illustration of a regenerative dye-sensitized electrochemical solar cell. CB, conduction band; VB, valence band of the semiconductor; D, dye molecule; Red/Ox, redox couple electrolyte; C, counterelectrode.

investigate those sensitized nanoporous TiO_2 electrodes, only a very limited number of dyes were found to act as good sensitizers [72, 74, 83–85]. Energy-conversion efficiencies on the order of 10% are reported; reliable information about the long-term stability of such solar cells is, however, hardly available.

We do not want to discuss these cells here in detail but want to focus on investigations closer related to small-particle research. Similar to dye molecules, Q-particles should also be able to sensitize oxide semiconductors, whereby the size-quantization effect could control the optimum bandgap and the absolute position of the electronic levels of the sensitizing particles. First results based on this idea were reported by Vogel et al. [86]. They embedded Q–CdS particles of different size into a highly porous polycrystalline TiO_2 electrode and reported photocurrent quantum yields of more than 70% under monochromatic illumination. The synthesis of the Q-particles in the pores of the matrix was remarkably easy. The surface of the oxide electrode was loaded with Cd^{2+} simply by dipping the electrode into a concentrated Cd^{2+} solution, then rinsed with water, and the absorbed ions were reacted with S^{2-} by dipping the electrode into a Na_2S solution. The strong adsorption of the metal ions on the negatively charged oxide electrode highly reduced the growth of the particles, leading to the formation of quantized CdS with diameters around 3 nm. The particle size could be increased by repetitive dip coating, whereby the originally formed nanoparticles acted as nucleation centers for the further deposited CdS molecules. During the last years, a series of

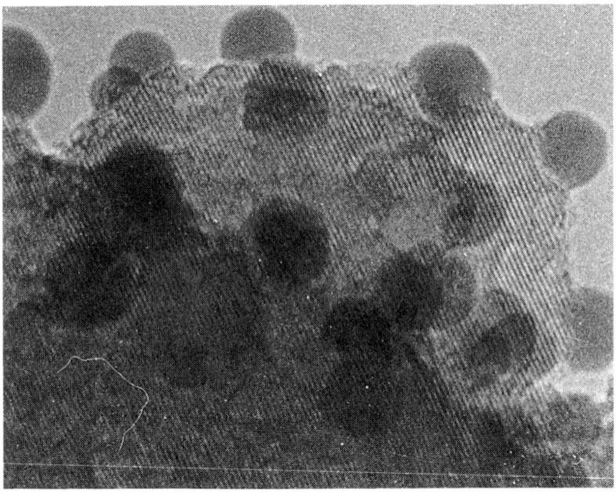

Figure 30. High-resolution transmission electron micrograph image of a Q–PbS-sensitized microporous TiO_2 electrode.

Q-particle-doped oxide electrodes have been described [86–91] including CdS–TiO_2, PbS–TiO_2, CdSe–ZnO, CdS–ZnO, and various combinations of CdS, PbS, Ag_2S, Sb_2S_3, and Bi_2S_3 with ZnO, TiO_2, Nb_2O_5, Ta_2O_5, and SnO_2.

A nice illustration of the morphology of such electrodes is seen in Figure 30, where a high-resolution electron microscopy image of a polycrystalline TiO_2 electrode after coating with Q–PbS particles is shown. The individual Q-particles of almost uniform size (6 nm) are clearly visible. They are well separated from each other and tightly embedded in the matrix (the lattice planes of which are also seen). In more recent papers, the substrate electrodes were usually prepared from colloidal oxide solutions similar to Spanhels preparation of nanoporous ZnO [73]. The nanopores of these electrodes act as a host for the embedded Q-particles. Those Q-particle doped nanoporous electrodes are perfectly transparent for light, and thus optical spectroscopy can easily be applied for photoelectrochemical investigations. The absorption spectra of a nanoporous TiO_2 electrode after embedding PbS particles of different sizes are shown in Figure 31. As expected for quantized particles, the absorption edge shifts to longer wavelengths with increasing particle size; for a comparison: bulk PbS exhibits a bandgap of approximately 0.4 eV, corresponding to an absorption edge at 3100 nm (i.e., even the largest particles of Figure 6 are in the quantization regime). The corresponding photocurrent spectra are displayed in Figure 32. The shape of these spectra is similar to the absorption spectra

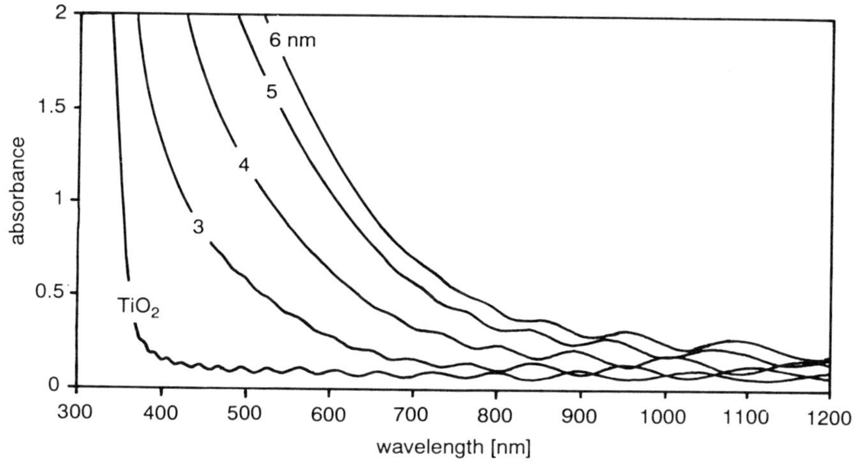

Figure 31. Absorption spectra of a pure nanoporous TiO_2 electrode and after embedding Q-PbS particles. The numbers at the spectra indicate the approximate mean size of the PbS particles.

(i.e., the onset of the photocurrent is shifted to longer wavelengths with increasing particle size). The ordinate scale in this figure is expressed as *incident photon to current efficiency* (IPCE), which gives the number of photoelectrons detected in the outer current circuit per incident photon of the respective wavelength.

It is readily seen that the IPCE values for electrodes with small particles are very high and that the IPCE decreases strongly for larger particles. This decrease in efficiency was shown to be a direct consequence of the diminishing size quantization effect with increasing particle size [89]. This is illustrated in the energy schemes of Figure 33. In small PbS particles (*a*, 3 nm), the size quantization effect drives the energetic level of electrons above the flatband potential of the TiO_2 substrate, leading to an efficient electron transfer from the particles into the substrate. With increasing particle size, the quantized state of the electron is shifted to more positive and the hole state to more negative potentials (e.g., Figure 33*b*, 7 nm), and finally, approach the bulk band edges. It is obvious that electron injection into the substrate cannot occur in the case of 7-nm particles or bulk PbS. Instead, it is expected that electrons from the TiO_2 matrix would be injected into PbS in the case of particles with diameters of 7 nm or more.

Vogel et al. [88] used PbS and CdS Q-particles of different sizes for the sensitization of various oxide substrate materials which differ in the absolute position of the flatband potentials. They showed that Q-particles of a characteristic minimal size were required to inject electrons efficiently in

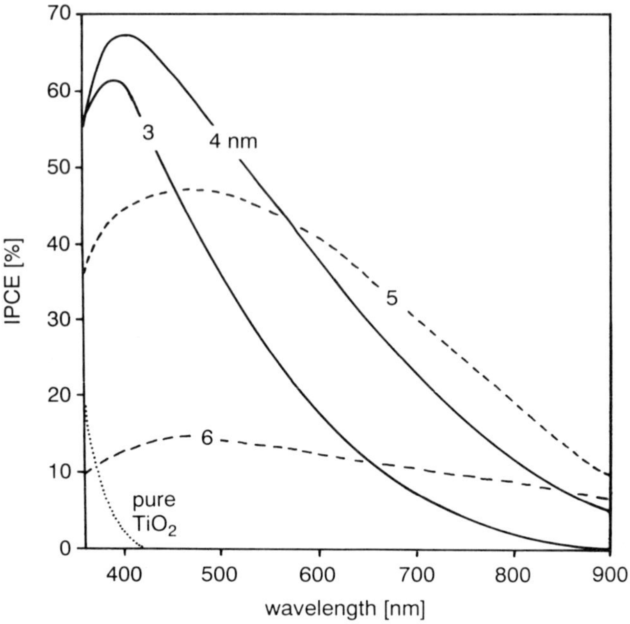

Figure 32. Photocurrent spectra of the electrodes of Figure 31. IPCE, incident photon to current efficiency, sulfide electrolyte.

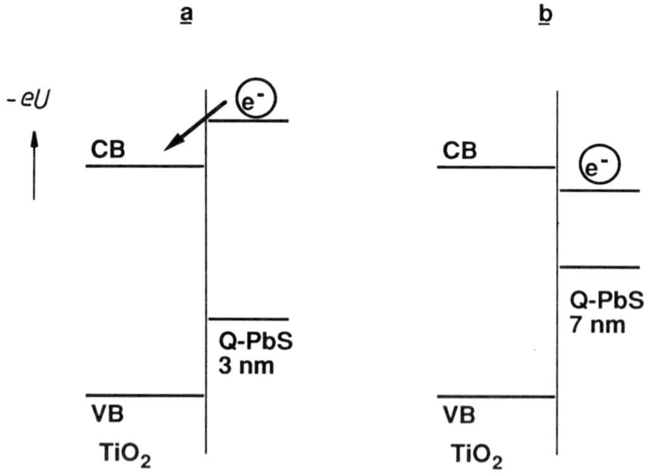

Figure 33. Energy scheme of the TiO_2–Q–PbS junction: (*a*) small PbS particles (3 nm); (*b*) large PbS particles (7 nm).

TiO_2, ZnO, Nb_2O_5, and Ta_2O_5, respectively. Thus the size quantization effect had to become more or less operative, depending on how negative the flatband potential of the substrates were.

Semiconductor heterojunctions are established in many modern electronic devices. The material combinations are highly restricted since, on the one hand, the lattice constants of the materials must be very similar to achieve epitaxial growth of the layers and, on the other hand, the energetics of the materials are given by the chemical composition and can only be altered with respect to the desired application. Small-particle research on the basis of colloid chemistry opens exciting perspectives. As seen in the case of TiO_2–Q–PbS, large lattice mismatches may be tolerated in a nanoheterojunction, and moreover, the size quantization effect can be utilized to determine the properties of the device.

At the present state of the art, however, the entire field of nanochemistry is still at a level of basic research. Despite the relatively high photocurrent quantum yields, solar applications did not arise until now, since the Q-particle doped electrodes decompose under illumination within several days or weeks. Keeping in mind, however, the complicated manner in which the photochemistry and photophysics of Q-particles is determined by the interplay of particle size and surface chemistry, it appears reasonable that a lot of detailed investigations have to be carried out to reliably judge the potential of Q-particle materials.

ACKNOWLEDGMENTS

We express our gratitude to all members of our group who contributed to this project. We dedicate this article to our academic teachers Dr. K.-H. Grellmann of the Max-Planck-Institute for Biophysical Chemistry in Göttingen and Prof. A. Henglein of the Hahn-Meitner-Institute in Berlin on the occasion of their retirement in 1994. We express our gratitude for their support, their ongoing interest in our work, and their never-ending patience.

REFERENCES

1. G. Schmid, *Polyhedron* **7**, 2321 (1988).
2. D. Fenske, and J.-C. Steck, *Angew. Chem.* **105**, 254 (1993).
3. N. Herron, J. C. Calabrese, W. E. Farneth, and Y. Wang, *Science* **259**, 1426 (1993).

4. A. Henglein, *Topics Current Chem.* **143**, 113 (1988).
5. A. Henglein, *Chem. Rev.* **89**, 1861 (1989).
6. A. Henglein, *J. Phys. Chem.* **97**, 5457 (1993).
7. Y. Wang and N. Herron, *J. Phys. Chem.* **95**, 525 (1991).
8. L. E. Brus, *Appl. Phys. A* **53**, 465 (1991).
9. T. Kataoka, T. Tokizaki, and A. Nakamura, *Phys. Rev. B* **48**, 2815 (1993).
10. A. Nakumura, H. Yamada, and T. Tokizaki, *Phys. Rev. B* **40**, 8585 (1989).
11. H. Weller, *Angew. Chem. Int. Ed. Engl.* **32**, 41 (1993).
12. H. Weller, *Adv. Mater.* **5**, 88 (1993).
13. A. Chemseddine and H. Weller, *Ber. Bunsenges. Phys. Chem.* **97**, 636 (1993).
14. C. B. Murray, D. J. Norris, and M. G. Bawendi, *J. Am. Chem. Soc.* **115**, 8706 (1993).
15. I. G. Dance, A. Choy, and M. L. Scudder, *J. Am. Chem. Soc.* **106**, 6285 (1984).
16. T. Vossmeyer, L. Katsikas, M. Giersig, A. Chemseddine, K. Diesner, I. G. Popovic, A. Eychmüller, and H. Weller, *J. Phys. Chem.* **98**, 7665 (1994).
17. T. Takagahara, *Phys. Rev. B* **36**, 9293 (1987).
18. Y. Kayanuma, *Phys. Rev. B* **38**, 9797 (1988).
19. T. Rajh, O. I. Micic, and A. J. Nozik, *J. Phys. Chem.* **97**, 11999 (1993).
20. A. Fojtik and A. Henglein, *Chem. Phys. Lett.* **221**, 363 (1994).
21. A. Fojtik, M. Giersig, and A. Henglein, *Ber. Bunsenges. Phys. Chem.* **97**, 1493 (1993).
22. W. L. Wilson, P. F. Szajowski, and L. E. Brus, *Science* **262**, 1242 (1993).
23. J. R. Heath, *Science* **258**, 1131 (1992).
24. Y. Maeda, N. Tsukamoto, Y. Yazawa, Y. Kanemitsu, and Y. Masumoto, *Appl. Phys. Lett.* **59**, 3168 (1991).
25. A. Kornowski, M. Giersig, R. Vogel, A. Chemseddine, and H. Weller, *Adv. Mater.* **5**, 634 (1993).
26. L. T. Canham, *Appl. Phys. Lett.* **57**, 1046 (1990).
27. L. E. Brus, *J. Phys. Chem.* **98**, 3575 (1994).
28. J. J. Ramsden and M. Grätzel, *J. Chem. Soc. Faraday Trans.* 1 **80**, 919 (1984).
29. H. Weller, U. Koch, M. Gutierrez, and A. Henglein, *Ber. Bunsenges. Phys. Chem.* **88**, 649 (1984).
30. N. Chestnoy, T. D. Harris, R. Hull, and L. E. Brus, *J. Phys. Chem.* **90**, 3393 (1986).
31. Y. Wang, and N. Herron, *J. Phys. Chem.* **92**, 4988 (1988).
32. Y. Wang, A. Suna, J. McHugh, E. F. Hilinski, P. A. Lucas, and R. D. Johnson, *J. Chem. Phys.* **92**, 6927 (1990).
33. A. I. Ekimov, I. A. Kudryavtsev, M. G. Ivanov, and A. L. Efros, *J. Lumin.* **46**, 83 (1990).
34. K. Misawa, H. Yao, T. Hayashi, and T. Kobayashi, *J. Lumin.* **48/49**, 269 (1991).

35. K. Misawa, H. Yao, T. Hayashi, and T. Kobayashi, *Chem. Phys. Lett.* **183**, 113 (1991).
36. M. O'Neil, J. Marohn, and G. McLendon, *J. Phys. Chem.* **94**, 4356 (1990).
37. A. Eychmüller, A. Hässelbarth, L. Katsikas, and H. Weller, *Ber. Bunsenges. Phys. Chem.* **95**, 79 (1991).
38. L. Spanhel, M. Haase, H. Weller, and A. Henglein, *J. Am. Chem. Soc.* **109**, 5649 (1987).
39. M. G. Bawendi, P. J. Carroll, W. L. Wilson, and L. E. Brus, *J. Chem. Phys.* **96**, 946 (1992).
40. A. Hässelbarth, A. Eychmüller, and H. Weller, *Chem. Phys. Lett.* **203**, 271 (1993).
41. T. Miyoshi, *Jpn. Appl. Phys.* **31**, 375 (1992).
42. T. Rajh, O. I. Micic, D. Lawless, and N. Serpone, *J. Phys. Chem.* **96**, 4633 (1992).
43. M. Nirmal, C. B. Murray, D. J. Norris, and M. G. Bawendi, *Z. Phys. D* **26**, 361 (1993).
44. A. V. Barzykin and M. A. Fox, *Isr. J. Chem.* **33**, 21 (1993).
45. A. Ueno, N. Kakuta, K. H. Park, M. F. Finalayson, A. J. Bard, A. Campion, M. A. Fox, S. E. Webber, and J. M. White, *J. Phys. Chem.* **89**, 3828 (1985).
46. L. Spanhel, H. Weller, A. Fojtik, and A. Henglein, *Ber. Bunsenges. Phys. Chem.* **91**, 88 (1987).
47. H.-C. Youn, S. Baral, and J. H. Fendler, *J. Phys. Chem.* **92**, 6320 (1988).
48. A. Henglein, M. Gutierrez, H. Weller, A. Fojtik, and J. Jirkovsky, *Ber. Bunsenges. Phys. Chem.* **93**, 593 (1989).
49. A. R. Kortan, R. Hull, R. L. Opila, M. G. Bawendi, M. L. Steigerwald, P. J. Carrol, and L. E. Brus, *J. Am. Chem. Soc.* **112**, 1327 (1990).
50. C. F. Hoener, K. A. Allan, A. J. Bard, A. Campion, M. A. Fox, T. E. Mallouk, S. E. Webber, and J. M. White, *J. Phys. Chem.* **96**, 3812 (1992).
51. A. Eychmüller, A. Hässelbarth, and H. Weller, *J. Lumin.* **53**, 113 (1992).
52. A. Hässelbarth, A. Eychmüller, R. Eichberger, M. Giersig, A. Mews, and H. Weller, *J. Phys. Chem.* **97**, 5333 (1993).
53. H. S. Zhou, I. Honma, H. Komiyama, and J. W. Haus, *J. Phys. Chem.* **97**, 895 (1993).
54. J. W. Haus, H. S. Zhou, I. Honma, and H. Komiyama, *Phys. Rev. B* **47**, 1359 (1993).
55. A. Eychmüller, A. Mews, and H. Weller, *Chem. Phys. Lett.* **208**, 59 (1993).
56. A. Mews, A. Eychmüller, M. Giersig, D. Schooss, and H. Weller, *J. Phys. Chem.* **98**, 934 (1994).
57. D. Schooss, A. Mews, A. Eychmüller, and H. Weller, *Phys. Rev. B* **49**, 17072 (1994).
58. A. Eychmüller, T. Vossmeyer, A. Mews, and H. Weller, *J. Lumin.* **58**, 223 (1994).
59. H. Zajicek, P. Juza, E. Abramof, O. Pankratov, H. Sitter, M. Helm, G. Brunthaler, W. Faschinger, and K. Lischka, *Appl. Phys. Lett.* **62**, 717 (1993).

60. I. Bedja, S. Hotchandani, and P. V. Kamat, *J. Phys. Chem.* **97**, 11064 (1993).
61. P. Hoyer and H. Weller, *Chem. Phys. Lett.* **221**, 379 (1994).
62. M. Haase, H. Weller, and A. Henglein, *J. Phys. Chem.* **92**, 482 (1988).
63. D. W. Bahnemann, C. Kormann, and M. R. Hoffmann, *J. Phys. Chem.* **91**, 3789 (1987).
64. A. Henglein, A. Kumar, E. Janata, and H. Weller, *Chem. Phys. Lett.* **132**, 133 (1986).
65. E. Burstein, *Phys. Rev.* **93**, 632 (1954).
66. E. F. Hilinski, P. A. Lucas, and Y. Wang, *J. Chem. Phys.* **89**, 343 (1988).
67. Y. Z. Hu, S. W. Koch, M. Lindberg, and N. Peyghambarian, *Phys. Status Solidi B* **159**, 249 (1990).
68. V. L. Colvin and P. Alivisatos, *J. Chem. Phys.* **97**, 730 (1992).
69. A. S. Baranski and W. R. Fawcett, *J. Electrochem. Soc.* **131**, 2509 (1984).
70. G. Hodes, I. D. J. Howell, and L. M. Peter, *J. Electrochem. Soc.* **139**, 3136 (1992).
71. G. Hodes, A. Albu-Yaron, F. Decker, and P. Motisuke, *Phys. Rev. B* **36**, 4215 (1987).
72. B. O'Regan, J. Moser, M. Anderson, and M. Grätzel, *J. Phys. Chem.* **94**, 8720 (1990).
73. L. Spanhel and M. A. Anderson, *J. Am. Chem. Soc.* **113**, 2826 (1991).
74. B. O'Regan, M. Grätzel, and D. Fitzmaurice, *Chem. Phys. Lett.* **183**, 89 (1991).
75. P. Hoyer and H. Weller, *J. Phys. Chem.*, submitted.
76. P. Hoyer, R. Eichberger, and H. Weller, *Ber. Bunsenges, Phys. Chem.* **97**, 630 (1993).
77. G. Petermann, H. Tributsch, and R. Bogomolni, *J. Chem. Phys.* **57**, 1026 (1972).
78. K. Schwarzburg and F. Willig, *Appl. Phys. Lett.* **58**, 2520 (1991).
79. S. Hotchandani and P. V. Kamat, *J. Electrochem. Soc.* **139**, 1630 (1992).
80. M. T. Spitler and M. Calvin, *J. Phys. Chem.* **66**, 4294 (1977).
81. H. Gerischer and F. Willig, *Topics Current Chem.* **61**, 31 (1976).
82. H. Gerischer and H. Tributsch, *Ber. Bunsenges. Phys. Chem.* **72**, 437 (1968).
83. N. Vlachopoulos, P. Liska, J. Augustynski, and M. Grätzel, *J. Am. Chem. Soc.* **110**, 1216 (1988).
84. B. O'Regan and M. Grätzel, *Nature* **353**, 737 (1991).
85. A. Hagfeldt, U. Björkstén, and S.-E. Lindquist, *Solar Energy Mater.* **27**, 293 (1992).
86. R. Vogel, K. Pohl, and H. Weller, *Chem. Phys. Lett.* **174**, 241 (1990).
87. R. Vogel and H. Weller, *Proc. Europto Ser. SPIE* **1729**, 82 (1992).
88. R. Vogel, P. Hoyer, and H. Weller, *J. Phys. Chem.* **98**, 3183 (1994).
89. H. Weller, *Ber. Bunsenges, Phys. Chem.* **95**, 1361 (1991).
90. D. Liu and P. V. Kamat, *J. Phys. Chem.* **97**, 10769 (1993).
91. S. Hotchandani and P. V. Kamat, *J. Phys. Chem.* **96**, 6834 (1992).

ARTIFICIAL PHOTOSYNTHETIC TRANSFORMATIONS THROUGH BIOCATALYSIS AND BIOMIMETIC SYSTEMS

Itamar Willner and Bilha Willner
Institute of Chemistry, The Hebrew University of Jerusalem,
Jerusalem 91904, Israel

CONTENTS

Advances in Photochemistry, Volume 20, Edited by Douglas C. Neckers, David H. Volman, and Günther von Bünau
ISBN 0-471-11469-3

I. INTRODUCTION

Photosynthesis is nature's process responsible for the oxygenic atmosphere of living systems and for the fossil fuels and many other raw materials. Not surprisingly, a long-term scientific challenge is the possibility of mimicking the natural process by artificial means [1–4]. Artificial photosynthetic systems could then provide human-made solar light harvesting and fuel generation technological devices. Progress in the development of artificial photosynthetic systems is possible due to the enormous progress in various scientific disciplines. The availability of time-resolved spectroscopic and magnetic techniques has allowed physical chemists and biophysicists to follow the fast stepwise reactions involved in the natural process [5–8] and to characterize analogous reactions in artificial assemblies. The progress in biological research in understanding the bioreactions of the natural process, and the scientific breakthrough resolving the crystal structure of the photosynthetic reaction center have stimulated chemists with effective tools to mimic functions of the natural process by synthetic means [9]. In addition to the experimental progress, the theory of light-induced electron transfer reactions [10] and its conjugation to the photosynthetic reaction center [11, 12] gained substantial attention and provides a solid theoretical guideline in mimicking photosynthesis by artificial means.

There are several approaches to mimic functions of photosynthesis, and each has been reviewed extensively. These include the design of molecular assemblies duplicating functions of the photosynthetic reaction center [13, 14], development of photosensitive materials capable of light harvesting and inducing electron transfer processes [15–17] and tailoring multi-component systems where light-induced electron transfer processes drive endoergic chemical transformations (generation of fuel products) [18, 19]. The aims of this account are to review several aspects involved in assembling artificial photosynthetic systems and to exemplify the progress in the area. The methodologies of designing molecular assemblies mimicking the photosynthetic reaction center and tailoring of integrated photosynthetic systems are discussed.

The photosynthetic process involves photochemical reactions followed by sequential dark chemical reactions [20, 21] (Figure 1). The photochemical processes occur in two photoactive sites, photosystem I and photosystem II (PS-I and PS-II, respectively), where chlorophyll *a* and chlorophyll *b* act as light active compounds. Excitation of PS-I results in an electron transfer (ET) process to ferredoxin, acting as a primary electron acceptor. In this process, light energy is converted to chemical potential which is stored in the reduced ferredoxin and oxidized chlorophyll. Excitation of PS-II results in a similar ET process where a quinone acts as electron acceptor. The reduced photoproduct generated in PS-II transfers the electron across a chain of acceptors to the oxidized chlorophyll of PS-I, and as a result, the light-harnessing component of PS-I is recycled. Reduced ferredoxin formed in PS-I induces a series of ET reactions that ultimately reduce the natural cofactor $NADP^+$ to NADPH. This reduced cofactor provides the electron source for CO_2 fixation to carbohydrates in a complex sequence of enzyme-catalyzed transformations (Calvin cycle). The oxidized chlorophyll of PS-II mediates the biocatalyzed oxidation of water to molecular oxygen. This scheme emphasizes that the photosynthetic apparatus operates by a synchronous mechanism involving a photochemical part, where light energy is converted to chemical potential by ET reactions, followed by a set of dark reactions utilizing the chemical energy in driving CO_2 fixation and O_2 evolution.

The general scheme of the photosynthetic apparatus outlined in Figure 1 is certainly oversimplified. The fundamental photochemical steps leading to the charge separation in PS-I and PS-II are certainly the result of evolutionary optimization. The quantum efficiency of the charge-separated photoproducts exceeds the value $\phi = 0.95$, and the lifetime of the separated products corresponds to milliseconds [22]. This implies that the photosynthetic reaction centers are extremely effective in light conversion and energy storage. Insight into the performance of photosynthetic reaction center has become available from structural characterization of photosynthetic bacteria reaction centers and spectroscopic analysis of the sequential ET reactions in these assemblies [23, 24]. In the photosynthetic bacteria, in contrast to green plants, a single photosystem consisting of a protein matrix and a *donor–acceptor complex* is operative in the ET process. The

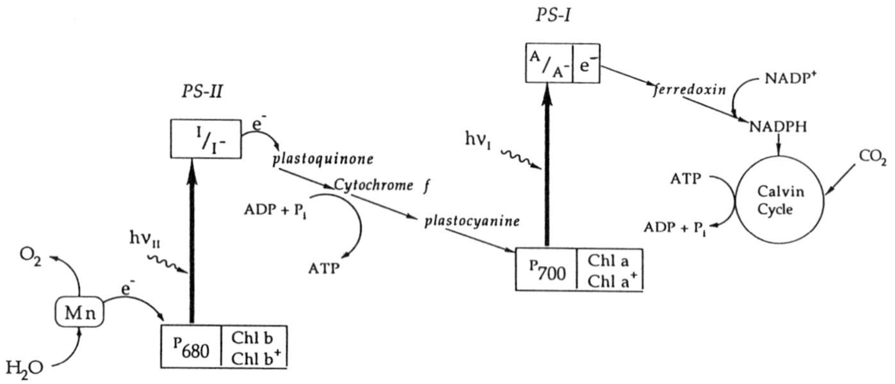

Figure 1. Sequence of photochemical reactions in the photosynthetic cycle (Z-scheme).

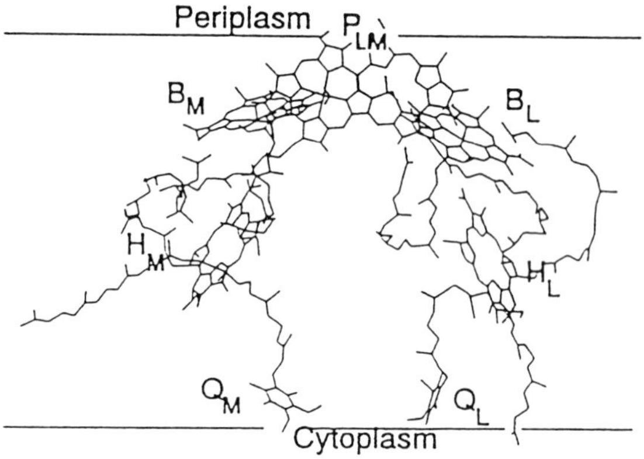

Figure 2. Donor–acceptor structure of the reaction center of *Rb. sphaeroides*.

donor–acceptor complex is embedded in the protein matrix that maintains the ET components in the proper spatial orientation and distances. The donor–acceptor complex structure of *Rb. sphaeroidis* bacteria is shown in Figure 2. It includes four bacteriochlorophylls (BChl), two bacteriopheophytins (BPheo), two quinones (Q), and a nonheme iron. These units are organized in an almost symmetrically spatial configuration consisting of two arms, L and M subunits. Two of the BChls in this assembly form the primary electron donor site (P) for ET in the reaction center. The remaining two BChls, two BPheos, and quinones occupy analogous positions in arms

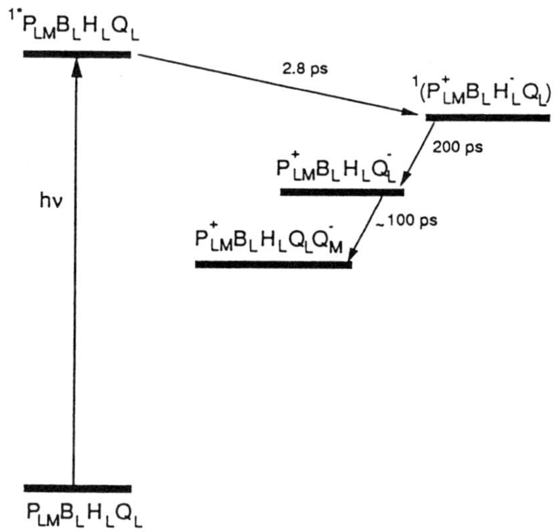

Figure 3. Charge-separated intermediates in the bacterial photosynthetic reaction center and their lifetimes.

L and M (and are represented in Figure 2 as B_L, B_M; H_L, H_M; and Q_L, Q_M, respectively). The photoinduced ET initiated by excitation of P proceeds only across arm L, and charge separation occurs in a sequential mechanism of ET reactions. The important charge-separated intermediates and their lifetime are represented in Figure 3. We realize that a vectorial electron transfer process proceeds in the photosynthetic reaction center and that the lifetime of the intermediate redox products is lengthened as the distance between the components increases.

II. ARTIFICIAL PHOTOSYNTHETIC MODEL SYSTEMS

Two general approaches of modeling photosynthesis can be considered, one involving mimicking the functions of the photosynthetic reaction center by means of synthetic analogs [25–27]. To this extent the synthesis of linked multicomponent donor–acceptor assemblies could lead to charge separation by means of sequential ET processes as outlined in Eqs. (1) to (4), where S is the light-active component and A and D represent electron acceptor and donor units, respectively.

$$*S{-}A_1{-}A_2 \rightarrow S^+{-}A_1^-{-}A_2 \rightarrow S^+{-}A_1{-}A_2^- \qquad (1)$$

$$D{-}S^*{-}A \rightarrow D{-}S^+{-}A^- \rightarrow D^+{-}S{-}A^- \qquad (2)$$

$$D_2{-}D_1{-}S^* \rightarrow D_2{-}D_1^+{-}S^- \rightarrow D_2^+{-}D_1{-}S^- \qquad (3)$$

$$D{-}S^*{-}A_1{-}A_2 \rightarrow D{-}S^+{-}A_1^-{-}A_2 \rightarrow D{-}S^+{-}A_1{-}A_2^- \rightarrow D^+{-}S{-}A_1{-}A_2^- \qquad (4)$$

A second approach of mimicking photosynthesis involves the design of integrated assemblies duplicating functions of the natural process to the extent that light is converted to redox energy that is stored in energy-rich chemical products [28–30]. A basic configuration of such integrated artificial photosynthetic device is shown schematically in Figure 4. Photoexcitation of the light-active component, S, results in ET to the electron acceptor, A. Subsequent oxidation of the electron donor by the intermediate oxidized species, S^+, recycles the light-active component, and light energy is converted into redox potential stored in the products D^+ and A^-. Subsequent dark reduction and oxidation processes of chemical substrates P_1 and P_2 yield the chemical products P_1^- and P_2^+, respectively. The overall process is considered as a photosynthetic device provided that the overall transformation, Eq. (5), is endoergic ($\Delta G^\circ > 0$) and light energy is converted and stored in the form of chemical energy in the final chemical products.

$$P_1 + P_2 \xrightarrow{h\nu} P_1^- + P_2^+ \qquad (\Delta G^\circ > 0) \qquad (5)$$

Several redox transformations that can be driven in such artificial photosynthetic systems are summarized in Table 1 and include photolysis of water, carbon dioxide fixation, and nitrogen fixation processes. The endoergicities of these transformations and the number of electrons involved in the reduction processes are also included in Table 1. It can be realized that the energy per electron required to drive the various transformations is available in visible-light quanta. It should be noted, however, that the overall

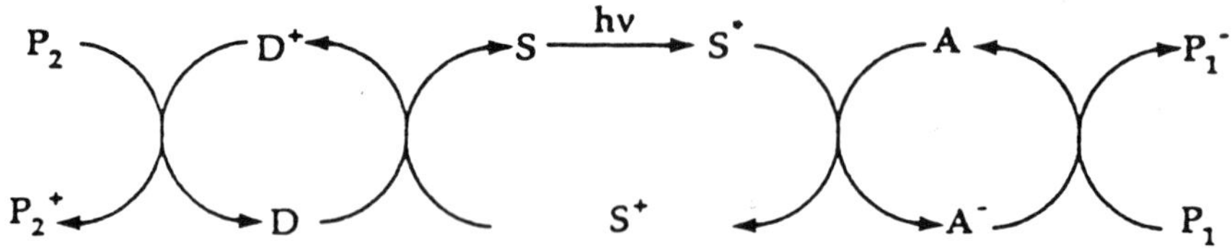

Figure 4. Scheme for an integrated artificial photosynthetic system. S, photosensitizer; A, electron acceptor; D, electron donor; P_1, P_2, reactants (substrates).

TABLE 1 Photosynthetic Transformations and Their Thermodynamics

S_1 S_2 P_1 P_2	ΔG° (kcal mol^{-1})	ΔG° [per e^- (eV)]	n (e^-)
$H^+ + OH^- \rightarrow H_2 + \frac{1}{2}O_2$	57	1.23	2
$H_2O(l) + CO_2(g) \rightarrow HCO_2H(l) + \frac{1}{2}O_2$	68	1.46	2
$H_2O(l) + CO_2(g) \rightarrow CH_2O(g) + O_2(g)$	125	1.35	4
$2H_2O(l) + N_2(g) \rightarrow N_2H_4(l) + O_2(g)$	181	1.95	4
$CO_2(g) + 2H_2O(l) \rightarrow CH_3OH(l) + \frac{3}{2}O_2(g)$	168	1.21	6
$N_2(g) + 3H_2O(l) \rightarrow 2NH_3(g) + \frac{3}{2}O_2(g)$	162	1.17	6
$CO_2(g) + 2H_2O(l) \rightarrow CH_4(g) + 2O_2(g)$	196	1.06	8

process includes kinetic limitations that were considered theoretically in several recent review articles and monographs [18, 29].

III. MOLECULAR ASSEMBLIES MIMICKING PHOTOSYNTHESIS

The theoretical grounds of ET processes are of basic importance to understand natural photosynthesis and to tailor artificial photosynthetic devices. Marcus theory [10, 31], summarized in Eq. (6), provides the essence of ET and charge separation. It states that the ET rate constant between a donor and an acceptor depends exponentially on the distance, d, separating the redox components, and on the free-energy change, ΔG°, and the reorganization energy, λ, associated with the ET process.

$$k_{et} \propto \exp[-\beta(d - d_0)] \exp\left[\frac{-(\lambda + \Delta G^\circ)^2}{4\lambda RT}\right] \tag{6}$$

The sequential ET in the photosynthetic reaction center (Figure 3) is accompanied by a free-energy drop and spatial distance separation of the redox intermediates. As a result, the ET intermediates are stabilized against back ET reactions in each of the sequential steps, and charge separation is assisted. Organization of rigid molecular assemblies capable of sequential ET, which involve a free energy drop and spatial removal of the redox intermediates, is the basis for tailoring photosynthetic model systems at the

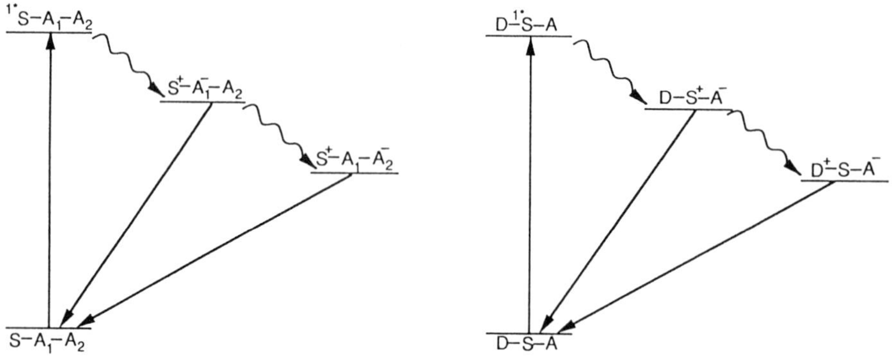

Figure 5. Artificial triad assemblies mimicking ET in the photosynthetic reaction center.

molecular level. Figure 5 shows these principles in two triad assemblies, $S-A_1-A_2$ and D–S–A.

The most simple kind of photosynthetic model systems consists of a chromophore linked to an electron acceptor or donor. For example, the porphyrinquinone diad **1** represents a model of the primary [32] photosynthetic reaction center. Various diads have been synthesized and their photophysical properties have recently been reviewed [33]. The general strategy of multistep ET in triad and multifunctionalized molecular assemblies provides a useful means for maximizing the quantum yields of charge-separated states and the lifetime of the redox intermediate states. Evidently, various molecular assemblies following this strategy have been exploited.

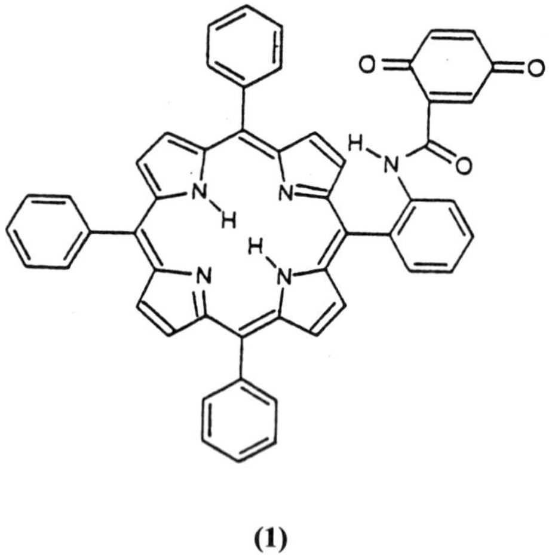

(1)

This specific subject has recently been reviewed [25–27], and in the following sections we exemplify this approach in the general context of recent progress in artificial photosynthetic systems.

A. Electron Donor/Photosensitizer/Electron Acceptor Systems (D–S–A)

Molecular assembly **2**, composed of a porphyrin chromophore covalently linked to a benzoquinone electron acceptor and a carotenoid polyene as electron donor provides an electron donor/photosensitizer/electron acceptor (D–S–A) configuration [34]. The sequential ET reactions occurring in the triad upon photoexcitation of the porphyrin center are outlined as follows (where C, P, and Q stand for the carotenoid polyene, porphyrin, and quinone components, respectively):

$$\text{C–P–Q} \xrightarrow{h\nu} \text{C–}^1\text{P*–Q} \longrightarrow \text{C–P}^{+\bullet}\text{–Q}^{-\bullet} \longrightarrow \text{C}^{+\bullet}\text{–P–Q}^{-\bullet} \qquad (7)$$

The lifetime of the final charge-separated ion pair, $C^{+\bullet}$–P–$Q^{-\bullet}$, is about 300 ns in a dichloromethane solution (and about 2 μs in a butyronitrile solution). This lifetime is substantially longer when compared to porphyrin–quinone ion pairs, $P^{+\bullet}$–$Q^{-\bullet}$ (ca. 100 to 200 ps). Other D–S–A assemblies containing porphyrins as chromophores are exemplified in structures **3** and **4**. In these triad assemblies an aniline residue provides the electron donor component [35]. System **3**, for example, exhibits structural rigidity, and thus defined spatial orientation between the donor and acceptor components is established. The sequence of ET reactions occurring upon excitation of the porphyrin is summarized by

$$\text{D–P–Q} \xrightarrow{h\nu} \text{D–}^1\text{P*–Q} \longrightarrow \text{D–P}^{+\bullet}\text{–Q}^{-\bullet} \xrightarrow{k = 1.4 \times 10^{10}\,\text{s}^{-1}} \text{D}^{+\bullet}\text{–P–Q}^{-\bullet} \qquad (8)$$

The primary ion pair decays to the charge-separated state $D^{+\bullet}$–P–$Q^{-\bullet}$ with a quantum efficiency corresponding to $\phi = 0.71$ and a rate constant of $1.4 \times 10^{10}\,s^{-1}$. The lifetime of the final charge separated ion pair is 2.45 μs, substantially longer than ordinary porphyrin quinone ion pairs, $P^{+\bullet}$–$Q^{-\bullet}$.

The excellent photophysical properties of Ru(II)–tris(bipyridine) complexes (see Section IV) have been applied to organize D–S–A triads, where a Ru(II)–tris(polypyridine) center acts as photoactive center, *N*,*N*′-trimethylene-2,2′-bipyridinium DQ^{2+}, or 4,4′-bipyridinium, MV^{2+}, are the electron acceptor units, and phenothiazine (PTZ), acts as electron donor component [36]. Two different D–S–A triads, formed by direct condensation of the three components (**5**), or organization of the components on a lysine residue,

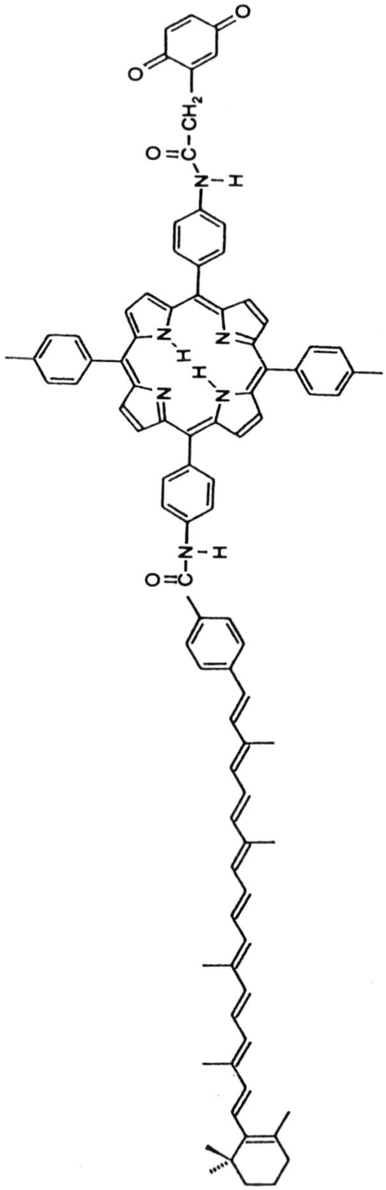

(2)

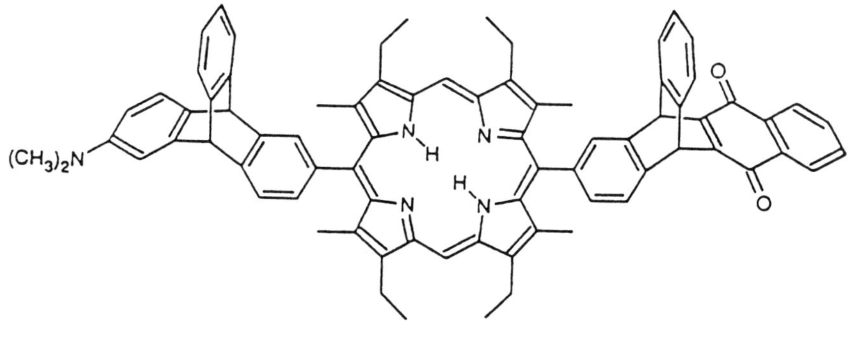

(3)

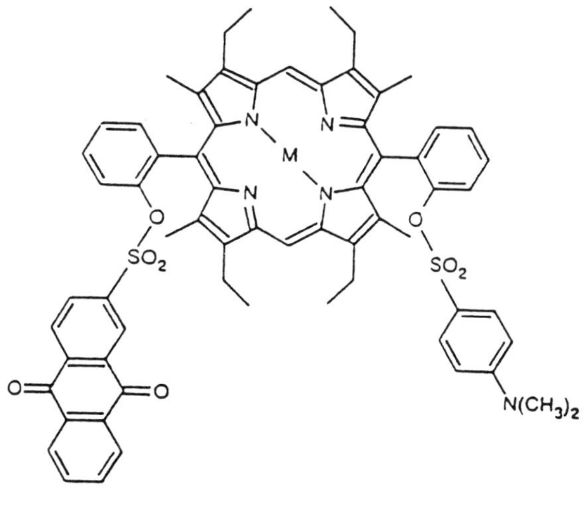

(4)

acting as a pivot site (**6**), have been prepared [37]. The lifetimes of the charge-separated ion pairs, formed upon excitation of the metal-to-ligand charge transfer state (MLCT) of the Ru(II)–polypyridine chromophores of **5** and **6**, $PTZ^{+\bullet}$–Ru^{2+}–$DQ^{+\bullet}$, are 165 ns and 146 ns, respectively. These lifetimes are significantly longer than those of the bifunctional ion pairs, $PTZ^{+\bullet}$–Ru^{+} or Ru^{3+}–$DQ^{+\bullet}$. The sequence of ET processes in triads **5** and **6** leading to the charge-separated state might proceed by two

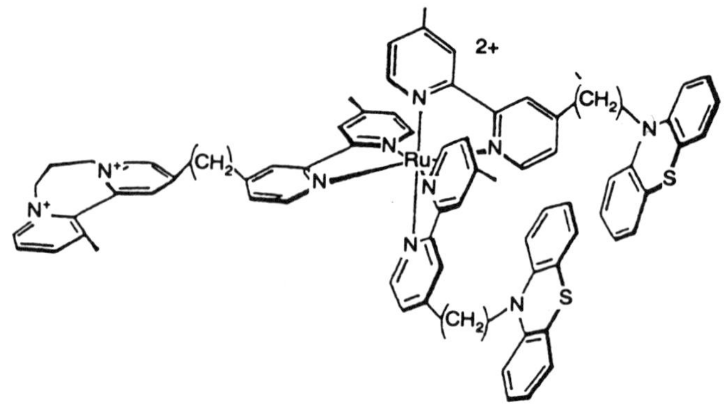

(5)

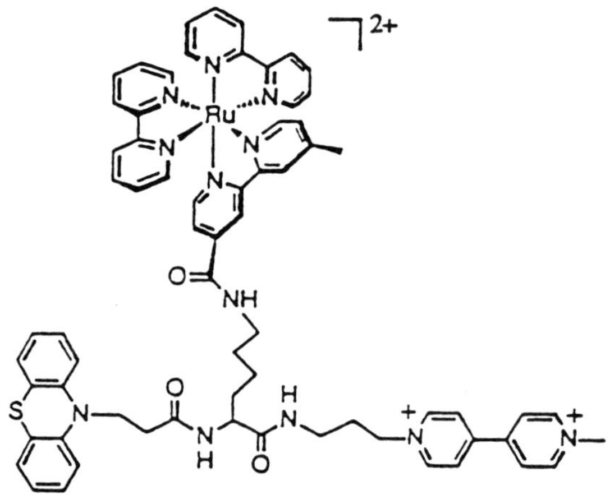

(6)

alternative routes:

$$PTZ\text{–}Ru^{2+}\text{–}DQ^{2+} \xrightarrow{h\nu} PTZ\text{–}{*}Ru^{2+}\text{–}DQ^{2+} \longrightarrow PTZ\text{–}Ru^{3+}\text{–}DQ^{+\cdot} \longrightarrow PTZ^{+\cdot}\text{–}Ru^{2+}\text{–}DQ^{+\cdot} \quad (9)$$

$$PTZ\text{–}Ru^{2+}\text{–}DQ^{2+} \xrightarrow{h\nu} PTZ\text{–}{*}Ru^{2+}\text{–}DQ^{2+} \longrightarrow PTZ^{+\cdot}\text{–}Ru^{+}\text{–}DQ \longrightarrow PTZ^{+\cdot}\text{–}Ru^{2+}\text{–}DQ^{+\cdot} \quad (10)$$

Evidently, sequential ET in tailored D–S–A assemblies leads to stabilization of the charge separation products at the molecular level.

B. Photosensitizer/Acceptor1/Acceptor2 Systems ($S-A_1-A_2$) and Hybrid $D-S-A_1-A_2$ Assemblies

The molecular assembly (7), consisting of a porphyrin chromophore linked to two quinones, which exhibit the proper ordering of redox potentials for stepwise ET, mimics the functionalities of the photosynthetic reaction center [38]. Transient absorption studies have revealed that upon excitation of the porphyrin, a charge-separated porphyrin–quinone ion pair with a lifetime of about 300 ps is formed. In a molecular assembly lacking the second quinone system, an ion pair with a lifetime of only 130 ps is observed. The increase in the lifetime of the triad ion pair has been attributed to the multistep ET, leading to spatial separation of the photoproducts, which

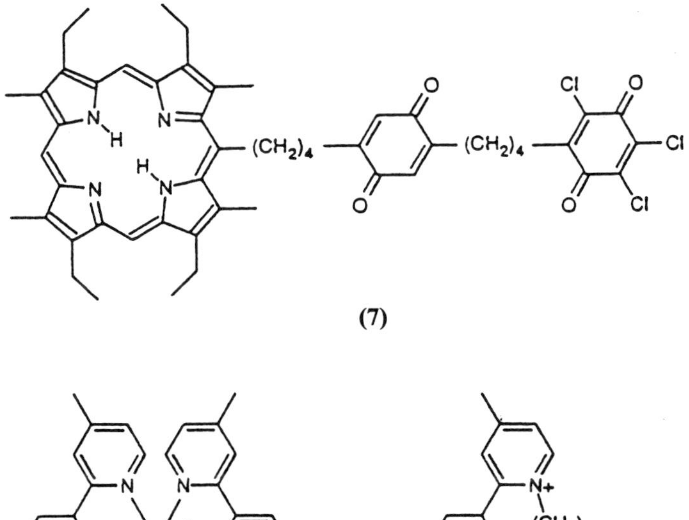

(7)

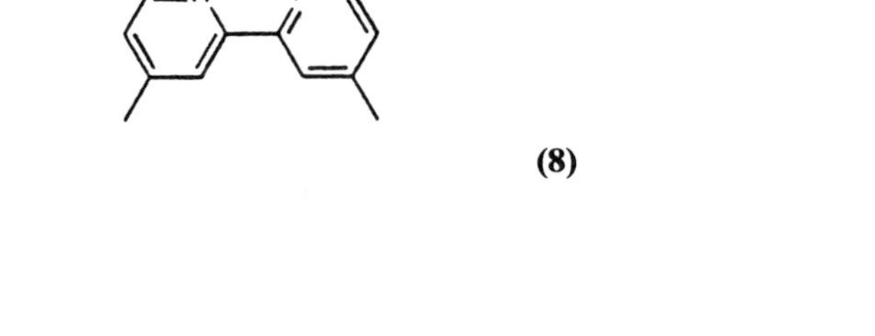

(8)

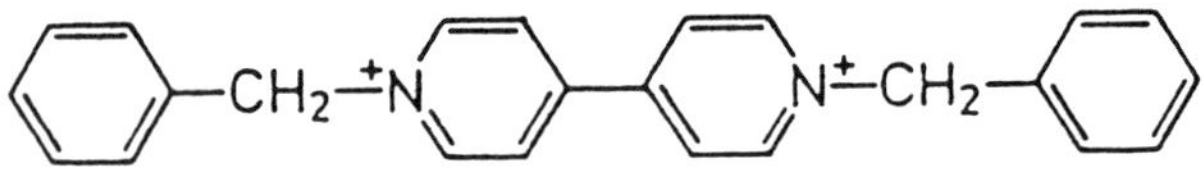

(9)

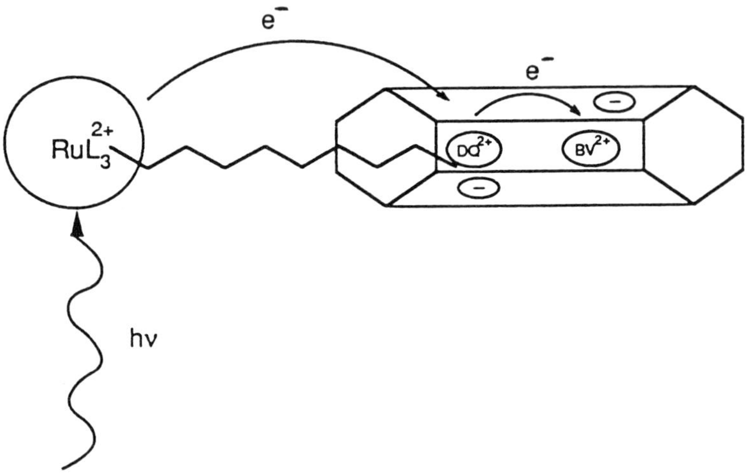

Figure 6. Photosensitized vectorial ET within an organized zeolite L assembly consisting of diad **8** and BV^{2+} (**9**).

increases their lifetime by retardation of the back ET:

$$P{-}Q_1{-}Q_2 \xrightarrow{h\nu} {}^1P^*{-}Q_1{-}Q_2 \longrightarrow P^{+\bullet}{-}Q_1^{-\bullet}{-}Q_2 \longrightarrow P^{+\bullet}{-}Q_1{-}Q_2^{-\bullet} \quad (11)$$

The flexibility of the quinone components in respect to the porphyrin in structure **7**, however, results in folding of the quinone units toward the porphyrin ring. As a result, the spatial separation of the charge-separated products formed in the multistep ET is not sterically optimized, and the extent of stabilization of the triad ion pair, $P^{+\bullet}{-}Q_1{-}Q_2^{-\bullet}$, is quite low.

An interesting $S{-}A_1{-}A_2$ triad assembly has been organized by coupling a diad $S{-}A_1$ to a secondary electron acceptor, A_2, on a solid support that orients the components in a triad array capable of vectorial ET. The covalently linked Ru(II)–tris(bipyridine) chromophore and *N*,*N*′-tetramethylene-2,2′-bipyridinium diad (**8**), $Ru^{2+}{-}DQ^{2+}$, has been exchanged onto zeolite L that is prefunctionalized with *N*,*N*′-dibenzyl-4,4′-

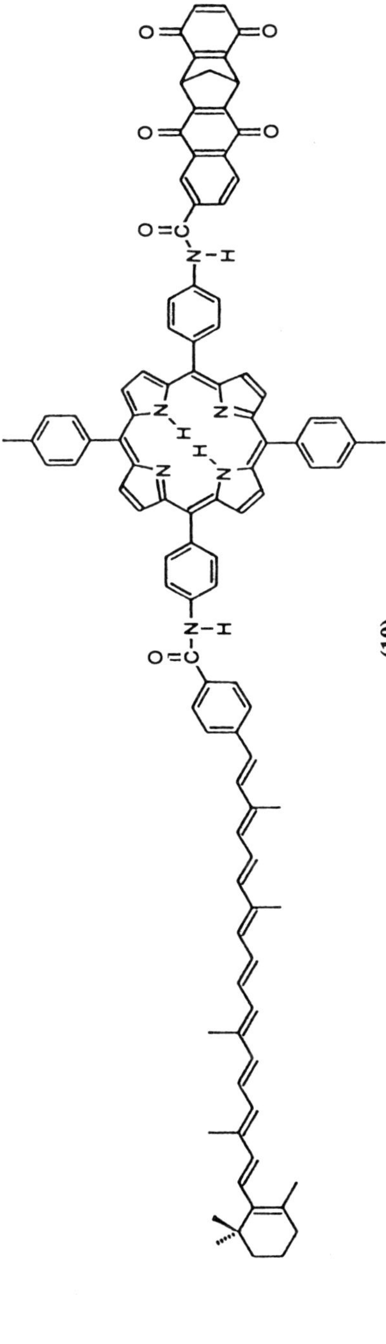

(10)

bipyridinium, BV^{2+} (**9**). The DQ^{2+} moiety of diad **8** penetrates into the pores of the solid zeolite support, but the large Ru^{2+} complex composing the diad is located externally to the zeolite pores [39]. As a result, an ordered three-component array is organized (Figure 6). The sequence of ET reactions following excitation of the Ru^{2+} complex is shown in Figure 6. The excited species decays within 5 ns to the primary redox products $Ru^{3+}–DQ^{+\bullet}$, which subsequently generate the solid immobilized charge-separated state, $[Ru^{3+}–DQ^{2+}, BV^{+\bullet}]$. The final charge-separated ion pair is formed with a quantum efficiency of about 0.17, and its lifetime corresponds to 37 μs.

Tetrad **10** illustrates the clear advantage of multistep ET pathways in enhancing the lifetime and quantum yield of the separated state at the molecular level [40]. The multifunctional tetrad **10** combines the functionalities of the two triads D–S–A and S–A_1–A_2 discussed earlier, and can be considered as a D–S–A_1–A_2 assembly. The porphyrin acts as the photo-

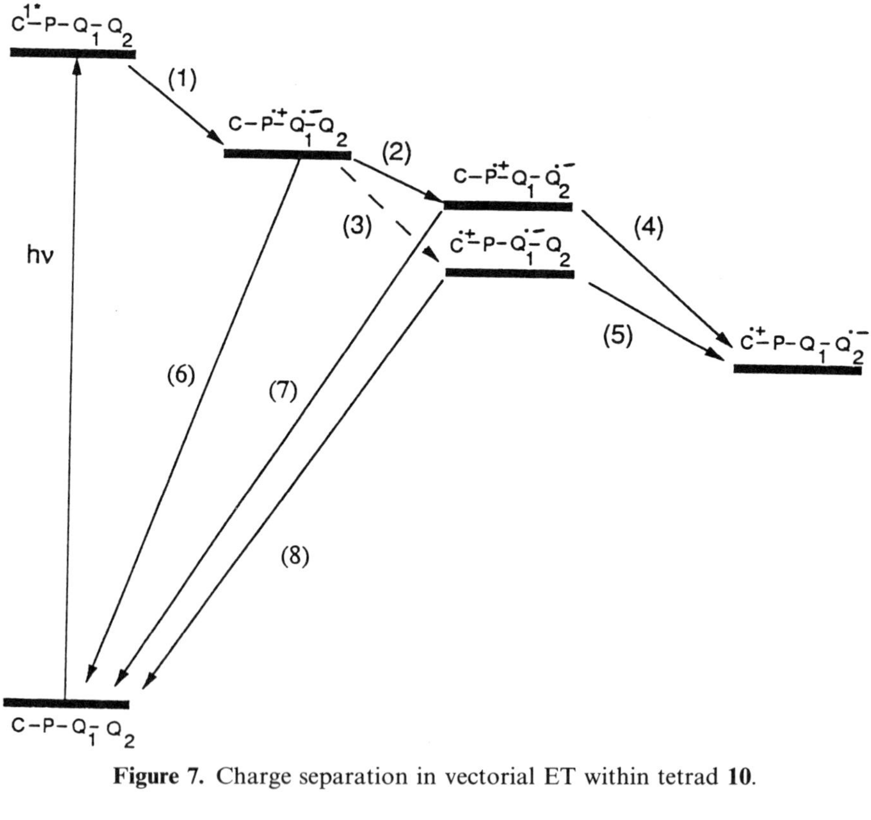

Figure 7. Charge separation in vectorial ET within tetrad **10**.

active site of the tetrad and the carotenoid polyene and two quinone units as the electron donor and series of electron acceptors, respectively, $C-P-Q_1-Q_2$. Tetrad **10** reveals rigidity, and therefore distance effects on charge separation are emphasized. In response to excitation of the porphyrin, the sequence of electron transfer reactions, illustrated in Figure 7, takes place. The singlet excited state of the porphyrin decays to the primary ion-pair state, $C-P^{+\bullet}-Q_1^{-\bullet}-Q_2$, with a rate constant of $6.7 \times 10^{10}\,s^{-1}$, corresponding to a quantum efficiency of unity. Two alternative ET reactions (steps 2 and 3) can follow and compete with the back ET of the primary redox ion pair (step 6). Each of these two routes generates new redox ion pairs $C-P^{+\bullet}-Q_1-Q_2^{-\bullet}$ and $C^{+\bullet}-P-Q_1^{-\bullet}-Q_2$, respectively. Both ion pairs are followed by ET reactions (steps 4 and 5) to generate the same charge-separated ion pair, $C^{+\bullet}-P-Q_1-Q_2^{-\bullet}$. The final charge-separated state competes with the back ET processes of the intermediate ion pairs (steps 7 and 8) and is formed with a quantum yield of 0.23 exhibiting a lifetime of 460 ns (in dichloromethane solution, 295 K).

The advantages of multistep ET pathways in stabilization of charge-separated states are well illustrated by comparing the photophysical properties of tetrad **10** with those of the structurally related D–S–A triads **11** and **12** [41]. Triad **11** lacks the secondary quinone component, and the resulting charge-separated ion pair, $C^{+\bullet}-P-Q_1^{-\bullet}$, is formed with a quantum efficiency of 0.04, exhibiting a lifetime of 70 ns only. The low quantum efficiency for formation of $C^{+\bullet}-P-Q_1^{-\bullet}$ in **11**, compared to the high quantum yield of the charge-separated state of **10**, has been attributed to the presence of a single inefficient ET pathway (analogous to step 3 in Figure 7) that competes with charge recombination of the primary ion pair, $C-P^{+\bullet}-Q_1^{-\bullet}$. In tetrad **10** two processes (steps 2 and 3) compete with charge recombination of the primary ion pair, and consequently, the formation of the redox products is assisted. The longer lifetime of the charge-separated state of tetrad **10**, $C^{+\bullet}-P-Q_1-Q_2^{-\bullet}$, originates mainly from the larger distance separating the donor–acceptor pair in respect to the charge recombination process. In triad **12**, the primary naphthoquinone component is replaced by a dimethoxynaphthalene unit that does not act as an electron acceptor. Excitation of the porphyrin site of **12** results in the charge-separated state $C^{+\bullet}-P-O-Q_2^{-\bullet}$ with a quantum yield of 0.11, exhibiting a lifetime of 1.9 μs. The primary step in the charge-separation pathway involves ET quenching of the excited state by the remote quinone unit (Q_2). This long-range ET is inefficient relative to array **10**, $k = 9.6 \times 10^8\,s^{-1}$, and does not compete effectively with the competitive decay of the excited species. Consequently, the quantum yield of the resulting ion pair is lower than for **10**. Nevertheless, the final charge-separated state of **12**, $C^{+\bullet}P-O-Q_2^{-\bullet}$, exhibits a spatial configuration similar to that of the charge-separated

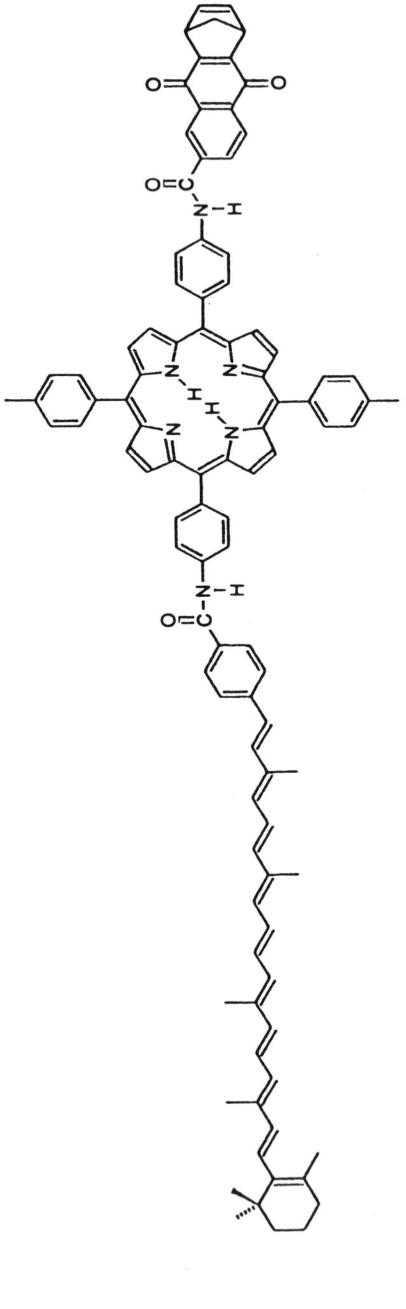

(11)

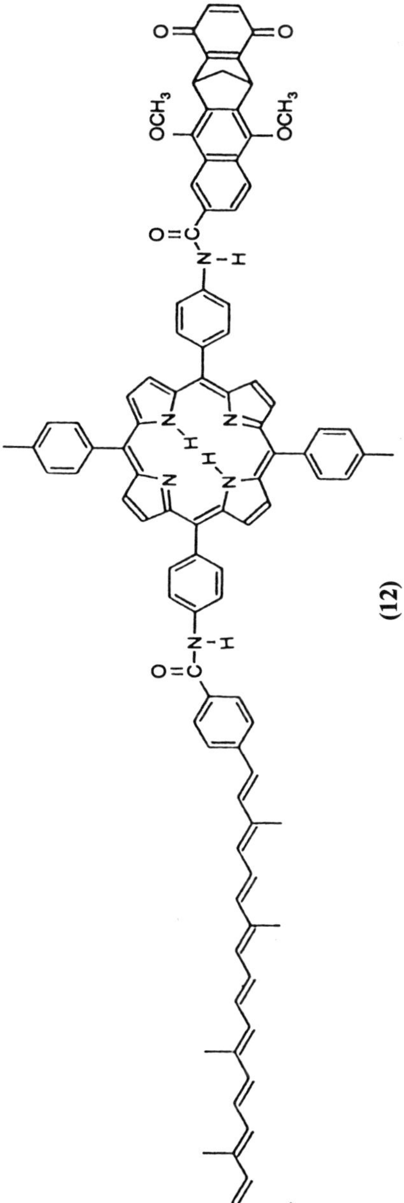

(12)

ion pair of **10**, $C^{+\bullet}{-}P{-}Q_1{-}Q_2^{-\bullet}$, and therefore both redox products reveal similar stabilization against charge recombination (long lifetimes). It is clear that molecular arrays that allow stepwise vectorial ET pathways mimic the functionalities of the photosynthetic reaction center and lead to enhancement of quantum yields and longer lifetimes of charge-separated states.

IV. INTEGRATED ARTIFICIAL PHOTOSYNTHETIC SYSTEMS

Construction of integrated artificial photosynthetic systems (Figure 4) involves various basic aspects of consideration. The primary event upon photoexcitation of the light-active component involves ET quenching of the excited species. This process should compete effectively with all other relaxation pathways of the excited species (i.e., radiative and nonradiative decay processes). Two general ET quenching pathways of the excited species, S*, are known: One pathway involves charge separation through reductive ET quenching of the excited photosensitizer by an electron donor, D [Eq. (12)], and a second route involves oxidative ET quenching of the excited species by an electron acceptor, A [Eq. (13)].

$$S^* + D \rightarrow S^- + D^+ \tag{12}$$

$$S^* + A \rightarrow S^+ + A^{-\bullet} \tag{13}$$

The overall ET processes outlined in Eqs. (12) and (13) proceed by a sequential pathway that is accompanied by limiting back ET reactions of the intermediate redox products. Figure 8 shows the stepwise charge separation process in oxidative ET quenching of an excited photosensitizer [42]. ET quenching results in an encounter cage complex of redox products

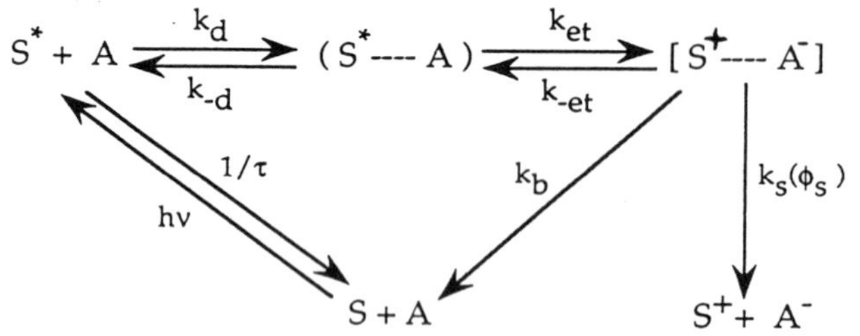

Figure 8. Scheme of reactions involved in photoinduced charge separation.

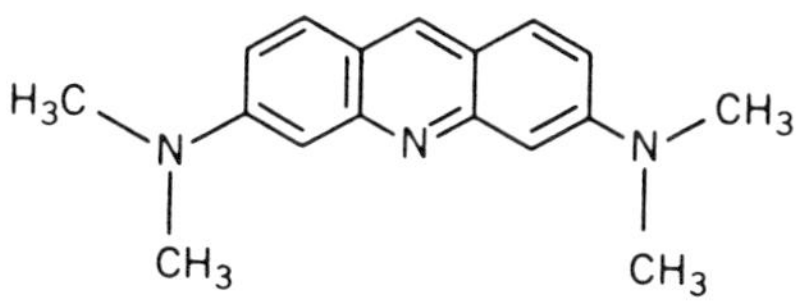

(13)

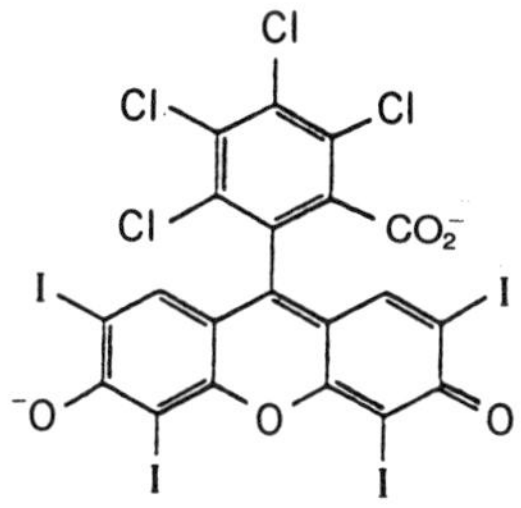

(14)

that can either recombine to the ground-state reactants or undergo dissociation to the charge-separated products. Thus an effective artificial photosynthetic device requires the application of light-active components of distinct properties: The excited photosensitizer should be efficiently quenched by the electron acceptor, and the primary encounter cage complex of photoproducts should effectively dissociate to the charge-separated products. The quantum efficiency for formation of the separated redox products

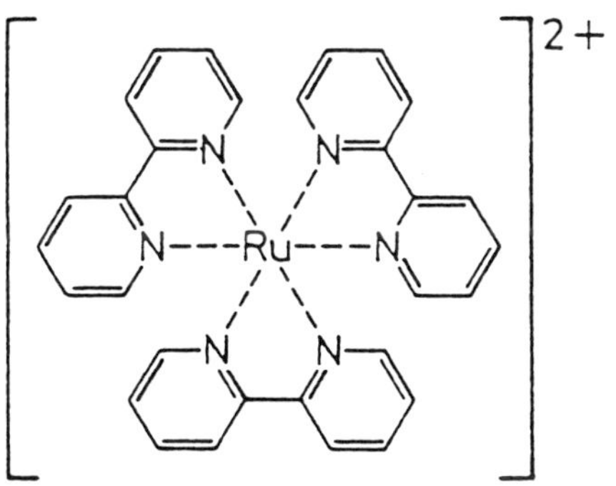

(15)

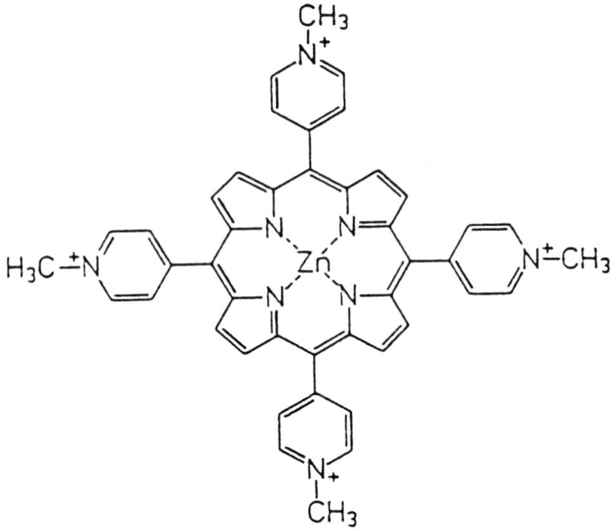

(16)

in view of the various degradative routes outlined in Figure 8 is given by

$$\phi_s = \frac{k_s}{k_b + k_s} \frac{k_{et} \tau_0 [A]}{1 + k_{et} \tau_0 [A]} \tag{14}$$

Different classes of light-active materials have been applied to trigger photoinduced ET reactions. These include organic dyes [43, 44] such as acridine orange (**13**) or rose bengal (**14**), and transition metal complexes [45–47] such as Ru(II)–tris(bipyridine) (**15**) or Zn(II)–*meso*-(*N*-tetramethylpyridinium) porphyrin (**16**). Inorganic semiconductor materials [48, 49] such as TiO_2 or CdS also reveal photoactivity and upon excitation, generate electron–hole pairs in the bulk inorganic material.

In this chapter we emphasize the use of homogeneous photosensitizers as light-harvesting components in artificial photosynthetic systems. Other reviews and monographs highlight the application of semiconductor materials in photosynthetic devices [48–50]. Nevertheless, several aspects of ET and charge separation are common to homogeneous photosensitizers and semiconductor photoactive materials, and these interrelations will be mentioned in later sections.

A. Control of Photosensitized ET Reactions in Organized Media

Figure 8 reveals that charge separation of ET transfer products is affected by the effectiveness of the ET quenching of the excited state and by the charge separation yield of the primary encounter complex of the photoproducts. As the photoproducts are designed as the driving intermediates for the endoergic photosynthetic process, they exhibit powerful reductive and oxidative properties. Consequently, the back ET of the products in the encounter cage complex is thermodynamically favored [51]. Similar back ET is also characteristic of the charge-separated photoproducts:

$$[S^{+} \cdots A^{-}] \rightarrow S + A \qquad (15)$$

$$S^{+} + A^{-} \rightarrow S + A \qquad (16)$$

or

$$D^{+} + A^{-} \rightarrow D + A$$

The design of vectorial ET processes might be accomplished at the molecular level (see Section III) or by the organization of microheterogeneous systems or microenvironments capable of controlling vectorial ET processes in the photochemical assembly [52, 53]. Two general interactions (i.e., electrostatic and hydrophobic–hydrophilic) can be utilized as driving forces for vectorial ET [54]. Figure 9*a* outlines the use of a negatively charged interface in controlling a photoinduced ET process: By predesign of the electrical properties of the photosensitizer–electron acceptor couple, the photoproducts of the primary encounter cage complex are tailored to include two oppositely charged species after ET. While the positive component is attracted by the interface, the negatively charged component is repelled by it. Consequently, charge separation is assisted and diffusional recombination is retarded by localization of the oxidized photoproduct at the charged interfacial region and concomitant repulsion of the reduced photoproduct. Utilization of hydrophobic–hydrophilic interactions in controlling photoinduced ET processes is exemplified in Figure 9*b*. The photosystem is assembled in an interfacial medium consisting of hydrophilic and hydrophobic microenvironments. The photoreactants are confined to one microenvironment (i.e., hydrophilic medium), yet upon photoinduced ET the products are predesigned to exhibit opposite hydrophobic and hydrophilic properties. As a result, the hydrophobic photoproduct is extracted into the hydrophobic microenvironment, and the charge separation is maintained by means of the two-phase system.

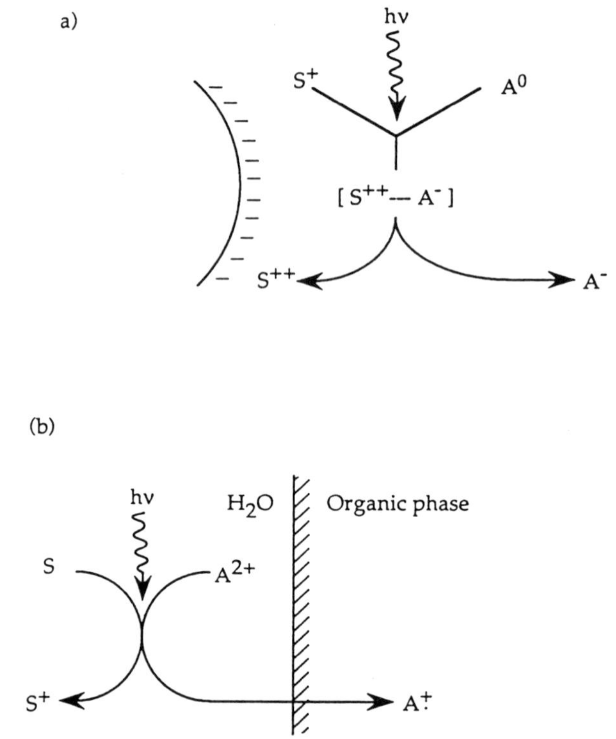

Figure 9. Control of charge separation by means of organized microenvironments: (*a*) application of charged interfaces as a means to induce charge separation and retard back electron transfer reactions; (*b*) application of a water–oil two-phase system for photoinduced charge separation and retardation of back reactions.

For effective charge separation by means of such two-phase systems the organization of a highly dispersed multiphase system is a prerequisite in order to maintain efficient extractability at the interface boundary.

Various charged microheterogeneous assemblies that control photochemically induced ET processes by means of electrostatic interactions have been developed. These include micelles [55–57], polyelectrolytes [58], inorganic colloids [59, 60], clays [61], and dendrides [62]. Hydrophilic–hydrophobic organized microenvironments that control photoinduced ET transformations include micelles [63], microemulsions [64, 65], vesicles [66], and cyclodextrin receptors [67]. The effects and functions of these organized systems in controlling vectorial ET reactions have been considered in detail in various reviews [52, 53]. For example, the effects of charged SiO_2 colloids on the photosensitized reduction of *N*,*N*′-(3-propylsulfon-

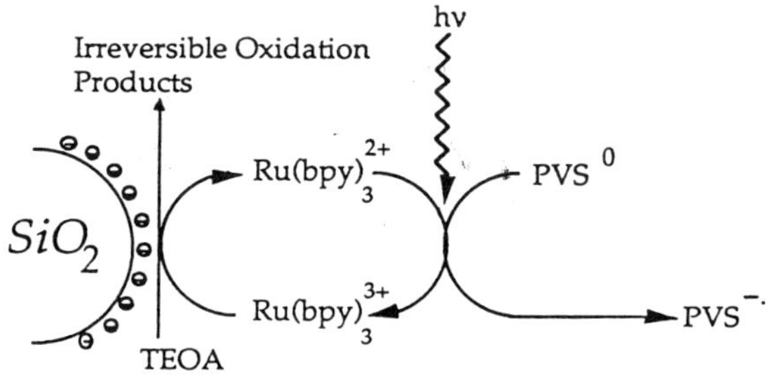

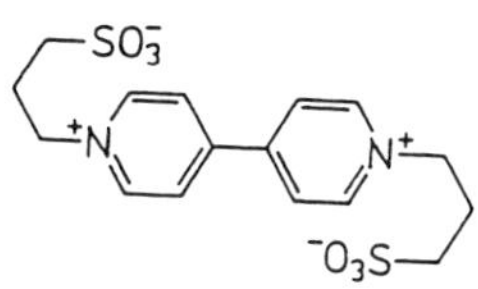

(17)

Figure 10. Control of photosensitized reduction of **17** in a microheterogeneous SiO_2 colloid using $Ru(bpy)_3^{2+}$ as photosensitizer.

ate)-4,4′-bipydirinium (**17**), PVS°, using Ru(II)–tris(bipyridine), $Ru(bpy)_3^{2+}$, have been characterized [59]. Upon excitation of the photosensitizer the respective oppositely charged photoproducts are generated [Eq. (17)]. In a homogeneous aqueous phase, the photoproducts recombine [Eq. (18)] with a diffusion-controlled bimolecular rate constant, $k_b = 7.9 \times 10^9\ M^{-1}\,s^{-1}$. In the presence of the SiO_2 colloid, the recombination rate is, however, substantially lower, $k_b = 5.7 \times 10^7\ M^{-1}\,s^{-1}$. This retardation in the back ET process is attributed to the electrostatic functions of the charged colloid (Figure 10). Photoinduced ET results in oppositely charged photoproducts: while the oxidized species is associated with the colloid interface, the negatively charged electron acceptor is repelled by the colloid interface. As a result, the topological dislocation of the photoproducts in the microheterogeneous system effects undirectional ET.

The vectorial ET process and stabilization of the intermediate photoproducts against recombination as compared to the homogeneous phase significantly affects the quantum efficiency of ET products under continuous illumination. In the presence of a sacrificial electron donor (i.e., triethanolamine), irreversible reduction of the oxidized photoproduct occurs [Eq. (19)] and the reduced product, $PVS^{-\bullet}$, is accumulated in the system. The quantum efficiency for $PVS^{-\bullet}$ formation in a homogeneous aqueous system corresponds to $\phi = 0.04$. In the presence of the SiO_2 colloid a substantially higher quantum efficiency for $PVS^{-\bullet}$ formation under steady-state illumination is observed, $\phi = 0.32$. This higher quantum yield originates from the retardation of back ET reactions, allowing improved irreversible oxidation of the electron donor and accumulation of the photoproduct.

$$^{*}Ru(bpy)_3^{2+} + PVS \xrightarrow{k_q} Ru(bpy)_3^{3+} + PVS^{-\bullet} \quad (17)$$

$$Ru(bpy)_3^{3+} + PVS^{-\bullet} \xrightarrow{k_b} Ru(bpy)_3^{2+} + PVS \quad (18)$$

$$Ru(bpy)_3^{3+} + TEOA \longrightarrow Ru(bpy)_3^{2+} + TEOA^{+\bullet} \quad (19)$$

$\quad\quad\quad\quad\quad\quad\quad\quad\quad \hookrightarrow$ decomposition products

The effects of hydrophilic–hydrophobic interactions on photochemically induced ET are exemplified in the micellar assembly composed of cetyl trimethylammonium chloride micelles and Zn(II)–*meso*-(*N*-tetramethylpyridinium)porphyrin (**16**), $Zn{-}TMPyP^{4+}$, as photosensitizer and *N*-methyl-*N*′-dodecadecyl-4,4′-bipyridinium (**18**), $C_{14}MV^{2+}$, as electron acceptor [63]. Photoinduced ET from the excited photosensitizer to (**17**) yields the photoproducts [Eq. (20)] that subsequently recombine in a homogeneous aqueous phase [Eq. (21)] with a diffusion-controlled back ET rate constant.

$$^{*}Zn{-}TMPyP^{4+} + C_{14}MV^{2+} \longrightarrow Zn{-}TMPyP^{5+\bullet} + C_{14}MV^{+\bullet} \quad (20)$$

$$Zn{-}TMPyP^{5+\bullet} + C_{14}MV^{+\bullet} \xrightarrow{k_b} Zn{-}TMPyP^{4+} + C_{14}MV^{2+} \quad (21)$$

In the micellar assembly, the recombination rate is about 500-fold slower, due to dislocation of the reactive photogenerated intermediates into confined microenvironments (Figure 11). Prior to light absorption the photosystem components exhibit hydrophilic character and are confined to the aqueous phase. Upon ET, the reduced photoproduct, $C_{14}MV^{+\bullet}$, reveals its hydrophobic nature due to removal of one positive charge and the long alkyl chain substituent. As a result, the photoproduct is entrapped in the micellar hydrophobic core, and the oxidized photoproduct, $Zn{-}TMPyP^{5+\bullet}$, is repelled electrostatically by the charged micellar interface. Thus separation of the photoproducts by two distinct phases affects stabilization of the reactive intermediates against back ET.

Semiconductor particles provide alternative photoactive materials for the formation of charge-separated products [48, 49]. Bandgap photochemical excitation of a semiconductor particle promotes an electron from the valence band to the conduction band, thus generating an electron–hole pair (Figure 12*a*). Recombination of the electron–hole pair represents an analogous degradative pathway to that of diffusional back ET of photoproducts.

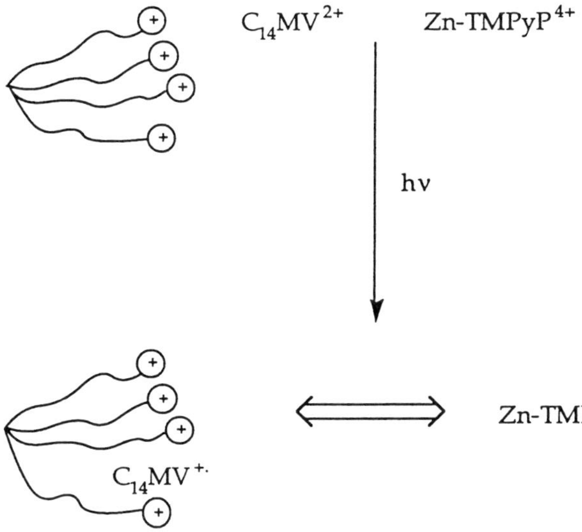

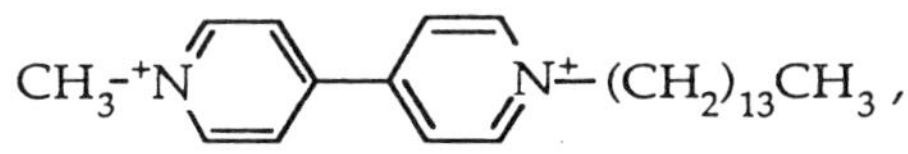

, $C_{14}MV^{2+}$, **(18)**

Figure 11. Control of photosensitized reduction of $C_{14}MV^{2+}$ by means of a hydrophilic–hydrophobic micellar system.

Trapping the conduction band electrons by ordered assemblies provides a vectorial pathway for charge separation. Immobilization of semiconductor TiO_2 particles onto a highly negatively charged SiO_2 matrix provides an organized assembly to control electron–hole recombination processes (Figure 12*b*) [68]. Photoexcitation of the semiconductor in the presence of PVS **(17)** as electron relay results in trapping of the conduction band electrons and concomitant electrostatic repulsion of the reduced photoproduct by the SiO_2 interface. Consequently, recombination of the trapped electrons in $PVS^{-\bullet}$ with valence-band holes is eliminated. A second organized semiconductor assembly where electron–hole recombinations are prevented is shown in Figure 12*c*. TiO_2 colloids are stabilized by a *N*-methyl-*N*′-carboxypropyl-4,4′-bipyridinium-functionalized polyethylenimine polymer [69]. Photoreduction of the polymer-anchored bipyridinium relay component is about 100-fold more efficient than that of a diffusional bipyridinium salt. This is attributed to the close proximity of the relay component to the semiconductor surface, where static (nondiffusional) uptake of conduction band electrons effectively competes with the electron–hole recombination.

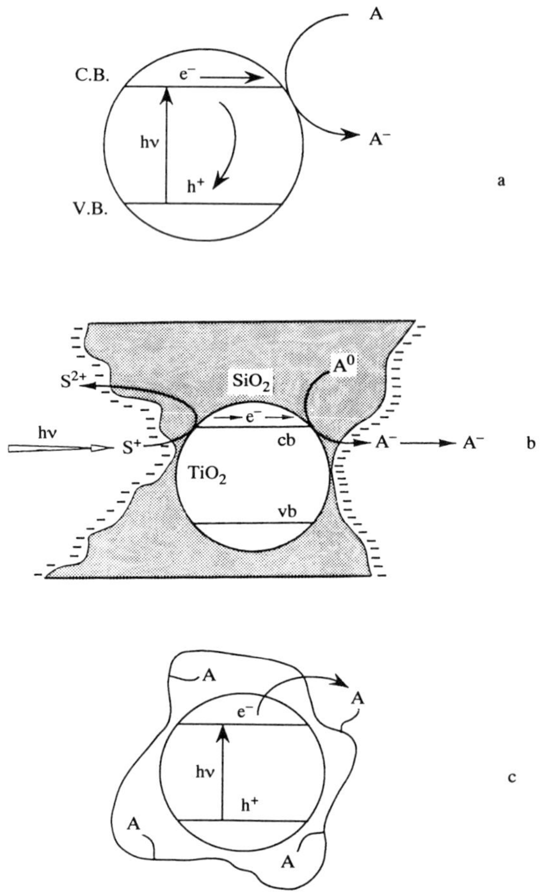

Figure 12. Control of electron transfer reactions at semiconductor solution interfaces: (*a*) electron transfer at a semiconductor solution interface and degradative recombination processes; (*b*) application of charged colloid as a supporting agent of semiconductor particles; (*c*) application of a redox functionalized polymer as a supporting matrix for stabilization of semiconductor colloids and effective trapping of conduction band electrons.

B. Catalysis in Artificial Photosynthetic Systems

The utilization of photogenerated ET in fuel production pathways is the ultimate goal of artificial photosynthesis. Several basic features need consideration upon coupling the electron transfer products to fuel generating routes:

1. The ET products must meet, as prerequisite, the thermodynamic redox potentials to drive the respective dark fuel production processes.
2. The raw materials acting as substrates for ET products should be abundant.
3. The fuel products generated in the photosynthetic device should generate in the energy release process the original substrates of the system. Under such circumstances the formation of waste products is eliminated.

With these considerations in mind, water and carbon dioxide provide the most attractive substrates for conjugation to the intermediate ET products [28]. Figure 13 shows the schematic configuration of artificial photosynthetic devices, where water or carbon dioxide are reduced to hydrogen or C_1-fuel products. Nevertheless, reduction of carbon dioxide (or its hydrated forms, bicarbonate and carbonate), as well as reduction of protons to hydrogen are accompanied by sluggish kinetics as reflected by their high overpotential in electroreduction processes. Thus catalysts are essential ingredients of photosynthetic systems, and they act in lowering the activation barrier of the coupled dark reactions by activation of the substrates. Although the need for catalysts is, in part, attributed to the inherent stability of the parent substrates, the detailed electronic and thermodynamic balance of CO_2 reduction processes and water photolysis respectively, Eqs. (22) to (30), reveal that the complementary functions of catalysts as multi-electron transfer mediators could facilitate these photosynthetic transformations [28].

$$CO_2 + 1e^- \rightarrow CO_2^{-\bullet} \qquad E^\circ = -1.9\ V \qquad (22)$$

$$CO_2 + 2H^+ + 2e^- \rightarrow CO + H_2O \qquad E^\circ = -0.53\ V \qquad (23)$$

$$CO_2 + 2H^+ + 2e^- \rightarrow HCO_2H \qquad E^\circ = -0.61\ V \qquad (24)$$

$$CO_2 + 4H^+ + 4e^- \rightarrow CH_2O + H_2O \qquad E^\circ = -0.48\ V \qquad (25)$$

$$CO_2 + 6H^+ + 6e^- \rightarrow CH_3OH + H_2O \qquad E^\circ = -0.38\ V \qquad (26)$$

$$CO_2 + 8H^+ + 8e^- \rightarrow CH_4 + 2H_2O \qquad E^\circ = -0.24\ V \qquad (27)$$

$$2H^+ + 2e^- \rightarrow H_2 \qquad E^\circ = -0.41\ V \qquad (28)$$

$$H_2O \rightarrow {\cdot}OH + H^+ + 1e^- \qquad E^\circ = 2.38\ V \qquad (29)$$

$$2H_2O \rightarrow O_2 + 4H^+ + 4e^- \qquad E^\circ = 0.82\ V \qquad (30)$$

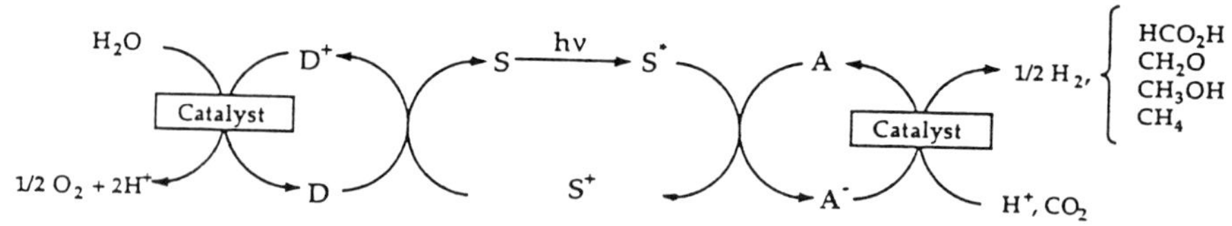

Figure 13. Photolysis of water and CO_2 fixation in an artificial photosynthetic system.

It is evident that the thermodynamic redox potential for the generation of C_1-fuel products becomes more positive as the number of electrons participating in the reduction process increases. Similarly, the multielectron oxidation of water to molecular oxygen is thermodynamically favored over the one-electron oxidation process. Thus multielectron redox reactions are thermodynamically favored over single-electron redox processes. The primary photosensitized ET process is, however, a single ET process, and thus transformation of the ET products to multielectron charge relays is essential. Catalysts could provide charge storage entities [28] and bridge the photogenerated single ET products with multielectron transfer reactions that proceed at the catalyst site.

Furthermore, it can be seen that reduction of CO_2 can lead to various C_1-fuel products, and that carbon dioxide fixation in aqueous media is accompanied by the competitive H_2-evolution process. Thus the developed catalysts must exhibit selectivity toward the desired product and respective substrate. In addition to carbon dioxide and water, nitrogen-containing compounds can be used as substrates for the reduced photoproduct. Several reduction processes of nitrogen-containing substrates are summarized as follows:

$$N_2 + 6e^- + 6H^+ \rightarrow 2NH_3 \tag{31}$$

$$N_2 + 4e^- + 4H^+ \rightarrow N_2H_4 \tag{32}$$

$$NO_3^- + 2e^- + 2H^+ \rightarrow NO_2^- + H_2O \tag{33}$$

$$NO_2^- + 6e^- + 8H^+ \rightarrow NH_4^+ + 2H_2O \tag{34}$$

Thus the designed catalysts included in artificial photosynthetic systems should involve three complementary functions: (1) The catalyst should activate the substrate toward the specific chemical transformation; (2) the catalyst should act as a multielectron charge storage intermediate, and (3) the catalyst should induce selectivity and specificity toward the designed photosynthetic process. Nature has resolved these catalytic requisites by utilizing enzymes and cofactors as biocatalysts that interlink the single ET process of the photosynthetic reaction center with the CO_2-fixation cycle. While the enzymes act as the substrate activation site, the cofactors accumulate the electrical charges. Recent developments reveal that natural bioactive materials (enzymes and cofactors) can be utilized as catalytic assemblies in artificial photosynthetic systems [70]. In addition, modification of enzymes by chemical means provides catalytic entities that activate the substrates and transduce the electrical charges. Pure chemical approaches involving homogeneous and heterogeneous catalysts duplicate functions of the bioactive enzyme–cofactor assembly and can be utilized similarly in artificial photosynthetic systems.

C. Enzymes as Catalysts in Artificial Photosynthetic Systems

Redox enzymes could provide the catalytic active site for activation of the selected substrate and exchange of electrons between an electron carrier (or hole carrier). Numerous redox enzymes are coupled to nicotinamide cofactors [i.e., nicotinamide adenine dinucleotide, NAD^+ (**19**), or nicotinamide adenine dinucleotide phosphate, $NADP^+$ (**20**)] acting as two-electron (or hole) carriers [71,72]:

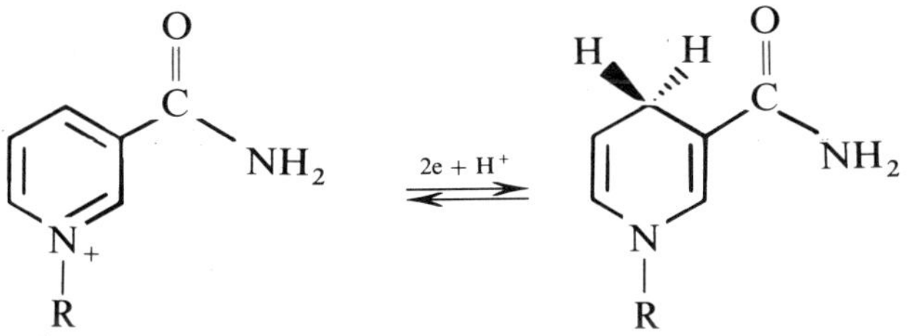

(35)

Thus utilization of the photogenerated ET reduced product in regeneration of NAD(P)H could provide a general means to utilize enzymes as catalysts (Figure 14*a*). Alternatively, biocatalyzed oxidation of substrates could proceed by regeneration of the oxidized cofactor, $NAD(P)^+$, using the oxidized intermediate formed upon photoinduced ET (Figure 14*b*).

A further possibility of applying enzymes in artificial photosynthetic systems involves the direct interaction between photogenerated charge carriers and the active site. Figure 15*a* shows the schematic interaction of a photogenerated reduced product, A^-, with the enzyme redox site. To maintain electron exchange within the protein, aspects of ET in such assemblies need consideration. According to Marcus theory, the ET rate constant between a donor–acceptor pair is given by Eq. (6), where ΔG and λ are the free-energy charge and reorganization energy associated with the

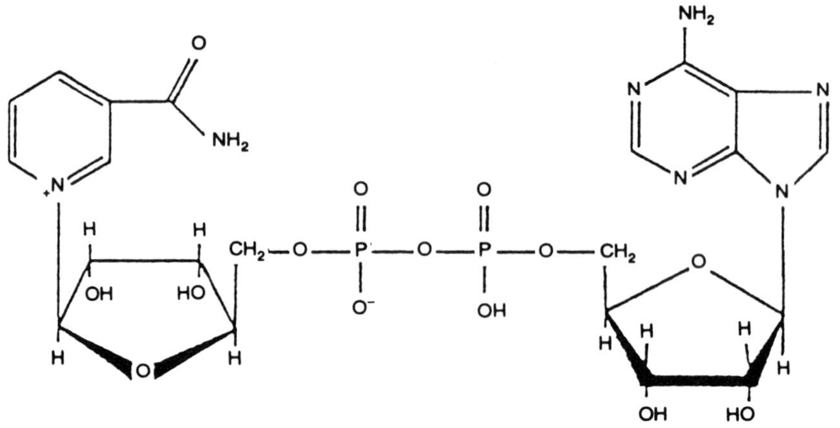

(19)

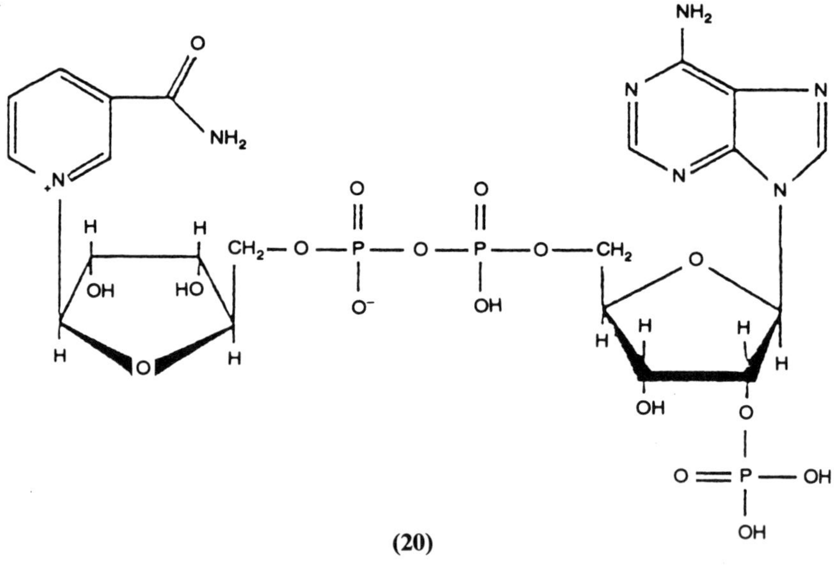

(20)

ET process, respectively, and *d* is the distance between the donor and the acceptor. The distance dependence of the electron transfer rate is extremely important in large molecular ensembles such as proteins, where the active redox site is shielded by the protein matrix [73–75]. As a result, electron exchange between the enzyme redox site and the artificial charge carrier is

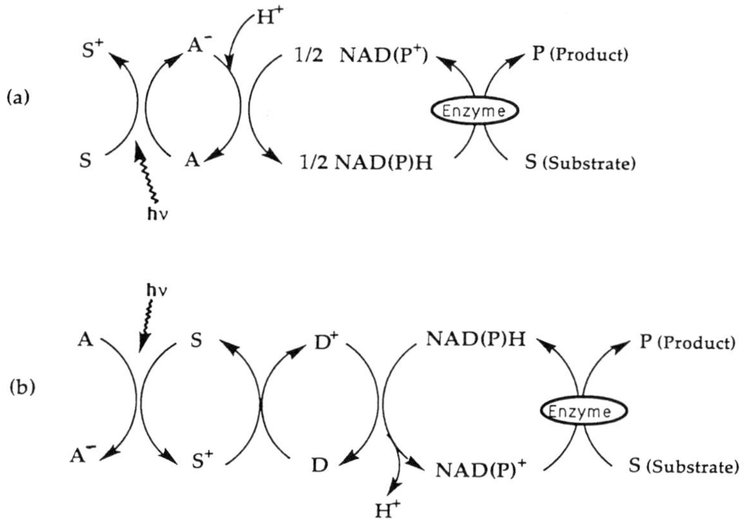

Figure 14. Photosensitized regeneration of cofactors as a means to drive photosynthetic biocatalyzed reactions: (*a*) photoinduced reductive transformations by NAD(P)H cofactors; (*b*) photosensitized oxidative transformation through $NAD(P)^+$ cofactors.

not feasible unless the charge carrier penetrates the protein backbone to a sufficiently close topological position relative to the active site. Thus the charge carrier must meet proper electrical and structural properties to allow its diffusion into the protein assembly.

An alternative approach to establish electrical communication between an enzyme redox site and a photochemical assembly involves "electrical wiring" of the protein by electron acceptor components that interact with the excited species (Figure 15*b*). Realizing that electron exchange between charge carriers and protein redox sites is limited by the protein backbone, chemical modification of protein amino acid residues by charge carrier units might resolve these difficulties. The protein-anchored charge carriers are distributed across the protein backbone, and several of the relay units are positioned close to the active site. Externally located charge carriers interact with the excited species, and electrons are trapped by these periphery anchored relay components. Subsequent to the photoinduced ET process, electrons are tunneled to intraprotein relay components that ultimately transfer the electrons to the enzyme active site. Thus chemical modification of the protein by charge carriers converts the protein into an electrically wired assembly and communicates the shielded redox site with an external photoactive species [76].

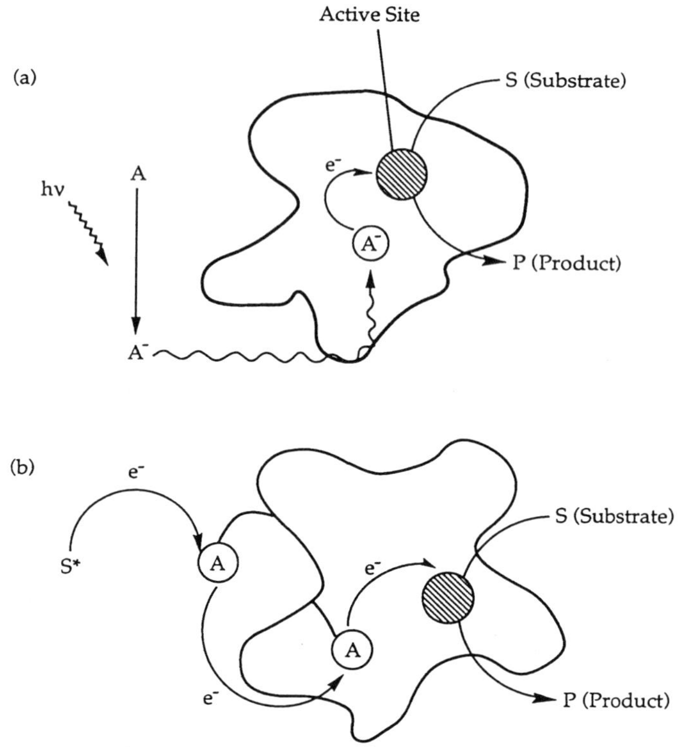

Figure 15. Electrical communication of enzyme redox sites with photoexcited species as a means to drive biocatalyzed photosynthetic transformations: (*a*) electrical communication by a diffusional photogenerated electron mediator; (*b*) direct communication between an electrically wired redox enzyme and a photoexcited species.

1. Photosynthetic Transformations by Regeneration of Cofactors. The subject of photochemical regeneration of $NAD(P)^+/NAD(P)H$ cofactors has recently been reviewed [70]. Photosystems for light-induced regeneration of NADH and NADPH cofactors have been developed using photogenerated *N,N'*-dimethyl-4,4'-bipyridinium radical cation, $MV^{+\bullet}$, as ET mediator [77, 78]. A photosystem composed of $Ru(bpy)_3^{2+}$ as photosensitizer, *N,N'*-dimethyl-4,4'-bipyridinium as primary electron acceptor, and $(NH_4)_3EDTA$ as sacrificial electron donor has been applied for regeneration of NADPH using the enzyme ferredoxin-$NADP^+$ reductase, FDR. In this photosystem, oxidative ET quenching of the excited species yields $MV^{+\bullet}$ that mediates reduction of $NADP^+$ using FDR as biocatalyst (Figure 16*a*). The quantum efficiency for NADPH formation corresponds to $\phi = 1.9 \times 10^{-2}$. A closely

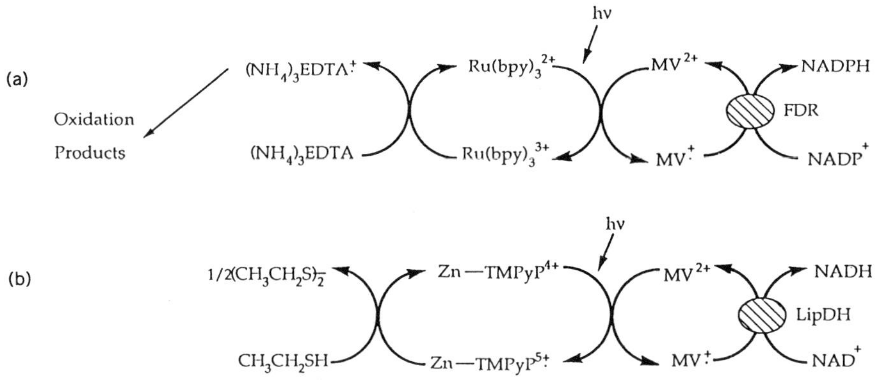

Figure 16. Photosensitized regeneration of NAD(P)H cofactors: (*a*) photosensitized regeneration of NADPH using $Ru(bpy)_3^{2+}$ as photosensitizer; (*b*) photochemical regeneration of NADH using Zn–TMPyP^{4+} as photosensitizer.

related photosystem is applied for the regeneration of NADH. Here Zn(II)–*meso*-(*N*-tetramethylpyridinium)porphyrin, Zn-TMPyP^{4+}, is used as photosensitizer and mercaptoethanol acts as sacrificial electron donor (Figure 16*b*). Photoinduced ET results in $MV^{+\bullet}$ that mediates the reduction of NAD^+ to NADH in the presence of lipoamide dehydrogenase, LipDH. Other light-active compounds, such as Ru(II)–tris(bipyrazine), $Ru(bpz)_3^{2+}$, or semiconductor particles have been used to generate $MV^{+\bullet}$ and subsequently utilize the charge carrier for NAD(P)H regeneration in the presence of FDR and LipDH, respectively.

Nonenzymatic photosensitized regeneration of NADH cofactor proceeds [79, 80] in the presence of various rhodium complexes such as Rh(III)–tris-(bipyridine), $Rh(bpy)_3^{3+}$, or Rh(I)–tris(triphenylsulfonatophosphine), $RhCl(dpm)_3^{3-}$, as catalysts. Single ET reduction of rhodium complexes is often accompanied by subsequent disproportionation of the product, resulting in a two-electron reduced product in the form of a hydridorhodium complex (see Section IV.E). The latter hydridorhodium complex can provide the hydride equivalent for reduction of NAD^+ to NADH (by hydride transfer or stepwise electron–proton–electron transfer). In a photosystem [79] composed of $Ru(bpy)_3^{2+}$ as photosensitizer, $Rh(bpy)_3^{3+}$ as charge acceptor and triethanolamine, TEOA, as sacrificial electron donor, NAD^+ is reduced to NADH by the stepwise mechanism outlined in Figure 17. ET quenching of excited $Ru(bpy)_3^{2+}$ by $Rh(bpy)_3^{3+}$ yields the metastable single-electron reduction product that undergoes disproportionation and ligand dissociation to generate the intermediate

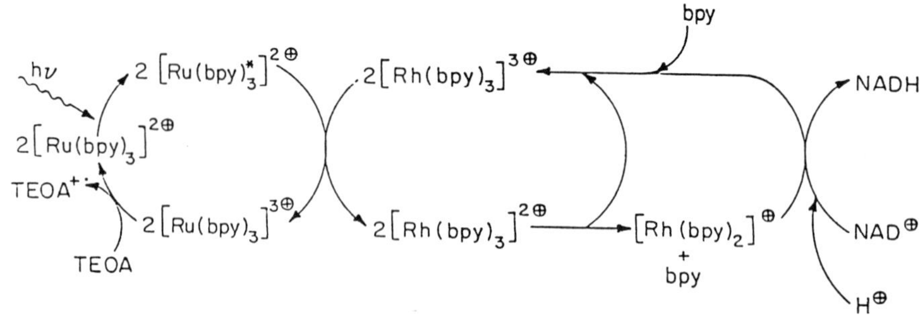

Figure 17. Photosensitized regeneration of NADPH using $Rh(bpy)_3^{3+}$ as electron transfer mediator.

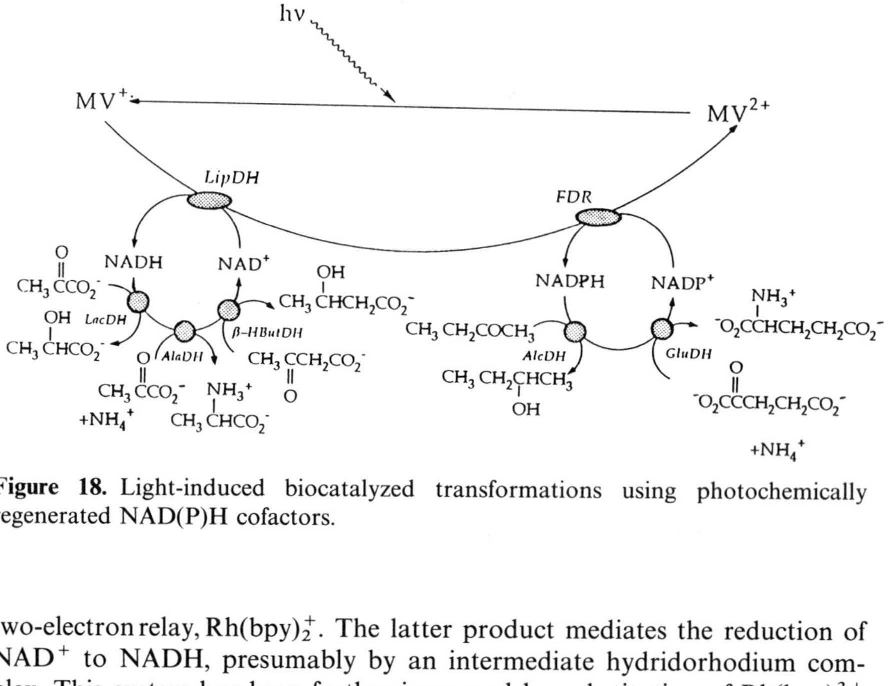

Figure 18. Light-induced biocatalyzed transformations using photochemically regenerated NAD(P)H cofactors.

two-electron relay, $Rh(bpy)_2^+$. The latter product mediates the reduction of NAD^+ to NADH, presumably by an intermediate hydridorhodium complex. This system has been further improved by substitution of $Rh(bpy)_3^{3+}$ with rhodium(III)–tris(2,2′-bipyridyl-5-sulfonic acid) as electron relay and hydride transfer catalyst [80].

Photosensitized regeneration of the biologically active NAD(P)H cofactors allows their subsequent coupling to cofactor-dependent enzymes. As a result, a series of photosynthetic transformations have been driven [78, 81] using various enzymes and their respective substrates by the general scheme outlined in Figure 18. Table 2 summarizes various photosynthetic

TABLE 2 Turnover Numbers of the Components in the Different Enzyme-Catalyzed Reactions

	$Ru(bpy)_3^{2+}$	$ZnTMPyP^{4+}$	MV^{2+}	$NAD(P)^+$	Enzymes		Percent Conversion	Percent Optical Purity	Final Product Concentration (*M*)	Initial Substrate Concentration (*M*)
Butan-2-ol	530		40	40	FDR[a] 2.4×10^4	AlDH[b] 6×10^3	27	100 ± 10	4.05×10^{-2}	0.15
Glutamic acid	225		15	15	FDR 9.75×10^3	GluDH[c] 1.26×10^5	13	99 ± 7	1.3×10^{-2}	0.1
Lactic acid		4500	13.5	27	LipDH[d] 5.67×10^4	LDH[e] 4.15×10^4	27	94 ± 7	2.7×10^{-2}	0.1
Alanine		7900	19	27	LipDH 1.7×10^4	AlaDH[f] 3.2×10^6	38	78 ± 7	3.8×10^{-2}	0.1
3-Hydroxybutyric acid		1900	2.4	10	LipDH 5.2×10^3	β-HBuDH[g] 8.3×10	21	105 ± 10	9.4×10^{-3}	4.52×10^2

[a]Formula weight (FW) ≈ 40,000 [M. Shin, *Methods Enzymol.* **23**, 441 (1971)].
[b]FW ≈ 40,000.
[c]FW ≈ 2,200,000 [H. Sund and W. Burchard, *Eur. J. Biochem.* **6**, 202 1968)].
[d]FW ≈ 70,000 [F. B. Straub, *Biochem. J.* **33**, 787 (1939)].
[e]FW ≈ 140,000 [R. Jaenicke and S. Knof, *Eur. J. Biochem.* **4**, 157 (1968)].
[f]FW ≈ 228,000 [A. Yoshida and G. Freese, *Biochim. Biophys. Acta* **92**, 33 (1964)].
[g]FW ≈ 850,000 [H. U. Bergmeyer, *Biochem. J.* **102**, 423 (1967)].

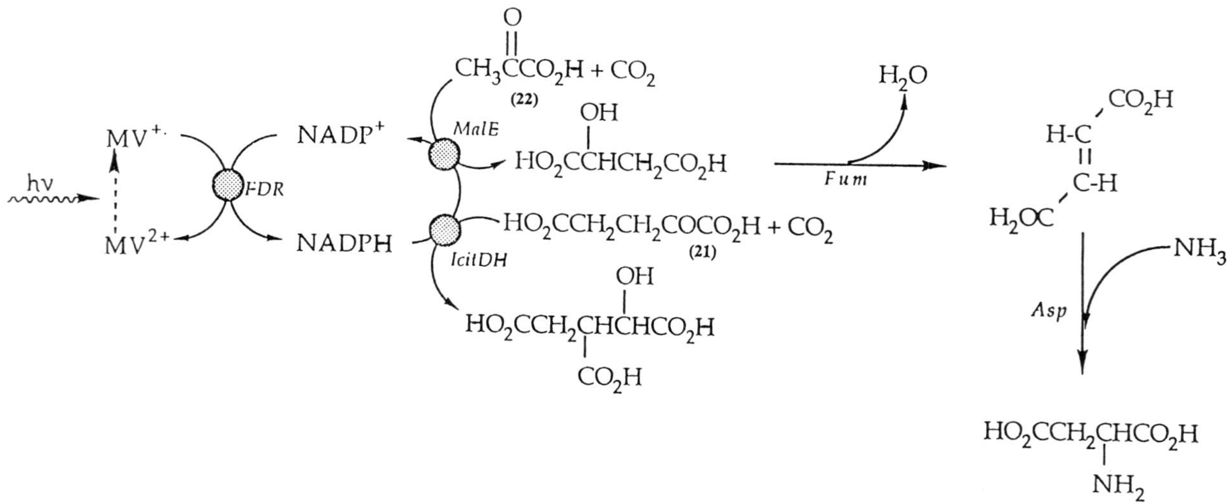

Figure 19. Photochemical-biocatalyzed CO_2-fixation processes and secondary biocatalyzed transformations using the carboxylation products by primary light-induced regeneration of NADPH.

transformations that were applied in the light-induced regeneration cycles of NAD(P)H cofactors. The substrates, products, and nature of enzymes coupled to the cofactors are detailed. Also, the turnover numbers of the enzymes and conversion yields to products are included.

Interestingly, the photoinduced regeneration system of NADPH has also been applied in photosynthetic carbon dioxide fixation processes [82]. Insertion of CO_2 into a carbon backbone represents a two-electron fixation of carbon dioxide:

$$\mathrm{R{-}CH_2{-}\overset{\overset{\displaystyle O}{\|}}{C}{-}CO_2^- + CO_2 + 2e^- + H^+ \rightarrow R{-}CH\overset{\overset{\displaystyle OH}{|}}{C}H{-}CO_2^-} \quad \text{with } \mathrm{CO_2^-} \text{ on the first CH} \qquad (36)$$

Accordingly, photogenerated NAD(P)H provides a two-electron relay for the insertion of CO_2 into α-ketoglutaric acid (**21**) and pyruvic acid (**22**), in the presence of isocitrate dehydrogenase, IcitDH, and the malic enzyme, MalE, respectively (Figure 19). In these photosystems, $Ru(bpy)_3^{2+}$ acts as photosensitizer, MV^{2+} as primary electron acceptor, and 2-mercaptoethanol is used as the sacrificial electron donor. Photogenerated $MV^{+\bullet}$ mediates regeneration of NADPH in the presence of FDR, and the resulting cofactor provides the electrons for the dark fixation of CO_2 into the keto acids. The net processes accomplished in these transformations correspond to the reductive insertion of CO_2 into keto acids via oxidation of the sacrificial thiol donor. Photochemical carboxylation of pyruvic acid by this process,

$$\mathrm{CH_3COCO_2^- + 2HOCH_2CH_2SH + CO_2 \rightarrow {}^-O_2CCH_2\overset{\overset{\displaystyle OH}{|}}{C}HCO_2^- + (HOCH_2CH_2S)_2 + H^+} \qquad \Delta G^\circ = 11.5\ \mathrm{kcal\ mol^{-1}} \qquad (37)$$

is endoergic by about $\Delta G^\circ = 11.5\ \mathrm{kcal\ mol^{-1}}$ and represents a true "uphill" photosynthetic pathway. The carbon dioxide fixation product can then act as the source substrate for subsequent biocatalyzed transformations. For example, Figure 19 shows that photogenerated malic acid acts as the source substrate for aspartic acid [82]. In this case, malic acid is dehydrated by fumarase, Fum, and the intermediate fumaric acid is aminated in the presence of aspartase, Asp, to form aspartic acid. The latter multienzyme photosynthetic assembly, composed of several biocatalysts, a cofactor, and an artificial photosystem, exhibits high stability and operates as an integrated system.

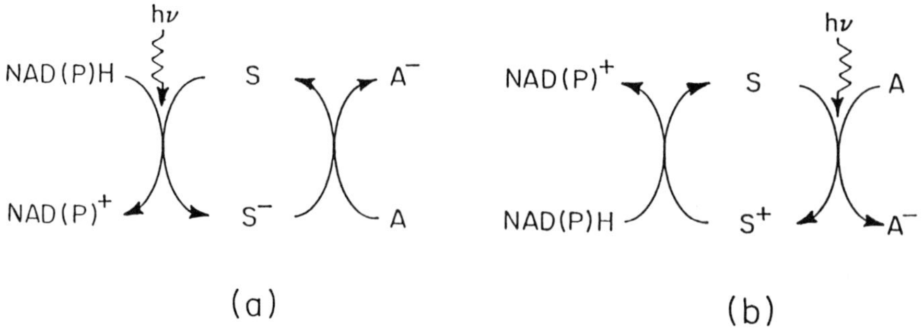

Figure 20. Photosensitized oxidative regeneration of $NAD(P)^+$ cofactors: (*a*) by reductive electron transfer quenching mechanism; (*b*) by an oxidative electron transfer quenching mechanism; (*b*) by an oxidative electron transfer quenching mechanism followed by dark oxidation of NADPH.

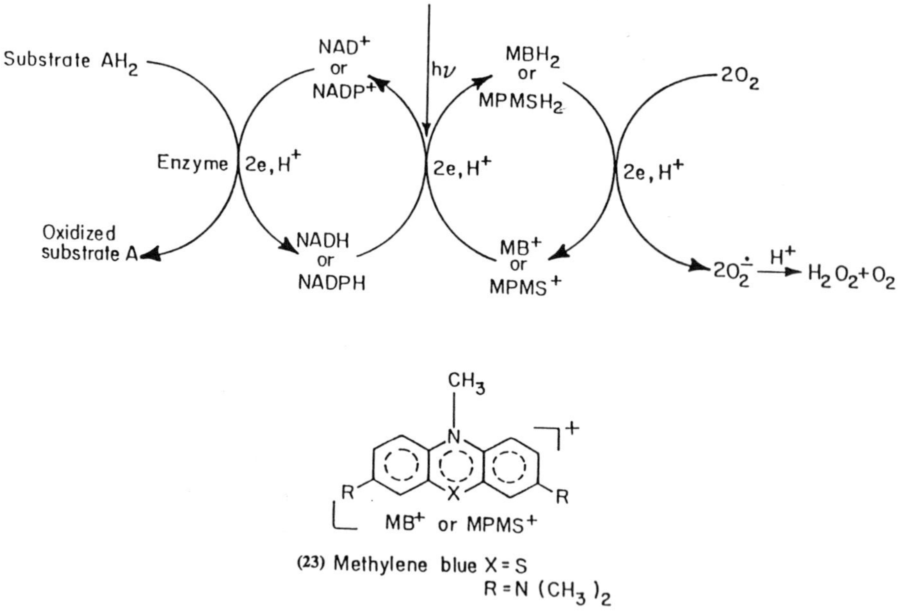

(23) Methylene blue X = S
R = N $(CH_3)_2$

(24) N-methylphenazonium methyl sulfate X = N
R = H

Figure 21. Photosensitized regeneration of $NADP^+$ cofactors using **23** and **24** as photosensitizers.

Photochemical regeneration of $NAD(P)^+$ cofactors provides an oxidant for uptake of two electrons from substrates. Photoinduced regeneration of $NAD(P)^+$ cofactors proceeds by two alternative routes (Figure 20). NAD(P)H can act as electron donor for reductive ET quenching of the excited species, thereby generating the reduced photosensitizer (Figure 20*a*). The second route involves oxidative ET quenching of the excited photosensitizer followed by dark oxidation of NAD(P)H by the photogenerated oxidized photoproduct (Figure 20*b*). Organic dyes such as methylene blue (**23**) or *N*-methyl phenazonium methyl sulfonate (**24**) have been used as light-active compounds for $NAD(P)^+$ regeneration by the reductive ET mechanism [83, 84] (Figure 21). The quantum efficiencies for NAD^+ formation using **23** and **24** correspond to $\phi = 3 \times 10^{-1}$ and $\phi = 2.8 \times 10^{-1}$, respectively. Oxidation of ethanol and lactic acid to acetaldehyde and pyruvic acid, using alcohol dehydrogenase, AlcDH, or lactate dehydrogenase, LacDH, is mediated by the photoregenerated NAD^+. Other dyes, such as rose bengal, eosin, or thionine, have been applied successfully as photosensitizers in similar photosystems.

Photolysis of organic compounds to molecular hydrogen and the complementary oxidized organic products has been reported by photoregeneration of NAD^+. The photochemical assembly includes Sn(IV)–*meso*-(*N*-tetramethylpyridinium)porphyrin, $Sn\text{-}TMPyP^{4+}$, as photosensitizer, a Pt colloid as a heterogeneous catalyst, the reduced cofactor NADH, an organic photolysable substrate, and the respective NAD^+-dependent enzyme (alcohol or lactic acid and AlcDH or LacDH, respectively) [85]. Photolysis of the organic material proceeds by the sequence of reactions outlined in Eqs. (38) to (44).

$$^*Sn\text{–}TMPyP^{4+} + NADH \longrightarrow Sn\text{–}TMPyP^{3+\cdot} + NAD^{\cdot} + H^+ \quad (38)$$

$$2Sn\text{–}TMPyP^{3+\cdot} + H^+ \rightleftharpoons Sn\text{–}TMPyP^{4+} + Sn\text{–}TMPyPH^{3+} \quad (39)$$

$$Sn\text{–}TMPyP^{3+\cdot} + NAD^{\cdot} \longrightarrow Sn\text{–}TMPyP^{3+} + NAD^+ \quad (40)$$

$$Sn\text{–}TMPyP^{4+} + NAD^{\cdot} \longrightarrow Sn\text{–}TMPyP^{3+\cdot} + NAD^+ \quad (41)$$

$$Sn\text{–}TMPyPH^{3+} + H^+ \xrightarrow{Pt} Sn\text{–}TMPyP^{4+} + H_2 \quad (42)$$

$$NAD^+ + CH_3CH_2OH \xrightarrow{AlcDH} NADH + CH_3CHO + H^+ \quad (43)$$

$$CH_3CH_2OH \xrightarrow[NAD^+/NADH]{Pt,\ AlcDH} CH_3CHO + H_2$$

$$\Delta G^\circ = 10\,\text{kcal mol}^{-1} \quad (44)$$

The primary step involves reductive ET of the metalloporphyrin. The resulting ET photoproducts undergo redox transformations that ultimately lead to NAD^+ and the hydrogenated porphyrin. These two photoproducts drive the oxidation and reduction of the organic material: while H_2 evolution is effected by the hydro metalloporphyrin in the presence of the Pt catalyst acting as charge storage site [Eq. (42)] (see Section IV.C.1), NAD^+ mediates the oxidation of the alcohol using AlcDH as biocatalyst [Eq. (43)]. All of the photosystem components are recycled within the photocycle and the net transformation corresponds to the endoergic cleavage (photolysis) of ethanol, Eq. (44). Thus this energy storage transformation represents a photosynthetic process in which H_2 fuel is produced. In view of the broad research activities in biomass utilization, photosynthetic production of hydrogen from biomass seems an attractive route for future developments.

Photolysis of organic materials has also been accomplished by the oxidative ET quenching route of excited states [86]. A photosystem that includes $Ru(bpy)_3^{2+}$ as photosensitizer, *N*,*N'*-dimethyl-4,4'-bipyridinium, MV^{2+}, and NADH as electron donor, leads to $MV^{+\cdot}$ and NAD^+ through the sequence of reactions outlined in Eqs. (45) to (47). The reduced photoproduct, $MV^{+\cdot}$, mediates H_2 evolution at a heterogeneous catalytic site of Pt [Eq. (48)], while NAD^+ mediates the two electron oxidation of ethanol [Eq. (43)] or lactic acid [Eq. (49)].

$$^*Ru(bpy)_3^{2+} + MV^{2+} \longrightarrow Ru(bpy)_3^{3+} + MV^{+\cdot} \qquad (45)$$

$$Ru(bpy)_3^{3+} + NADH \longrightarrow Ru(bpy)_3^{2+} + NAD^{\cdot} \qquad (46)$$

$$2NAD^{\cdot} + H^+ \longrightarrow NADH + NAD^+ \qquad (47)$$

$$MV^{+\cdot} + H^+ \xrightarrow{\text{Pt}} \tfrac{1}{2}H_2 \qquad (48)$$

$$CH_3CH(OH)CO_2H + NAD^+ \xrightarrow{\text{LacDH}} CH_3COCO_2H + NADH \qquad (49)$$

2. Biocatalyzed Photosynthetic Transformations Using Artificial Electron Mediators. A variety of NAD(P)H-dependent enzymes recognize artificial electron carriers. With such enzymes, the natural cofactors can be substituted by a nonnative photogenerated electron relay. Photogenerated *N*,*N'*-dimethyl-4,4'-bipyridinium radical cation (Figure 22) is recognized by various enzymes such as hydrogenase, Hyd [87,88], formate dehydrogenase, ForDH [82], nitrate and nitrite reductases [89,90], NitraR and NitriR, and

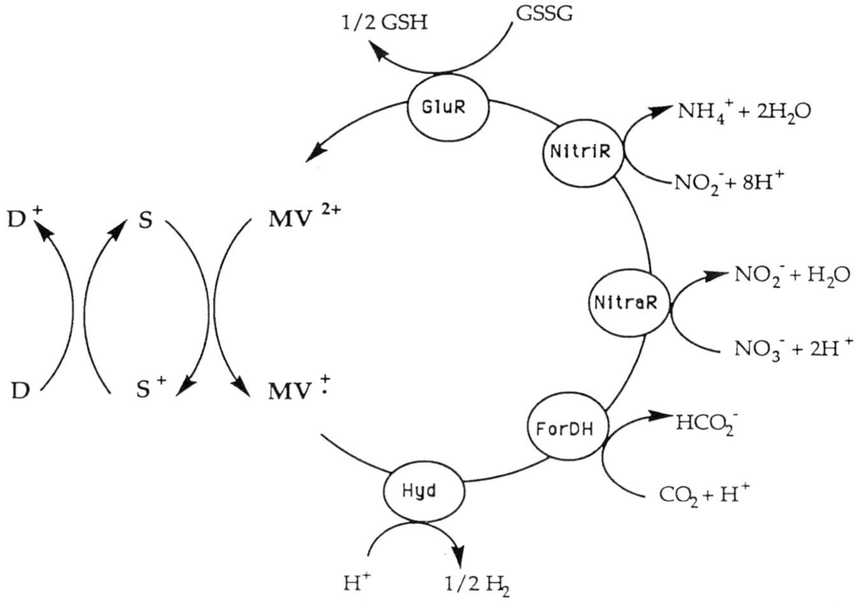

Figure 22. Biocatalyzed photosynthetic system mediated by *N*,*N'*-dimethyl-4,4'-bipyridinium radical cation, $MV^{+\cdot}$.

glutathione reductase [91], GluR. Various biocatalyzed photosynthetic systems using light-induced regeneration of $MV^{+\cdot}$ by Zn-porphyrins or $Ru(bpy)_3^{2+}$ as photosensitizers and different sacrificial electron donors such as EDTA, cysteine, and mercaptoethanol have been developed [82, 87–90]. The photobiocatalyzed transformations effected by the artificial electron mediator $MV^{+\cdot}$ are summarized as follows:

$$2e^- + 2H^+ \xrightarrow{\text{Hyd}} H_2 \quad (50)$$

$$CO_2 + 2e^- + H^+ \xrightarrow{\text{ForDH}} HCO_2^- \quad (51)$$

$$NO_3^- + 2e^- + 2H^+ \xrightarrow{\text{NitraR}} NO_2^- + H_2O \quad (52)$$

$$NO_2^- + 6e^- + 8H^+ \xrightarrow{\text{NitriR}} NH_4^+ + 2H_2O \quad (53)$$

$$GSSG + 2e^- + 2H^+ \xrightarrow{\text{GluR}} 2GSH \quad (G = \text{Glu–Cys–Gly}) \quad (54)$$

Carbon dioxide reduction to formate in the presence of ForDH represents a photochemical CO_2-fixation process. The enzyme recognizes in addition to $MV^{+\bullet}$ other 2,2-bipyridinium radical cations [82]. Reduction of nitrate to NO_2^- and reduction of nitrite to ammonia in the presence of NitraR and NitriR, respectively, allows the sequential $8e^-$ reduction of nitrate to ammonia,

$$NO_3^- + 10H^+ + 8e^- \rightarrow NH_4^+ + 3H_2O \tag{55}$$

in a photosystem that includes the two enzymes. In a photosystem including $Ru(bpy)_3{}^{2+}$ as photosensitizer, the fixation of $NO_3{}^-$ to ammonia was driven with a quantum efficiency corresponding to $\phi = 3 \times 10^{-2}$.

3. Electrical Communication in Biocatalyzed Artificial Photosynthetic Systems. The various biocatalyzed photosynthetic systems discussed in Sections IV.C.1 and IV.C.2 have applied diffusional electron carrier (native cofactors or synthetic electron mediators). For practical application of biocatalyzed photosynthetic systems, it is important to construct photosystems where the components are organized in an immobilized configuration that allows continuous regeneration of photochemical assembly. Yet the redox sites of enzymes do not communicate electrically with electron sources (or holes) located externally to the protein backbone (see Section IV.C). Thus construction of immobilized biocatalytic assemblies requires the development of nondiffusional means for electrical communication [92] between the enzyme redox site and the externally located excited species.

Means to establish such electrical communication are presented schematically in Figure 23. In Figure 23*a* the enzyme is chemically modified by relay components that shorten the ET distances within the protein backbone and convert the protein into a conductive matrix. Also, the anchored relay component acts as electron quencher of the excited species. As a result, the enzyme acts as the active quencher of the excited state and as biocatalyst. Figure 23*b* shows a second approach where the enzyme active site communicates electrically with an excited state by the intermediate participation of a redox functionalized polymer. Here, relay components are linked to the polymer backbone by long-chain flexible bridging groups ("arms"). These redox groups act as quenchers of the excited state and mediate ET to the enzyme redox site by penetration of the redox groups into the protein backbone. The linkage of the enzyme to the functionalized redox polymer can be tailored by physical encapsulation of the protein in polymer gels, covalent linkage, or electrostatic attractions.

Photoreduction of nitrate to nitrite, Eq. (52), was accomplished by physical encapsulation of nitrate reductase, NitraR, in a redox functional-

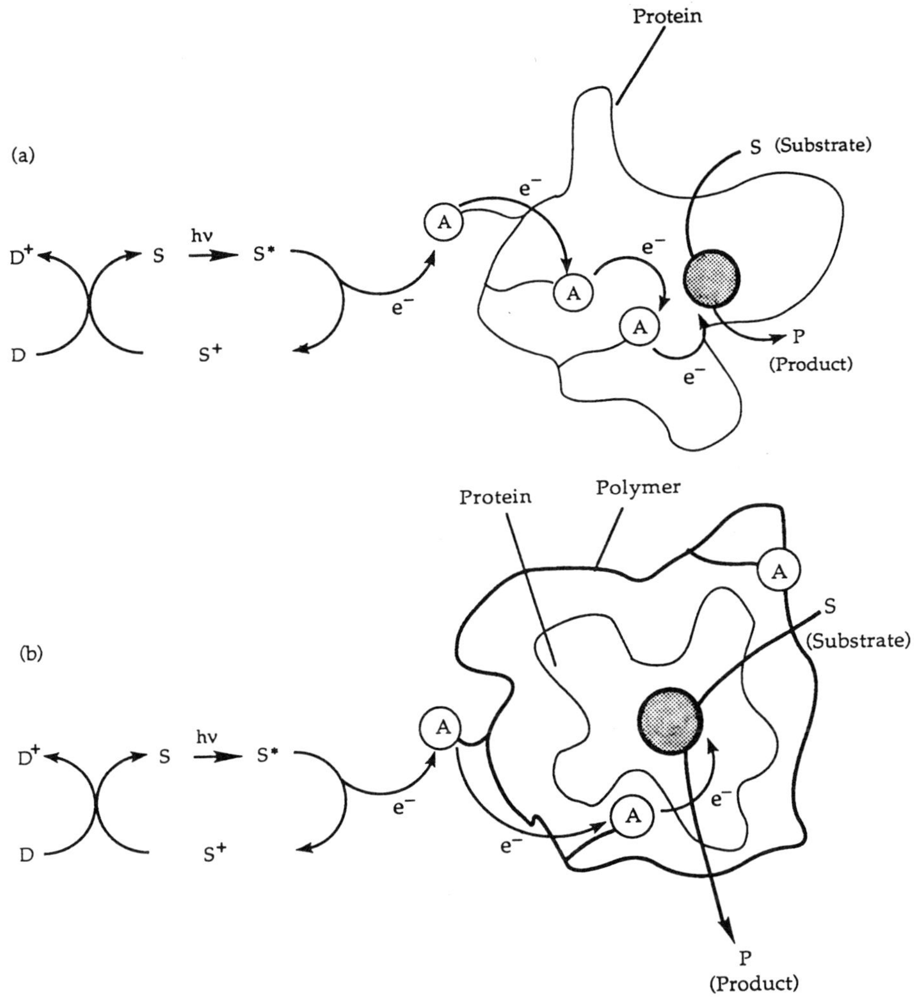

Figure 23. Design of electrical communication between an enzyme redox center and a photoexcited species to attain light-induced biocatalyzed transformations: (*a*) direct electrical wiring of the protein by its chemical modification with electron relay components; (*b*) electrical communication by immobilization of the protein into a redox functionalized polymer matrix.

ized 4,4′-bipyridinium acrylamide copolymer [93]. In this photosystem, $Ru(bpy)_3^{2+}$ was used as photosensitizer and EDTA as sacrificial electron donor. Electron transfer from the excited photosensitizer results in ET to the polymer, and the redox sites on the polymer mediate ET to the protein redox center, where the biocatalyzed transformation occurs (Figure 24*a*).

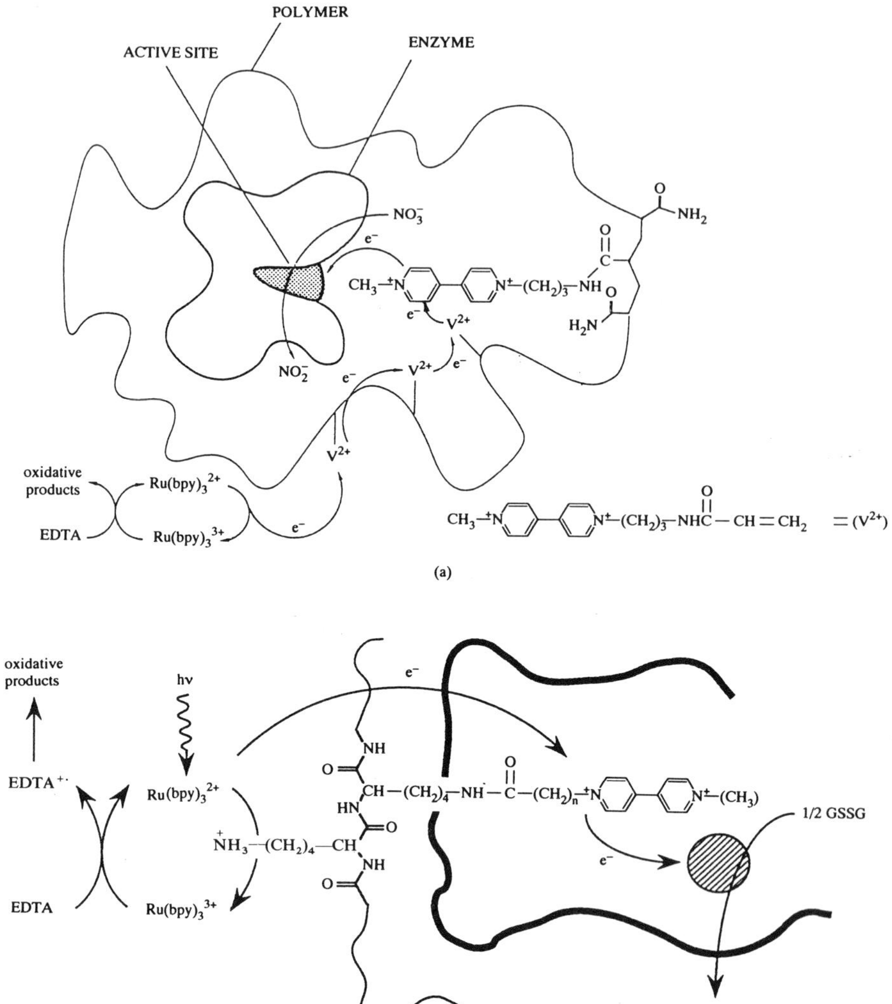

Figure 24. Photosensitized biocatalyzed transformations in electrically communicated redox functionalized polymers: (*a*) reduction of nitrate in a bipyridinium–acrylamide-cross-linked polymer; (*b*) reduction of oxidized glutathione using a redox functionalized bipyridinium–polylysine polymer.

The rate constant for ET from the redox polymer functionalities to the protein active site is $k_{et} = (9 \pm 3) \times 10^5\ s^{-1}$. Similarly, the enzyme glutathione reductase was electrically wired by interacting the protein with a redox polymer composed of polylysine modified by *N*-methyl-*N'*-carboxyalkyl-4,4'-bipyridinium. The photosensitized reduction of oxidized gluta-

TABLE 3 Electrical Communication of Glutathione Reductase by *N*-Methyl-*N'*-carboxyalkyl Bipyridinium Functionalized Polylysine[a]

Electron Mediator[b]	v (M h^{-1})[c]	k_{et} (s^{-1})[d]
$n = 2$	0.12	490
$n = 5$	0.16	1090
$n = 8$	—	1700
$n = 12$	0.50	4350

[a]The photosystem is composed of $Ru(bpy)_3^{2+}$ as photosensitizer, EDTA as electron donor, 4,4'-bipyridinium-modified polylysine, and GluR as biocatalyst.
[b]Polylysine-4,4'-bipyridinium-modified polymer has the structure

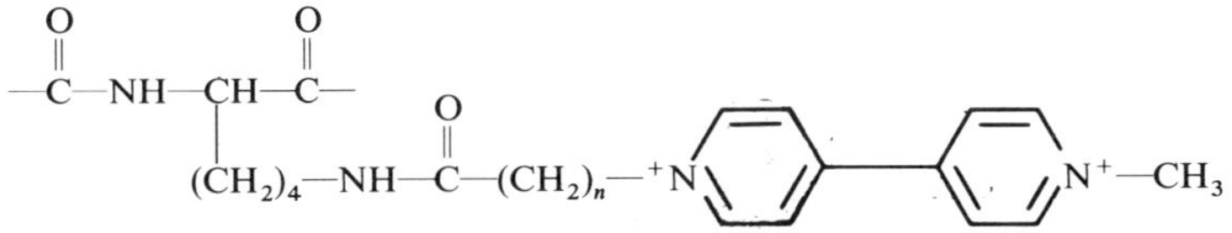

The loading degree of lysine residues is 1:68 ± 5.
[c]Rate of reduced glutathione formation under steady-state illumination.
[d]ET rate constant from polymer redox component to protein redox site (determined by flash photolysis).

thione, Eq. (54), was examined in a photosystem composed of $Ru(bpy)_3^{2+}$ as sensitizer, EDTA as sacrificial electron donor, the polylysine redox-modified polymer as electron wiring matrix, and the enzyme glutathione reductase, GluR. The effectiveness of electrical communication between the excited species and the redox center was established as a function of the alkyl chain length bridging the bipyridinium salt to the polymer backbone [94].

In this photosystem, ET from excited $Ru(bpy)_3^{2+}$ results in polymer-anchored 4,4'-bipyridinium radical cation that mediates a long-range ET to the protein redox center, where the biocatalyzed transformation proceeds (Figure 24*b*). The effectiveness of the biocatalyzed reduction of oxidized glutathione is controlled by the alkyl chain length bridging the bipyridinium radical cation to the polymer backbone. As the chain length increases, the electrical communication between the redox polymer and protein active site is improved. Table 3 summarizes the rates of reduced glutathione formation in the photosystem under steady-state illumination, using different chain-length-bridged bipyridinium-modified polylysine polymers as electrical communicating matrices. For comparison, the ET rate constants from the polymer-anchored bipyridinium radical cation redox components to the protein active site are included. Thus, as the flexible alkyl chain, linking the bipyridinium radical cation to the polymer, is longer, the redox mediating

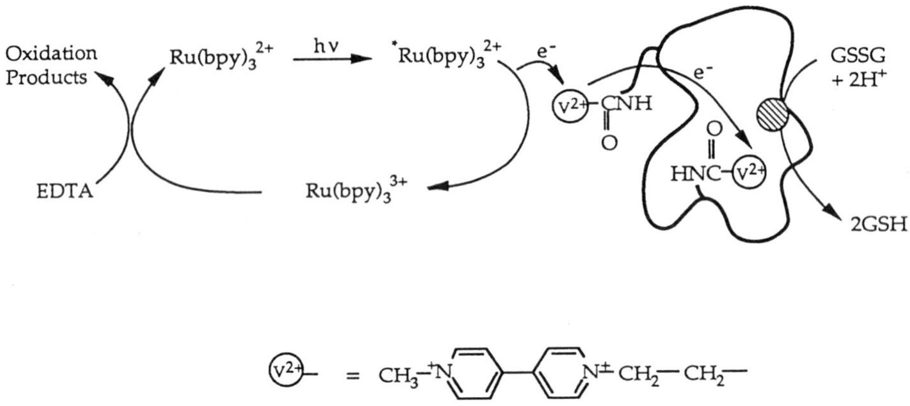

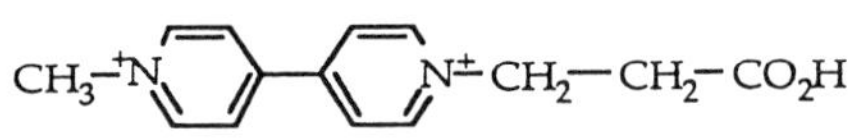

(25)

Figure 25. Photosensitized reduction of oxidized glutathione using bipyridinium-modified nitrate reductase as electrically wired assembly for direct interaction with the excited species.

group reaches a closer distance to the protein active center, and electrical communication is improved.

The enzyme glutathione reductase, GluR, was electrically wired also by the alternative approach involving chemical modification of the protein backbone by electron relay components [95]. *N*-Methyl-*N*′-propionic acid-4,4′-bipyridinium (**25**) was coupled to lysine residues of GluR. In a photosystem composed of $Ru(bpy)_3^{2+}$ as photosensitizer, EDTA as electron donor and the bipyridinium-modified GluR, reduction of glutathione proceeds. In this system the bipyridinium relay components act as ET quenchers of the excited species and as electrical wiring components tunneling the electron to the biocatalyst active site (Figure 25).

D. Heterogeneous Metal Colloids Mimicking Cofactor–Enzyme Functionalities in Artificial Photosynthetic Systems

Heterogeneous catalysts provide catalytically active solid interfaces in conjunction with liquid, gaseous, or other solid phases [96, 97]. A variety of metal catalysts, such as Ni, Pd, Pt, and metal oxides, such as NiO, RuO_2, and Fe_2O_3, influence various chemical transformations, including oxidations, hydrogenations, isomerizations, and dehydrogenation processes. The activities of heterogeneous catalysts are determined by their bulk chemical composition and surface properties. Comprehensive studies have revealed that surface structural properties of the solid, such as steric order and distance of atoms, crystal defects, steps and kinks, porosity, and surface area, control its catalytic behavior [98–100]. In general terms, the basic step of heterogeneous catalysis involves adsorption of the substrate and its activation toward the respective chemical transformation. Adsorption may proceed by dissociative mechanism (i.e., hydrogen adsorption), where chemical bonds are cleaved in the activation process. Alternatively, adsorption of a substrate to the solid interface can occur by an associative pathway, where the substrate adsorption includes specific bond deformation or bond isomerization, leading to the activation of the substrate. Various studies revealed that activation of the substrates on solid surfaces proceeds at distinct sites and that the modes of substrate activation and its orientation lead to product specificity and stereospecificity. To this extent, the heterogeneous catalysts resemble the functions of active sites in proteins.

Metal particles also behave as microelectrodes, and electron exchange between the microelectrode material and redox species present in the solution occurs. Upon conjugation of the metal particle with a reduced species, the electrode is charged and in the presence of an oxidant the electrode material is discharged. The ET exchange rate between a metal particle and a redox relay is given by

$$k_{\mathrm{et}} = 4\pi N D r^2 \tag{56}$$

where N is the Avogadro number, D the diffusion coefficient of the redox component, and r the radius of the metal particle [101]. Thus coupling of metal particles with photogenerated reduced or oxidized species provides a route to charge or discharge the heterogeneous catalysts by electrons (or "holes"), respectively. The effectiveness of charging relates to the particle size, and colloid particles ($r = 20$ to $100\,\text{Å}$) are anticipated to interact rapidly with molecular electron (or "holes") carriers. Such electrical interaction between metal particles and solution redox substrates provides a pathway to generate charge reservoirs that can subsequently be coupled to chemical transformations utilizing surface-activated substrates. Various methods for preparing metal colloids are known, including chemical means [102–104], photochemical [105] and radiolytic methods [106], and vacuum evaporation techniques [107]. The various preparations lead to colloids of different sizes, surface structures, and porosity and, consequently, to different catalytic activities.

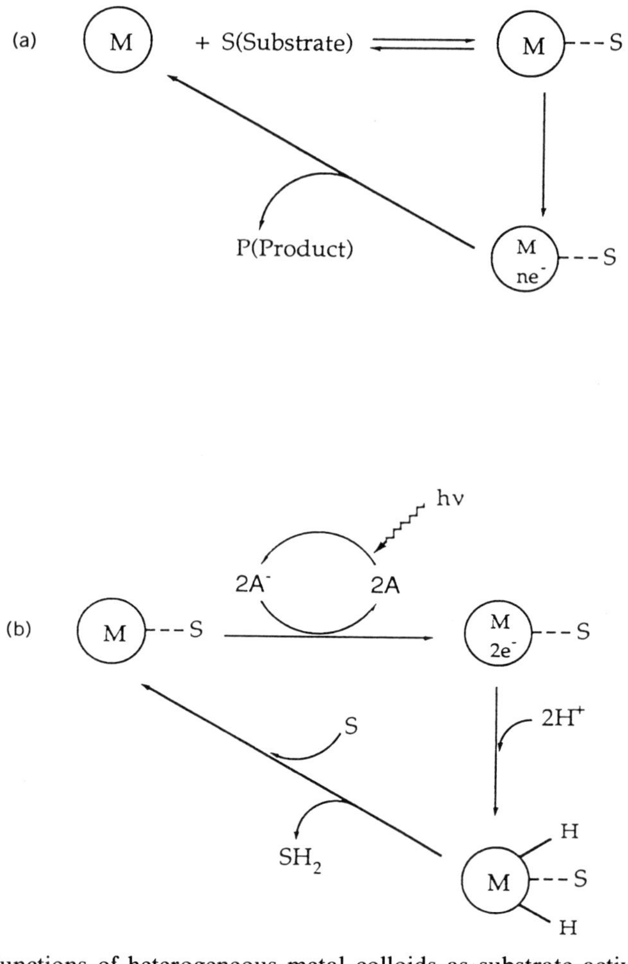

Figure 26. Functions of heterogeneous metal colloids as substrate activation sites and electron accumulation centers: (*a*) photoreduction of a metal-activated substrate by metal-accumulated electrons; (*b*) photoreduction of a metal-activated substrate by in situ–generated hydrogen atoms and a subsequent hydrogenation process.

The two complementary functions, substrate activation and charge accumulation of metal colloids, are utilized in artificial photosynthetic devices (Figure 26). Upon charging of the metal colloid with electrons (Figure 26*a*), multi-ET reduction processes of metal-activated substrates can be driven. Alternatively, electrical charging of the colloid could effect in aqueous media the generation of surface-bound hydrogen atoms (Figure 26). This surface-active H-atom species can then mediate the hydrogenation of surface-activated substrates (in analogy with hydrogenation of substrates by molecular hydrogen on heterogeneous catalysts). The functions of metal colloids in accumulating electrical charges or generating surface-bound hydrogen (or hydride) species resemble the activities of natural cofactors in biological systems. Thus heterogeneous catalysts mimic functionalities of biological cofactor–enzyme assemblies to the extent that site-activated substrate moieties are conjugated to electron (or hydride) reservoirs.

1. Photosynthetic Transformations Utilizing Heterogeneous Catalysts. Photogeneration of metal-bound hydrogen-atom or hydride intermediates has been examined extensively in relation to photolysis of water [108–112]. The subject of water photocleavage and photochemical H_2-evolution systems has been reviewed extensively [113, 114], and only aspects related to the present discussion are mentioned here. The basic configuration of a hydrogen evolution system is outlined schematically in Figure 27*a*. It includes a photosensitizer, S, that upon excitation results in ET to an acceptor component, A, and the sacrificial electron donor recycles the light-active component. The reduced photoproduct, A^-, mediates, in the presence of a heterogeneous catalyst, hydrogen evolution. The catalytic steps involved with H_2 formation are detailed schematically in Figure 27*b* and include electron exchange between the reduced relay and the metal microelectrode, reduction of protons to generate surface-associated H atoms, and dimerization of the active intermediates to yield H_2 evolution. Various noble-metal colloids composed of Pt, Rh, Pd, Ag, and Ru have been applied as heterogeneous catalysts for H_2 evolution [108–114]. Photosensitizers that were used to induce the ET processes include polypyridine metal complexes [115] [i.e., $Ru(bpy)_3^{2+}$] (**15**), metal porphyrins such as Zn–*meso*-(*N*-tetramethylpyridinium)porphyrin (**16**) [116], organic dyes [44] (i.e., acridine orange) (**13**), and semiconductor particles or colloids [117]. Electron acceptors that were widely employed as ET mediators for H_2 evolution include *N,N'*-dialkyl-4,4'-bipyridinium [118] [i.e., (**26**)] and *N,N'*-dialkyl-2,2'-bipyridinium salts (**27**) such as Co(III)–sepulchurate, (**28**) [119], Rh(III)–tris(bipyridine), $Rh(bpy)_3^{3+}$ [120] and metal ions such as Eu^{3+} or V^{5+} [121]. Sacrificial electron donors that were broadly applied for regeneration of the light-active component include triethanolamine, TEOA, ethylenediaminetetraacetic acid, EDTA, ascorbic acid, and thiols. Figure 27*c* exemplifies H_2 evolution from a photochemical assembly that includes *N,N'*-dimethyl-4,4'-bipyridinium, MV^{2+} (**29**), as ET mediator.

To utilize heterogeneous particles as catalytic sites for hydrogenation of substrates, the directed photosynthetic transformation must kinetically

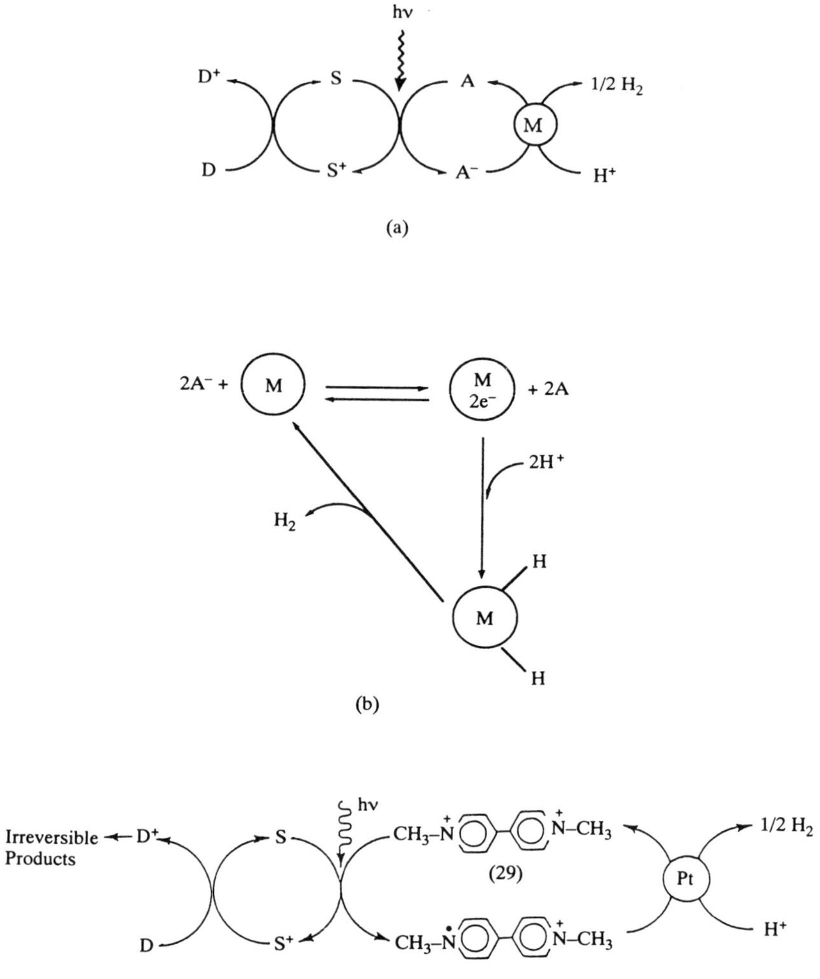

Figure 27. Schematic photosensitized hydrogen evolution: (*a*) sequence of electron transfer reactions involved in photosensitized hydrogen evolution; (*b*) catalyzed hydrogen formation at a metal–colloid interface; (*c*) photosensitized hydrogen evolution using **26** as electron transfer mediator.

compete with the H-atom dimerization process. Such competitive elimination of H_2 evolution can be achieved by two alternative routes: (1) By utilizing a substrate where its hydrogenation is favored over H-atom dimerization, and (2) by utilizing substrates exhibiting high association affinity to the catalyst surface. This leads to dilution of the active sites

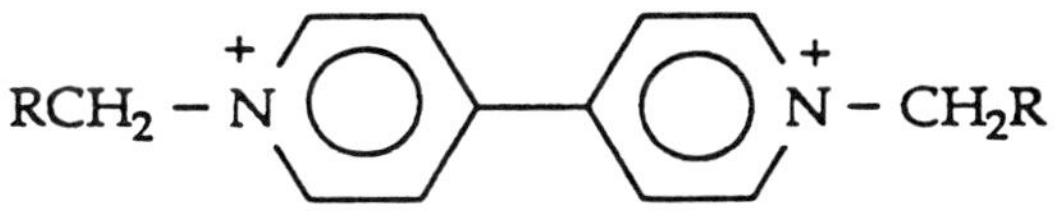

(26)

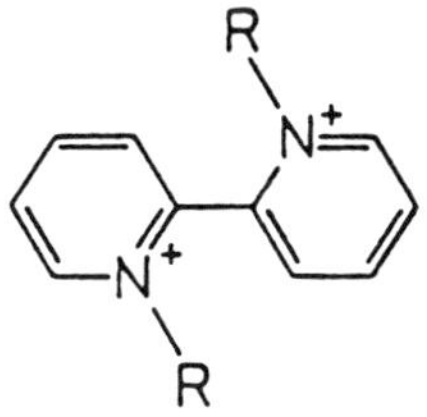

(27)

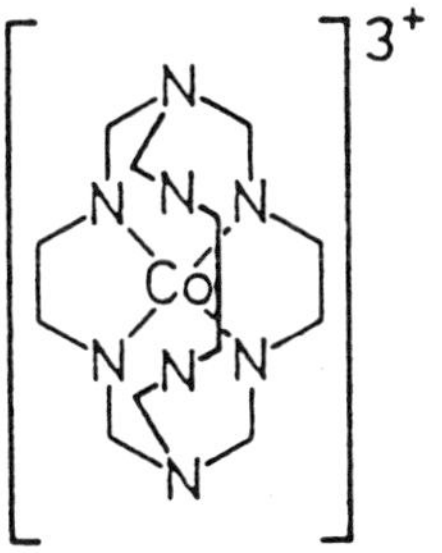

(28)

available for H-atoms formation and forms migration barriers for the dimerization of surface-bound hydrogen species. Consequently, the hydrogenation processes are kinetically favored over H_2 evolution.

Photosensitized hydrogenation of ethylene, C_2H_4, and acetylene, C_2H_2, in the presence of Pt, Rh, or Ru as heterogeneous catalysts represent photosynthetic transformations mimicking cofactor–enzyme systems [122]. In a photosystem composed of $Ru(bpy)_3^{2+}$ as photosensitizer, MV^{2+} (**29**) as electron acceptor, Na_2EDTA as sacrificial electron donor, and Pt as heterogeneous catalyst, H_2 evolution proceeds with a quantum efficiency corresponding to $\phi = 9.9 \times 10^{-2}$. With added ethylene, no H_2 evolution occurs and ethane, C_2H_6, is formed with a quantum yield corresponding to $\phi = 9.9 \times 10^{-2}$, identical to that of H_2-evolution. The formation of ethane has been attributed to the sequence of reactions outlined by

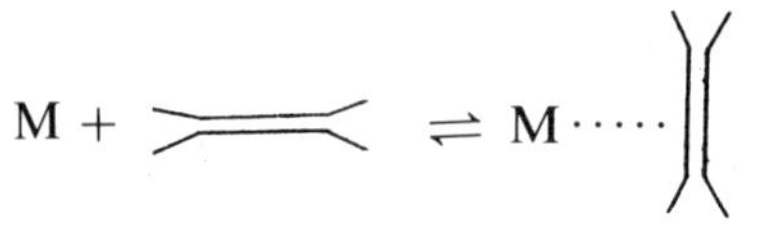
(57)

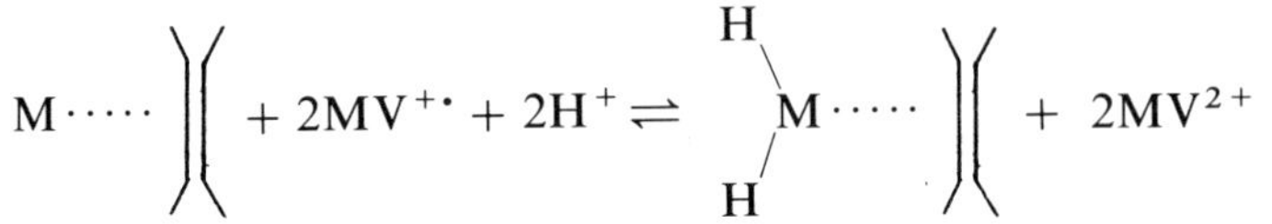
(58)

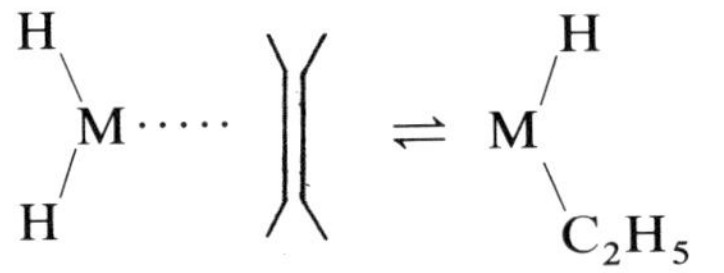
(59)

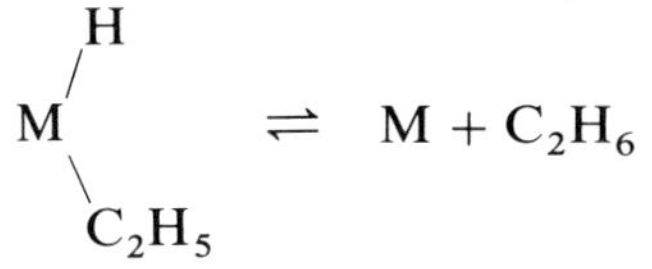
(60)

where all surface-bound H atoms are trapped by metal-activated ethylene.

Application of acetylene as substrate in the similar photosystem used for hydrogenation of ethylene leads, in the presence of Pt, to hydrogenation of acetylene to ethylene at a poor quantum efficiency, $\phi = 3.2 \times 10^{-3}$, but the competitive H_2-evolution process is blocked [122]. The low quantum yield was attributed to poor activation of the substrate by the metal catalyst, while inhibition of H_2 evolution was rationalized in terms of strong association of C_2H_2 onto the Pt surface, resulting in dilution of surface-bound H atoms that prevents their dimerization. With Pd as heterogeneous catalyst in the photosystem, a lower quantum yield for C_2H_2 hydrogenation is observed than for the Pt photosystem, although the Pd is a superior catalyst for acetylene activation toward hydrogenation. The poor efficiency of the Pd-containing photosystem was attributed to the inefficient activity of the catalyst in generating surface-bound H atoms. On the other hand, upon application of a bifunctional heterogeneous catalyst composed of Pt coated by islands of Pd in the photosystem, an impressive enhancement in the hydrogenation of acetylene to ethylene, $\phi = 4.4 \times 10^{-2}$, is observed [122]. The synergetic activity of the bifunctional Pt–Pd catalyst was attributed to two complementary functions of the composite colloid

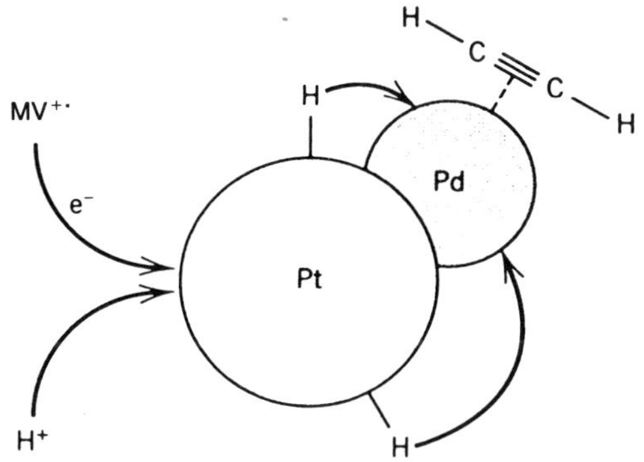

Figure 28. Synergetic activity of a Pt–Pd colloid in a photoinduced hydrogenation of acetylene.

(Figure 28): The Pt component acts as an effective microelectrode for acceptance of the electrons from the photoreduced electron carrier, $MV^{+\cdot}$, and provides the metal surface for the generation of active H-atom species. The Pd component provides catalytic sites for activation of acetylene, C_2H_2, and acts as a "hydrogen sponge" for uptake of Pt-generated H atoms. Hydrogenation of acetylene proceeds at the Pd active sites. The bifunctional bimetallic catalyst composite nicely simulates the cofactor–enzyme activities in natural systems. The Pt component mimics the cofactor functions in charge accumulation and generation of reducing equivalents, H atoms. The Pd component resembles the functions of an enzyme active site and provides the activation sites for acetylene. The improved catalytic properties are maintained only by immobilization of the two metals into a united catalytic composite. This resembles the activity of protein-associated cofactors (i.e., flavins), where cofactor–enzyme communication is established in highly organized assemblies.

Photohydrogenation of various other organic substrates (i.e., 3-hexyne, 1,1-diphenylethylene, 3-pentanone, and butyraldehyde) was accomplished in similar photosystems in the presence of different heterogeneous metal colloids [123]. Also, organization of photosynthetic multiphase assemblies for photoinduced hydrogenation of water-insoluble alkenes and alkanes have been reported using multifunctional heterogeneous catalysts [124].

Closely related photoinduced hydrogenation processes were accomplished by immobilization of noble metals such as Pt, Pd, and Rh onto CdS semiconductor particles [125]. Upon excitation of the semiconductor material, conduction band electrons are trapped by the noble metal surface, where

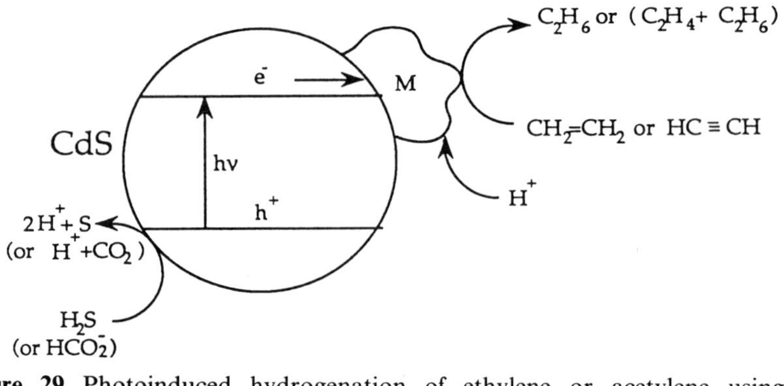

Figure 29. Photoinduced hydrogenation of ethylene or acetylene using CdS semiconductor photocatalytic particles.

adsorbed H atoms are formed. The photogenerated valence-band holes are utilized to oxidize a solution-solubilized electron donor, such as sulfide or formate (Figure 29). In the absence of ethylene or acetylene, H_2 evolution proceeds by dimerization of metal-bound H atoms. In the presence of ethylene or acetylene, however, hydrogen evolution is blocked and photocatalyzed hydrogenation of the substrates to ethane and ethylene, respectively, occurs:

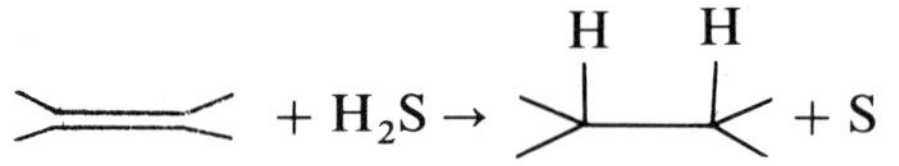

(61)

$$H{-}C{\equiv}C{-}H + H_2S \rightarrow C_2H_4 + S \tag{62}$$

Interestingly, the yield of the hydrogenated products is substantially higher than the yield of hydrogen evolved in the absence of the unsaturated compounds. This was attributed to the function of surface-activated ethylene or acetylene in scavenging metal-associated H atoms: In the absence of the unsaturated substrates, surface-associated H atoms recombine with valence-band holes. This recombination process competes with H-atom dimerization. In the presence of ethylene or acetylene, hydrogen atoms are trapped by the hydrogenation pathway, and consequently, the recombination process is diminished.

Photosensitized carbon dioxide fixation to formate,

$$HCO_3^- + 2H^+ + 2e^- \rightarrow HCO_2^- + H_2O \tag{63}$$

was accomplished by a hydrogenation pathway [126, 127]. In this photosystem a Pd colloid stabilized by cyclodextrin, a cyclic polyglucose macromolecule, acts as an effective and selective catalyst for the photosensitized reduction of carbon dioxide to formate in aqueous media, pH 7.5. In fact, experimental evidence revealed that bicarbonate, HCO_3^-, the hydrated product of CO_2 at pH 7.5, is hydrogenated rather than CO_2. The photosystem is composed of deazariboflavin, (**30**), as photosensitizer, *N*,*N*′-dimethyl-4,4′-bipyridinium, MV^{2+}, as electron acceptor, and sodium oxalate as sacrificial electron donor. Illumination of the photosystem in the presence of the Pd-β-cyclodextrin catalyst and of CO_2/HCO_3^- results in effective formation of formate, $\phi = 1.1$, and only traces of hydrogen are produced. In the absence of CO_2/HCO_3^-, the catalytic photosystem leads to effective H_2 evolution, implying that photoinduced formate production competes effectively with the hydrogen evolution pathway. The cyclic process leading to HCO_2^- formation or H_2 evolution is presented schematically in Figure 30*a*. Detailed mechanistic studies revealed important catalytic functions of the heterogeneous Pd catalyst. It has been shown that the rate of formate production as a function of HCO_3^- concentration shows a saturation behavior. This suggests that the Pd catalyst contains bicarbonate activation sites that are saturated upon increase of HCO_3^- concentration. Furthermore, the photocatalytic activity of the system is inhibited by added thiols acting as inhibitors. Kinetic analyses of the Pd-catalyst activities in the absence and presence of the added thiol reveal that the catalyst follows the Michaelis–Menten enzyme-active-site kinetic model, and the competitive inhibition pathway of active sites in proteins. Interestingly, when thiols are added to the photocatalytic assembly and formate production is inhibited, the system restores its activity toward H_2 evolution. In view of the mechanistic studies, the stepwise pathway shown in Figure 30*b* was suggested as the pathway for formate production. The photogenerated electron carrier, *N*,*N*′-dimethyl-4,4′-bipyridinium radical cation, $MV^{+\bullet}$, charges the Pd microelectrode and generates surface-associated H atoms (or hydrides). Bicarbonate, HCO_3^-, is activated at distinct sites on the Pd metal. Upon generation of surface-bound H atoms and in the presence of activated bicarbonate species, rapid hydrogenation of the activated substrate occurs. The hydrogenation process is fast and competes with the H_2-evolution process, being slow compared to the former hydrogenation route. The added inhibitor (thiol) selectively associates to the bicarbonate activation sites. As a result, only the slow H_2 evolution proceeds by surface dimerization of metal-bound H atoms. The activity of the heterogeneous Pd catalyst in the photosensitized CO_2/HCO_3^- reduction to formate simulates functionalities of biomaterials. The heterogeneous metal catalyst provides activation sites for HCO_3^-. These sites can be inhibited competitively by added ingredients similar to the

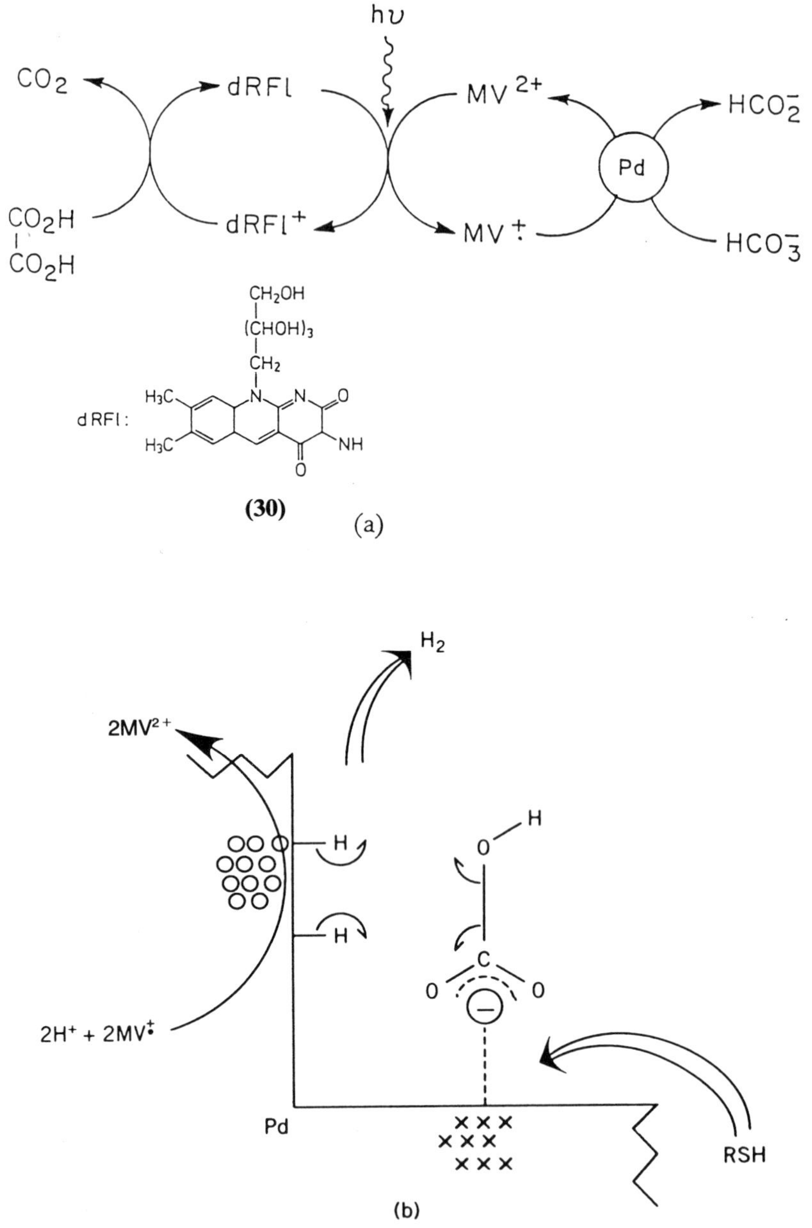

Figure 30. (*a*) Scheme of photochemical assembly for the reduction of CO_2/HCO_3^- to formate: (*b*) schematic functions of the Pd catalyst in activation of CO_2/HCO_3^- and hydrogenation of activated carbon substrate.

inhibition process of enzyme active sites. In addition, the metal cluster acts as a sink for electrons and as a reservoir of surface-bound H atoms (or hydrides) and thus resembles the functions of bioactive cofactors. In this way the particles provide the functionalities of a cofactor–enzyme assembly.

Photosensitized CO_2 reduction to methane, CH_4, through an eight-electron fixation process [Eq. (27)] was accomplished using Ru- or Os-metal colloids [128, 129]. Two different photosystems based on these heterogeneous catalysts were developed for the selective reduction of CO_2 to methane (Figure 31). One photosystem applies Ru(II)–tris(bipyrazine), $Ru(bpz)_3^{2+}$ (**31**), as photosensitizer, and triethanolamine, TEOA, as sacrificial electron donor [128]. Illumination of this photosystem in the presence of the Ru-colloid results in reduction of CO_2 to methane, $\phi = 4 \times 10^{-4}$, and to the higher hydrocarbons ethylene and ethane at quantum efficiencies corresponding to $\phi = 7.5 \times 10^{-5}$ and $\phi = 4 \times 10^{-5}$, respectively. In this photosystem, reductive ET quenching of the excited photosensitizer occurs (Figure 31*a*), and the reduced photoproduct, $Ru(bpz)_3^+$ ($E^\circ = -0.86$ V vs. SCE), mediates the reduction of CO_2 to methane and hydrocarbon oligomers, in the presence of the Ru colloid. The system lacks any activity toward H_2 evolution, although the reduced electron carrier, $Ru(bpz)_3^+$, is thermodynamically capable of mediating hydrogen evolution. Thus the formation of surface-bound H atoms in this photocatalytic assembly is eliminated, and electrical charging of the Ru microelectrode is followed by reduction of catalyst-activated CO_2. In a second photosystem composed of Ru(II)–tris(bipyridine), $Ru(bpy)_3^{2+}$, as photosensitizer, a bipyridinium electron acceptors of structures **32**–**34**, and TEOA as sacrificial electron donor, reduction of CO_2 to methane and ethylene and simultaneous H_2 evolution takes place in the presence of Ru or Os heterogeneous catalysts [129]. In these photosystems, oxidative ET quenching of the excited photosensitizer proceeds, and the resulting bipyridinium radical cation mediates CO_2 fixation and H_2 evolution in the presence of the metal catalysts (Figure 31*b*). Thus this system exhibits the catalytic property to generate surface-bound H atoms in contrast to the former photosystem that lacks this ability.

To account for this difference even though the same heterogeneous catalysts are applied in the two assemblies, the activity of the latter photosystem including $Ru(bpy)_3^{2+}$ as photosensitizer, was examined in the presence of added bipyrazine. It was found that the bipyrazine inhibited H_2 evolution, but did not affect the CO_2-reduction process. It was concluded that the Ru and Os catalysts included two specific and distinct activation sites: One class of sites generates surface-bound H atoms and leads to H_2 evolution, while the second type corresponds to CO_2-activation sites, where its reduction to methane proceeds. Bipyrazine is an effective and selective inhibitor of the H-atom binding sites. This leads to the selective reduction of

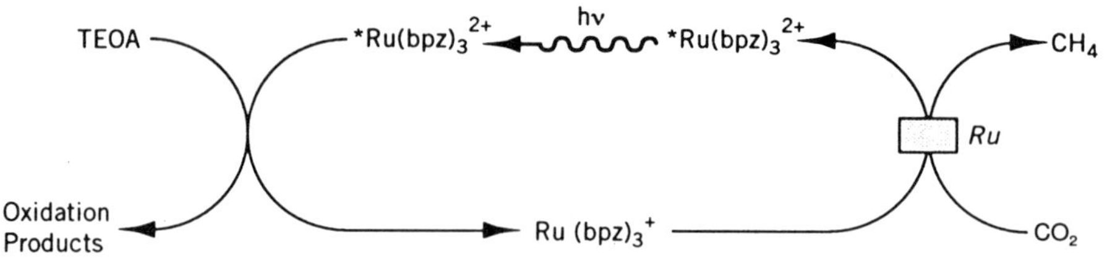

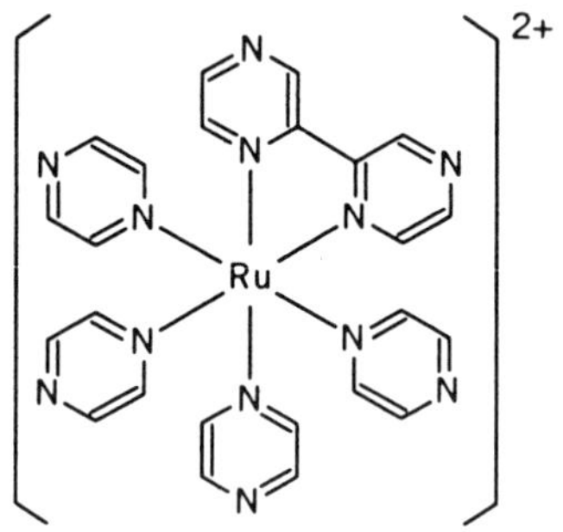

(31)

(a)

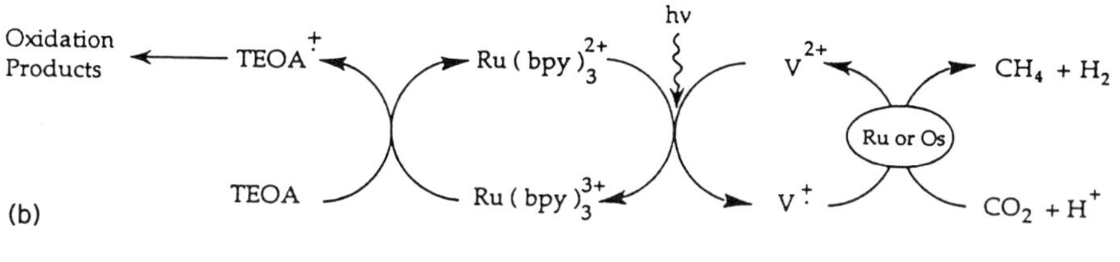

V^{2+} = (32) - (34)

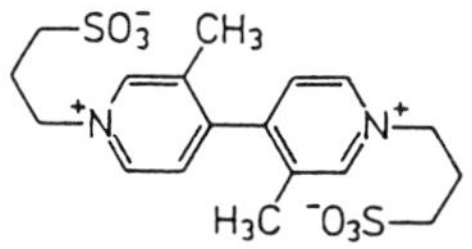

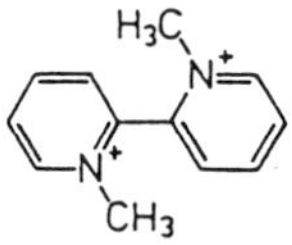

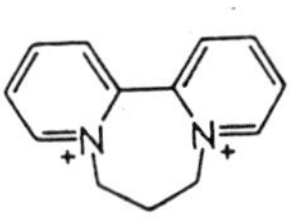

(32) **(33)** **(34)**

Figure 31. Photosynthetic systems for CO_2 fixation to methane and other hydrocarbon products using Ru and Os colloids as catalysts: (*a*) CO_2-fixation through reductive ET quenching of $Ru(bpz)_3^{2+}$ (**31**) as photosensitizer; (*b*) CO_2 fixation by oxidative ET quenching of $Ru(bpy)_3^{2+}$ and application of relays **32**–**34** as ET mediators.

CO_2 in the $Ru(bpz)_3^{2+}$-based system, where the inhibitor is a part of the photosensitizer backbone. Again, the photosystems for CO_2 fixation to methane reveal that heterogeneous catalysts provide means to accumulate electrons and activate the substrates toward photosynthetic transformations.

E. Homogeneous Catalysts Mimicking Cofactor–Enzyme Functionalities in Artificial Photosynthetic Systems

The central metal composing many transition metal complexes can exist in various oxidation states. Often, some of the oxidation states undergo spontaneous disproportionation, allowing the transformation of single-ET products into a multielectron charge relay. Two possible routes for the participation of transition metal complexes in photoinduced ET transformations are outlined in Eqs. (64) and (65), involving oxidative and reductive ET quenching mechanisms, respectively. The spontaneous disproportionation of a single-ET product to a multielectron charge relay is summarized in Eq. (66).

$$S^* + M^{n+1}L_x \rightarrow S^+ + M^nL_x \quad \text{(oxidative ET quenching)} \quad (64)$$

$$^*M^{n+1}L_x + D \rightarrow M^nL_x + D^+ \quad \text{(reductive ET quenching)} \quad (65)$$

$$2M^nL_x \rightleftharpoons M^{n-1}L_x + M^{n+1}L_x \quad (66)$$

A two-electron reduction product of a transition metal complex, formed by a single ET transformation followed by a disproportionation pathway, can act as a precursor for generating a hydride transition metal intermediate [130, 131] in the presence of a proton source (i.e., H_2O):

$$M^{n-1}L_x + H^+ \rightarrow H\text{—}M^{n+1}L_x \quad (67)$$

Hydridometal complexes are simple model systems for a cofactor–enzyme biological assembly. The metal hydride provides a two-electron center, and the reactivity of this species allows the insertion of various substrates into the metal–hydride bond. Thus this reactive species can be considered as an active site, where two ET transformations can proceed. Several transformations that can be driven by such a pathway are summarized as follows:

$$H{-}M^{n+1}L_x + H_2C{=}CH_2 \rightarrow CH_3CH_2{-}M^{n+1}L_x \tag{68}$$

$$H{-}M^{n+1}L_x + H_2C{=}CH_2 + CO \rightarrow CH_3CH_2\overset{\overset{\displaystyle O}{\|}}{C}{-}M^{n+1}L_x \tag{69}$$

$$H{-}M^{n+1}L_x + CO_2 \rightarrow H{-}\overset{\overset{\displaystyle O}{\|}}{C}O{-}M^{n+1}L_x \quad (\text{or } L_xM^{n+1}{-}\overset{\overset{\displaystyle O}{\|}}{C}{-}OH) \tag{70}$$

An alternative pathway for the participation of transition metal complexes in photosynthetic transformations involves activation of the substrates by their direct coordination to the metal center. In this respect the metal center acts as an active site, in addition to its redox functions. Numerous transition metal complexes activating inert molecules (i.e., carbon dioxide or nitrogen) by coordination to the metal center are known [132, 133].

A variety of transition metal complexes were applied as homogeneous catalysts for photosensitized CO_2 fixation. Typical products generated in such systems include carbon monoxide, formate, and molecular hydrogen. Photosensitized reduction of CO_2 to CO was achieved in a photosystem containing $Ru(bpy)_3^{2+}$ as the light-active component, Co(II)–tris(bipyridine), $Co(bpy)_3^{2+}$, as the homogeneous catalyst, and TEOA or ascorbic acid as the sacrificial electron donor [134–136]. Hydrogen evolution occurs from the systems and competes with the CO_2-fixation pathway. The active catalyst in these photosystems is a Co(III)–hydride species that is generated by protonation of a Co(I) complex formed by oxidative or reductive ET quenching processes:

$$^*Ru(bpy)_3^{2+} + Co(bpy)_3^{2+} \rightarrow Ru(bpy)_3^{3+} + Co(bpy)_3^{+} \tag{71}$$

$$Ru(bpy)_3^{3+} + TEOA \rightarrow Ru(bpy)_3^{2+} + TEOA^{+\bullet} \tag{72}$$

$$^*Ru(bpy)_3^{2+} + HA^- \rightarrow Ru(bpy)_3^{+} + HA^\bullet \quad (HA = \text{ascorbic acid}) \tag{72}$$

$$Ru(bpy)_3^{+} + Co(bpy)_3^{2+} \rightarrow Ru(bpy)_3^{2+} + Co(bpy)_3^{+} \tag{74}$$

Figure 32 outlines the suggested cycle leading to the formation of CO and hydrogen. Insertion of carbon dioxide into the Co(III)–H species yields a formate ligated complex that upon decarbonylation generates CO. Protonation of the hydride intermediate is a competitive pathway and yields H_2 evolution. Other Co(II) complexes, such as Co(II)–ethylene glycol dimethyl

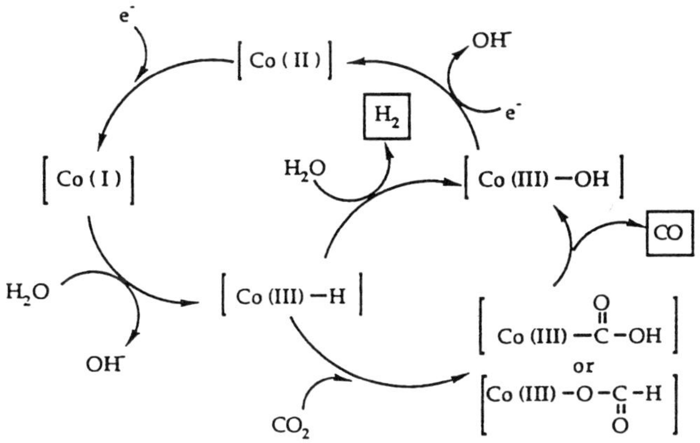

Figure 32. Schematic catalytic cycle for photoreduction of CO_2 and H_2 evolution using Co(II)–homogeneous complexes.

ether complexes [137] or various tetraazamacrocyclic Co(II) complexes [138, 139], such as **35** or **36**, can substitute $Co(bpy)_3^{2+}$ as a homogeneous catalyst for CO_2 reduction to CO in these photosystems. Also, the Ni(II)–tetraazamacrocyclic complex (**37**) acts as a homogeneous catalyst for CO_2 reduction to carbon monoxide in a photosystem composed of $Ru(bpy)_3^{2+}$ as photosensitizer and ascorbate as sacrificial electron donor [139]. The mechanism of the catalytic activity of the Ni(II) complex was suggested to be similar to that of the Co(II) complexes.

Direct excitation of the transition metal complex active in CO_2 reduction was demonstrated in a photosystem composed of tricarbonyl(2,2′-bipyridinium)rhenium(I), *fac*-$Re(bpy)(CO)_3X$ (X = Cl, Br) as a light-active compound and a homogeneous catalyst [140–142]. Photosensitized reduction of carbon dioxide to CO proceeds in nonaqueous solutions (i.e., dimethylformamide) including the rhenium(I) complex and TEOA as the sacrificial electron donor. The quantum efficiency for CO formation in the system corresponds to $\phi = 0.14$. Mechanistic studies show that the primary step in

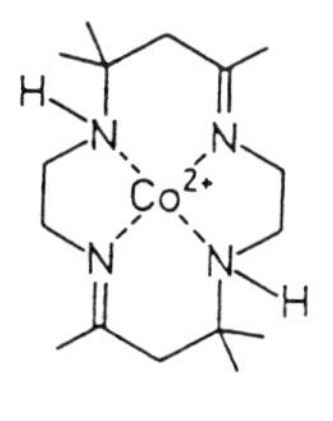

(35)

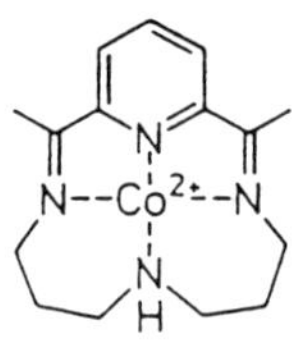

(36)

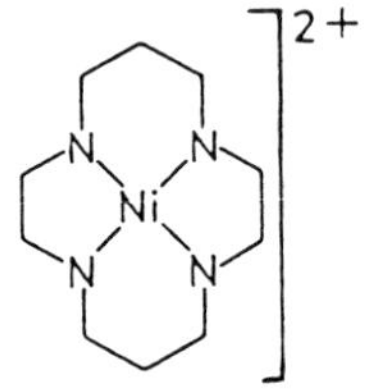

(37)

the photosensitized transformation involves reductive ET quenching of the excited species:

$$*[Re(bpy)(CO)_3X] + TEOA \rightarrow [Re(bpy)(CO)_3X]^- + TEOA^{+\cdot} \quad (75)$$

The resulting reduced photoproduct acts as the catalytic intermediate for CO_2 reduction (Figure 33). It has been suggested that carbon monoxide ligand dissociation, followed by formation of a rhenium–formate intermediate, leads to cyclic reduction of carbon dioxide to CO. Interestingly, a rhenium(I)–carboxylate complex, *fac*-Re(O—C(=O)—H)(bpy)$(CO)_3$, has been isolated as a by-product of the photosystem. The latter complex is inactive in CO_2 reduction to CO, and hence a rhenium–formate intermediate formed by CO_2 insertion into the hydride bond was suggested as the active intermediate for CO formation (Figure 33).

Selective photosensitized reduction of carbon dioxide to formate was accomplished in a photosystem composed of $Ru(bpy)_3^{2+}$ as photosensitizer, TEOA as electron donor, and dicarbonyl Ru(II)–bis(bipyridine), $Ru(bpy)_2(CO)_2^{2+}$, as homogeneous catalyst [143]. Formate, HCO_2^-, is generated in this photosystem with a quantum efficiency of $\phi = 0.14$. A possible mechan-

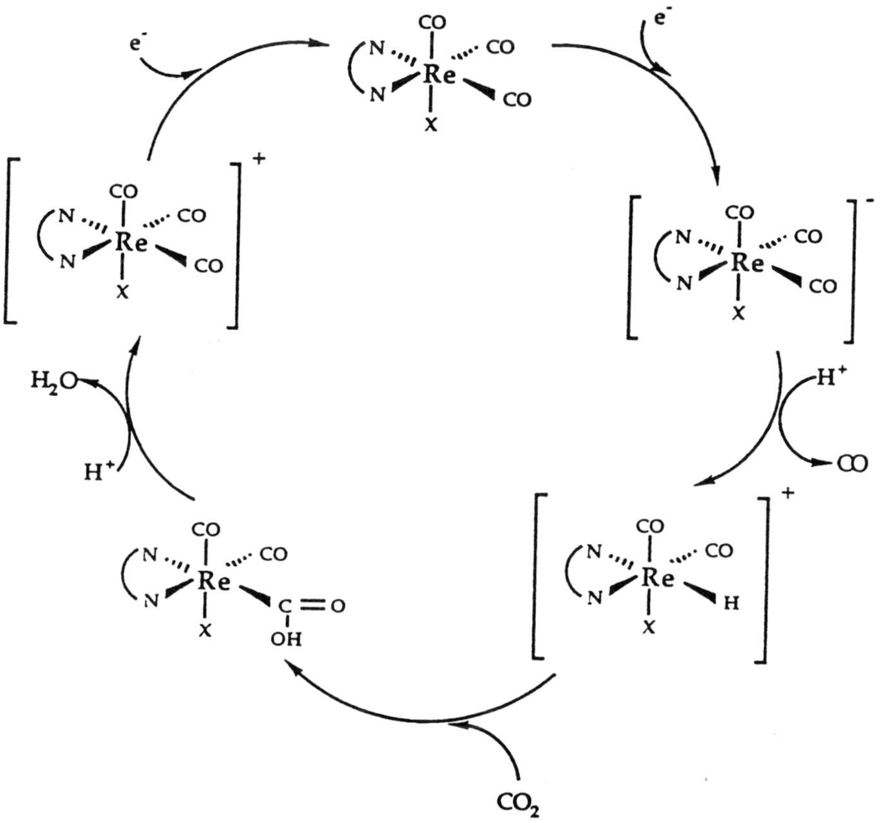

Figure 33. Photocatalytic cycle for CO_2 reduction to CO using *fac*-Re(bpy)$(CO)_3$X.

ism for the cyclic-catalyzed formation of formate is formulated as follows:

$$\mathrm{Ru(bpy)_3^{2+} + TEOA \rightarrow Ru(bpy)_3^{+} + \underset{\substack{\downarrow \\ \text{decomposition products}}}{TEOA^{+\cdot}}} \tag{76}$$

$$[\mathrm{Ru(bpy)_2(CO)_2}]^{2+} + 2e^- \rightarrow [\mathrm{Ru(bpy)_2(CO)_2}]^0 \tag{77}$$

$$\textit{or}\quad [\mathrm{Ru(bpy)_2(CO)L}]^{2+} \qquad \textit{or}\quad [\mathrm{Ru(bpy)_2(CO)L}]^{-1}$$

$$[\mathrm{Ru(bpy)_2(CO)_2}]^0 \rightarrow [\mathrm{Ru(bpy)_2(CO)}] + \mathrm{CO} \tag{78}$$

$$\textit{or}\quad [\mathrm{Ru(bpy)_2(CO)(L)}^{-1} \qquad \textit{or}\quad (\mathrm{Ru(bpy)_2(CO)}] + \mathrm{L}^-$$

$$[Ru(bpy)_2(CO)] + H^+ \rightarrow Ru(bpy)_2(CO)(H)]^+ \quad (79)$$

$$[Ru(bpy)_2(CO)(H)]^+ + CO_2 \rightarrow [Ru(bpy)_2(CO)(CO_2H)]^+ \quad (80)$$

$$[Ru(bpy)_2(CO)(CO_2H] + LH \rightarrow [Ru(bpy)_2(CO)(L)]^+ + HCO_2H$$
$$(LH = \text{solvent}) \quad (81)$$

Reductive ET quenching of $Ru(bpy)_3^{2+}$ results in the reduced photoproduct that provides the electrons for reduction of the homogeneous catalyst. The reduced catalyst undergoes CO-ligand dissociation, and the pentacoordinated intermediate generates the hydridoruthenium species. Insertion of CO_2 into the metal–hydride bond and subsequent protonation of the formate complex yields formate. Supportive evidence for this mechanism was obtained by direct application of the complex $[Ru(bpy)_2(CO)(H)]^+$ that was found to act as catalyst for reduction of CO_2 to formate [144]. Interestingly, it was found that illumination of a photosystem composed of $Ru(bpy)_3^{2+}$ and triethanolamine leads to the reduction of CO_2 to formate even in the absence of the homogeneous catalyst. This was attributed to in situ generation of the homogeneous catalyst by ligand dissociation from the photogenerated product:

$$Ru(bpy)_3^+ + 2L(\text{solvent}) \rightarrow Ru(bpy)_2L_2^+ + bpy \quad (82)$$

Hydridometal complexes were also applied in multielectron photoinduced hydrogenation [145] and hydroformylation [145] processes. The water-soluble Wilkinson-type catalyst chlorotris(diphenylphosphinobenzene-*m*-sulfonate)–rhodium(I), $Rh(dpm)_3^{3-}$ (**38**) acts as a homogeneous catalyst generating a hydridorhodium intermediate. In a photosystem composed of $Ru(bpy)_3^{2+}$ as photosensitizer and ascorbate as sacrificial electron donor, hydrogenation of ethylene and acetylene proceeds with quantum yields of $\phi = 1.8 \times 10^{-2}$ and $\phi = 6 \times 10^{-3}$, respectively. In the presence of ethylene or propene and carbon monoxide, the hydroformylation products propionaldehyde and butyraldehyde are photosynthesized. The primary photoinduced ET process in these systems involves the generation of $Ru(bpy)_3^+$ which mediates the reduction of $Rh(dpm)_3^{3-}$. The reduced homogeneous catalyst generates hydridorhodium intermediates, and in the presence of CO, hydridocarbonyl complexes are formed. These catalytic intermediates drive the hydrogenation and hydroformylation processes by the pathways suggested in Figure 34. We thus realize that homogeneous catalysts undergo light-induced redox transformations that provide active sites for photosynthetic chemical pathways.

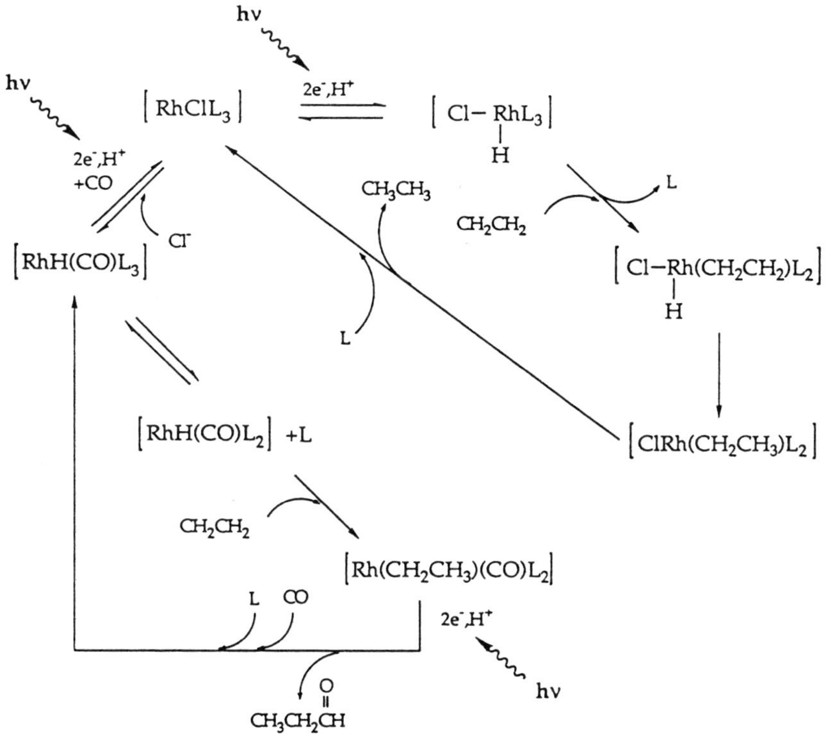

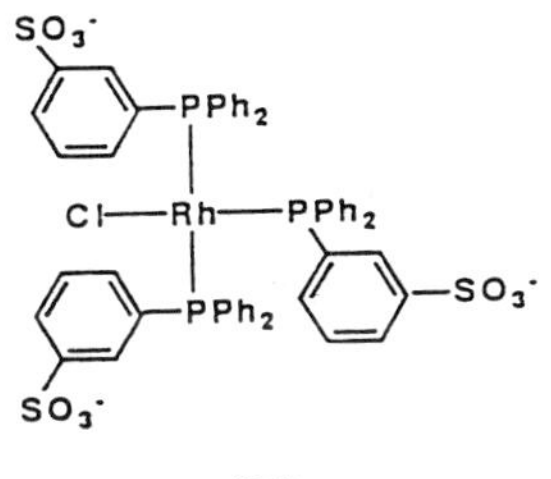

(38)

Figure 34. Photosensitized hydrogenation and hydroformylation of ethylene using $Rh(dpm)_3^{3-}$ (**38**) as homogeneous catalyst.

V. CONCLUSIONS AND OUTLOOK

In this chapter we have addressed biomimetic approaches to mimic natural photosynthesis. The photosynthetic apparatus is a complex system that allows us to extract selected native functionalities and duplicate them by artificial means. We have followed two general approaches to mimic photosynthesis: One approach involves the tailoring of molecular assemblies mimicking the photosynthetic reaction center to the extent that light energy is trapped in the form of redox intermediates within the organized molecular backbone. The second approach includes strategies to organize integrated multicomponent photosynthetic devices, where light energy is transformed into chemical potential and the redox intermediates are utilized in subsequent photosynthetic transformations. These integrated photosynthetic assemblies duplicate charge separation membrane functionalities by means of organized microenvironments and highlight the use of enzymes, homogeneous catalysts, and heterogeneous catalysts as catalytic entities that can be coupled to photogenerated redox products.

Recent advances in the development of artificial photosynthetic systems, where native enzymes are coupled to photoinduced ET products, have been reviewed. Chemical means to modify bioactive proteins and to establish electrical communication with photoexcited entities have been addressed. Such functionalized proteins can substitute complex cofactor–enzyme assemblies. Finally, the use of heterogeneous and homogeneous catalysts, and their functions as cofactor–enzyme models in activation of the substrates and accumulation of electrical charges, have been discussed.

Mimicking photosynthesis by artificial means is anticipated to receive further research efforts. Ingenious recent strategies to tailor molecular charge separation devices coupled to catalytic materials, and organization of chemical matrices for vectorial ET and charge separation, represent novel attitudes and future research directions [146, 147].

Artificial photosynthetic systems have important practical implications. The conversion of light energy to chemical potential allows the photosynthesis of new fuel products and valuable chemicals. In addition, the basic knowledge attained in this broad subject in relation to charge transport and photocatalysis reflects other research domains: Electrical wiring of proteins toward their macroscopic environment is the basis of many amperometric biosensors [148, 149]. The design of ordered donor–acceptor assemblies is important in the development of new nonlinear optical materials [150, 151]. Light-generated redox intermediates can be utilized in the removal of environmental pollutants [48, 152, 153], and light-induced ET processes can be applied for information storage and imaging [154].

ACKNOWLEDGMENT

The support of our research activities in artificial photosynthesis by the Bundesministerium für Forschung und Technologie, Germany, is gratefully acknowledged.

REFERENCES

1. J. R. Bolton, *Science* **202**, 705 (1978).
2. J. R. Bolton, S. J. Strickler, and J. S. Connolly, *Nature* **316**, 495 (1985).
3. J. R. Bolton, A. F. Haught, and R. T. Ross, in *Photochemical Conversion and Storage of Solar Energy*, J. S. Connolly, Ed., Academic Press, New York, 1981, p. 297.
4. J. O'M. Bockris, B. Dandapani, D. Cocke, and J. Ghoroghchian, *Int. J. Hydrogen Energy* **10**, 179 (1985).
5. G. R. Fleming, J. L. Martin, and J. Breton, *Nature* **333**, 190 (1988).
6. C. Kirmaier and D. Holten, *Photosynth. Res.* **13**, 225 (1987).
7. W. W. Parson, in *New Comprehensive Biochemistry: Photosynthesis*, J. Ames, Ed., Elsevier, New York, 1987, p. 43.
8. H. Levanon and J. R. Norris, *Chem. Rev.* **78**, 185 (1978).
9. J. Deisenhofer, O. Epp, K. Miki, R. Huber, and H. Michel, *Nature* **318**, 618 (1985).
10. R. A. Marcus and N. Sutin, *Biochim. Biophys. Acta* **811**, 265 (1985).
11. M. E. Michel-Beyerle, M. Plato, J. Deisenhofer, H. Michel, M. Bixon, and J. Jortner, *Biochim. Biophys. Acta* **932**, 52 (1988).
12. R. A. Marcus, in *Light-Induced Charge Separation in Biology and Chemistry*, H. Gerischer, and J. J. Katz, Eds., Verlag Chemie, Berlin, 1979, p. 15.
13. D. Gust and T. A. Moore, in *Topics in Current Chemistry: Photoinduced Electron Transfer III*, Vol. 159, J. Mattay, Ed., Springer-Verlag, Berlin, 1991, p. 103.
14. D. Gust and T. A. Moore, *Science* **244**, 35 (1989).
15. A. J. Bard, *Science* **207**, 139 (1980).
16. A. Heller, *Science* **223**, 1141 (1984).
17. R. Memming, in *Topics in Current Chemistry: Electrochemistry II*, Vol. 143, E. Steckhan, Ed., Springer-Verlag, Berlin, 1988, p. 79.
18. I. Willner and B. Willner, in *Topics in Current Chemistry: Photoinduced Electron Transfer III*, Vol. 159, J. Mattay, Ed., Springer-Verlag, Berlin, 1991, p. 153.
19. T. J. Meyer, *Acc. Chem. Res.* **22**, 163 (1989).
20. R. P. F. Gregory, *Biochemistry of Photosynthesis*, Wiley-Interscience, New York, 1977.
21. R. K. Clayton, *Photosynthesis: Physical Mechanism and Chemical Patterns*, Cambridge University Press, Cambridge, 1980.
22. C. A. Wraight, J. S. Leigh, P. L. Dutton, and R. K. Clayton, *Biochim. Biophys. Acta* **333**, 401 (1974).
23. C. H. Chang, D. M. Tiede, J. Tang, U. Smith, J. R. Norris, and M. Schiffer, *FEBS Lett.* **205**, 82 (1986).
24. M. R. Wasielewski, J. M. Fenton, and Govindjee, *Photosynth. Res.* **12**, 181 (1987).

25. D. Gust and T. A. Moore, in *Supramolecular Photochemistry*, V. Balzani, Ed., NATO ASI Ser. C214, D. Reidel, Dordrecht, The Netherlands, 1987, p. 267.
26. M. R. Wasielewski, in *Metal Ions in Biological Systems: Electron Transfer Reactions in Metalloproteins*, Vol. 27, H. Sigel and A. Sigel, Eds., Marcel Dekker, New York, 1991, p. 361.
27. M. R. Wasielewski, D. G. Johnson, and W. A. Svec, in *Supramolecular Photochemistry*, V. Balzani, Ed., D. Reidel, Amsterdam, 1987, p. 255.
28. M. Grätzel, Ed., *Energy Resources Through Photochemistry and Catalysis*, Academic Press, New York, 1983.
29. A. Harriman and M. E. West, Eds., *Photogeneration of Hydrogen*, Academic Press, London, 1983.
30. I. Willner and B. Willner, in *Photochemical Conversion and Storage of Solar Energy*, E. Pelizzetti, and M. Schiavello, Eds., Kluwer Academic Publishers, Dordrecht, The Netherlands, 1991, p. 151.
31. R. A. Marcus, *Annu. Rev. Phys. Chem.* **15**, 155 (1964).
32. I. Tabushi, N. Koga, and M. Yanagita, *Tetrahedron Lett.* 257 (1979).
33. J. S. Connolly and J. R. Bolton, in *Photoinduced Electron Transfer*, M. A. Fox, and M. Chanon, Eds., Elsevier, Amsterdam, 1988, part D, p. 303.
34. T. A. Moore, D. Gust, P. Mathis, J.-C. Mialocq, C. Chachaty, R. V. Bensasson, E. J. Land, D. Doizi, P. A. Liddell, W. R. Lehman, G. A. Nemeth, and A. L. Moore, *Nature* **307**, 630 (1984).
35. M. R. Wasielewski, M. P. Niemczyk, W. A. Svec, and E. B. Pewitt, *J. Am. Chem. Soc.* **107**, 5562 (1985).
36. J. Olmsted III, S. F. McClanahan, E. Danielson, J. N. Younathan, and T. J. Meyer, *J. Am. Chem. Soc.* **109**, 3297 (1987).
37. S. L. Mecklenburg, B. M. Peek, B. W. Erickson, and T. J. Meyer, *J. Am. Chem. Soc.* **113**, 8540 (1991).
38. N. Mataga, A. Karen, T. Okada, S. Nishitani, N. Kurata, Y. Sakata, and S. Misumi, *J. Phys. Chem.* **88**, 5138 (1984).
39. J. S. Krueger, J. E. Mayer, and T. E. Mallouk, *J. Am. Chem. Soc.* **110**, 8232 (1988).
40. D. Gust, T. A. Moore, A. L. Moore, G. Seely, P. Liddell, D. Barrett, L. O. Harding, X. C. Ma, S. J. Lee, and F. Gao, *Tetrahedron* **45**, 4867 (1989).
41. D. Gust, T. A. Moore, A. L. Moore, D. Barrett, L. O. Harding, L. R. Makings, P. A. Liddell, F. C. De Schryver, M. Van der Auweraer, R. V. Bensasson, and M. Rougee, *J. Am. Chem. Soc.* **110**, 321 (1988).
42. N. Sutin, *J. Photochem.* **10**, 19 (1979).
43. M.-P. Pileni and M. Grätzel, *J. Phys. Chem.* **84**, 2402 (1980).
44. I. Willner, Y. Eichen, and E. Joselevich, *J. Phys. Chem.* **94**, 3092 (1990).
45. K. Kalyanasundaram, *Coord. Chem. Rev.* **46**, 159 (1982).
46. A. Juris, V. Balzani, F. Barigelletti, S. Campagna, P. Belser, and A. von Zelewsky, *Coord. Chem. Rev.* **84**, 85 (1988).

47. J. R. Darwent, P. Douglas, A. Harriman, G. Porter, and M.-C. Richoux, *Coord. Chem. Rev.* **44**, 83 (1982).
48. E. Pelizzetti and N. Serpone, Eds., *Photocatalysis: Fundamentals and Applications*, Wiley, New York, 1989.
49. M. Grätzel, Ed., *Heterogeneous Photochemical Electron Transfer*, CRC Press, Boca Raton, Fla., 1989.
50. M. Schiavello, Ed., *Photocatalysis and Environment: Trends and Applications*, NATO Adv. Sci. Ser. C237, Kluwer Academic Publishers, Dordrecht, The Netherlands, 1988.
51. I. Willner, Z. Goren, D. Mandler, R. Maidan, and Y. Degani, *J. Photochem.* **28**, 215 (1985).
52. I. Willner, in *Organic Phototransformations in Nonhomogeneous Media*, M. A. Fox, Ed., ACS Symp. Ser. 278, American Chemical Society, Washington, D.C., 1985, p. 191.
53. M. A. Fox, in *Topics in Current Chemistry: Photoinduced Electron Transfer III*, Vol. 159, J. Mattay, Ed., Springer-Verlag, Berlin, 1991, p. 67.
54. I. Willner and B. Willner, in *Frontiers in Supramolecular Organic Chemistry and Photochemistry*, H.-J. Schneider and H. Dürr, Eds., VCH, Weinheim, Germany, 1991, p. 337.
55. N. J. Turro, M. Grätzel, and A. M. Braun, *Angew. Chem. Int. Ed. Engl.* **19**, 675 (1980).
56. T. Matsuo, K. Takuma, Y. Tsutsui, and T. J. Nishizima, *J. Coord. Chem.* **10**, 187 (1980).
57. J. K. Thomas, *Chem. Rev.* **80**, 283 (1980).
58. R. E. Sassoon and J. Rabani, *J. Phys. Chem.* **84**, 1319 (1980).
59. I. Willner, J. M. Yang, J. W. Otvos, and M. Calvin, *J. Phys. Chem.* **85**, 3277 (1981).
60. I. Willner, J. W. Otvos, and M. Calvin, *J. Am. Chem. Soc.* **103**, 3203 (1981).
61. R. A. DellaGuardia and J. K. Thomas, *J. Phys. Chem.* **88**, 964 (1984).
62. N. J. Turro, J. K. Barton, and D. A. Tomalia, *Acc. Chem. Res.* **24**, 332 (1991).
63. P.-A. Brugger, P. P. Infelta, A. M. Braun, and M. Grätzel, *J. Am. Chem. Soc.* **103**, 320 (1981).
64. D. Mandler, Y. Degani, and I. Willner, *J. Phys. Chem.* **88**, 4366 (1984).
65. S. S. Atik and J. K. Thomas, *J. Am. Chem. Soc.* **103**, 7403 (1981).
66. J. H. Fendler, *Chem. Rev.* **87**, 877 (1987).
67. E. Adar, Y. Degani, Z. Goren, and I. Willner, *J. Am. Chem. Soc.* **108**, 4696 (1986).
68. A. J. Frank, I. Willner, Z. Goren and Y. Degani, *J. Am. Chem. Soc.* **109**, 3568 (1987).
69. I. Willner, Y. Eichen, A. J. Frank, and M. A. Fox, *J. Phys. Chem.* **97**, 7264 (1993).
70. I. Willner and D. Mandler, *Enzyme Microb. Technol.* **11**, 467 (1989).

71. G. M. Whitesides and C.-H. Wong, *Angew. Chem. Int. Ed. Engl.* **24**, 617 (1985).
72. H. K. Chenault and G. M. Whitesides, *Appl. Biochem. Biotechnol.* **14**, 147 (1987).
73. G. McLendon, *Acc. Chem. Res.* **21**, 160 (1988).
74. H. B. Gray, *Chem. Soc. Rev.* **15**, 17 (1986).
75. S. L. Mayo, W. R. Ellis, Jr., R. J. Crutchley, and H. B. Gray, *Science* **233**, 948 (1986).
76. A. Heller, *J. Phys. Chem.* **96**, 3579 (1992).
77. D. Mandler and I. Willner, *J. Am. Chem. Soc.* **106**, 5352 (1984).
78. D. Mandler and I. Willner, *J. Chem. Soc. Perkin Trans. 2*, 805 (1986).
79. R. Wienkamp and E. Steckhan, *Angew. Chem. Int. Ed. Engl.* **22**, 497 (1983).
80. I. Willner, R. Maidan, and M. Shapira, *J. Chem. Soc. Perkin Trans. 2*, 559 (1990).
81. D. Mandler and I. Willner, *J. Chem. Soc. Chem. Commun.* 851 (1986).
82. D. Mandler and I. Willner, *J. Chem. Soc. Perkin Trans. 2*, 997 (1988).
83. M. Julliard and J. Le Petit, *Photochem. Photobiol.* **36**, 283 (1982).
84. R. P. Chambers, J. R. Ford, J. H. Allender, W. H. Baricos, and W. Cohen, *Enzyme Eng.* **2**, 195 (1974).
85. J. Handman, A. Harriman, and G. Porter, *Nature* **307**, 534 (1984).
86. I. Willner, R. Maidan, and B. Willner, *Isr. J. Chem.* **29**, 289 (1989).
87. I. Okura, *Coord. Chem. Rev.* **68**, 53 (1985).
88. Y. Kinumi and I. Okura, *J. Mol. Catal.* **52**, L33 (1989).
89. I. Willner, N. Lapidot, and A. Riklin, *J. Am. Chem. Soc.* **111**, 1883 (1989).
90. I. Willner, N. Lapidot, S. Rubin, A. Riklin and B. Willner, in *Biotechnology: Bridging Research and Applications*, D. Kamely, A. M. Chakrabarthy, and S. E. Kornguth, Eds., Kluwer Academic Publishers, Dordrecht, The Netherlands, 1991, p. 341.
91. I. Willner, N. Lapidot, A. Riklin, R. Kasher, E. Zahavy, and E. Katz, *J. Am. Chem. Soc.* **116**, 1428 (1994).
92. A. Heller, *Acc. Chem. Res.* **23**, 128 (1990).
93. I. Willner, N. Lapidot, and A. Riklin, *J. Am. Chem. Soc.* **112**, 6438 (1990).
94. I. Willner, R. Kasher, E. Zahavy, and N. Lapidot, *J. Am. Chem. Soc.* **114**, 10963 (1992).
95. I. Willner and N. Lapidot, *J. Am. Chem. Soc.* **113**, 3625 (1991).
96. G. C. Bond, *Heterogeneous Catalysis: Principles and Applications*, Oxford University Press, London, 1974.
97. R. Pearce and W. R. Patterson, Eds., *Catalysis and Chemical Processes*, Blackie, London, 1981.
98. G. A. Somorjai, *Chemistry in Two Dimensions: Surfaces*, Cornell University Press, Ithaca, N.Y., 1981.
99. G. A. Somorjai, *Pure Appl. Chem.* **60**, 1499 (1988).

100. J. M. Thomas, *Pure Appl. Chem.* **60**, 1517 (1988).

101. A. Harriman, *Platinum Met. Rev.* **27**, 102 (1983).

102. L. D. Rampino and F. F. Nord, *J. Am. Chem. Soc.* **63**, 2745 (1941).

103. J. Turkevich, R. S. Miner, Jr., and L. Babenkova, *J. Phys. Chem.* **90**, 4765 (1986).

104. A. Gutbier, *Z. Anorg. Chem.* **32**, 347 (1902).

105. H. Hirai, M. Ohtaki, and M. Komiyama, *Chem. Lett.* 269 (1986).

106. M. O. Delcourt, N. Keghouche, and J. Belloni, *Nouv. J. Chim.* **7**, 131 (1983).

107. K. Kimura and S. Bandow, *Bull. Chem. Soc. Jpn.* **56**, 3578 (1983).

108. K. I. Zamaraev and V. N. Parmon, *Catal. Rev.* **22**, 261 (1980).

109. A. J. Frank and K. L. Stevenson, *J. Chem. Soc. Chem. Commun.* 593 (1981).

110. R. Rafaeloff, Y. Haruvy, J. Binneboym, G. Baruch, and L. A. Rajbenbach, *J. Mol. Catal.* **22**, 219 (1983).

111. A. Henglein and J. Lilie, *J. Am. Chem. Soc.* **103**, 1059 (1981).

112. K. Kopple, D. Meyerstein, and D. Meisel, *J. Phys. Chem.* **84**, 870 (1980).

113. I. Willner and B. Steinberger-Willner, *Int. J. Hydrogen Energy* **13**, 593 (1988).

114. N. Serpone, in *Photochemical Energy Conversion*, J. R. Norris and D. Meisel, Eds., Elsevier, New York, 1989, p. 297.

115. J. Kiwi and M. Grätzel, *Nature (London)* **281**, 657 (1979).

116. M. C. Richoux and D. O. Hall, *Photobiochem. Photobiophys.* **1**, 375 (1980).

117. J. Kiwi, K. Kalyanasundaram, and M. Grätzel, *Struct. Bonding* **49**, 37 (1982).

118. P. Keller and A. Moradpour, *J. Am. Chem. Soc.* **102**, 7193 (1980).

119. P. P. Lay, A. W. H. Mau, W. H. F. Sasse, I. Creaser, L. R. Gahan and A. M. Sargeson, *Inorg. Chem.* **22**, 2347 (1983).

120. M. Kirch, J.-M. Lehn, and J.-P. Sauvage, *Helv. Chim. Acta* **62**, 1345 (1979).

121. B. V. Koryakin, T. S. Dzhabiev, and A. E. Shilov, *Dokl. Acad. Nauk SSSR* **233**, 620 (1977).

122. Y. Degani and I. Willner, *J. Chem. Soc. Perkin Trans. 2*, 37 (1986).

123. C.-K. Tan, V. Newberry, T. R. Webb, and C. A. McAuliffe, *J. Chem. Soc. Dalton Trans.* 1299 (1987).

124. D. Mandler and I. Willner, *J. Phys. Chem.* **91**, 3600 (1987).

125. A. J. Frank, Z. Goren, and I. Willner, *J. Chem. Soc. Chem. Commun.* 1029 (1985).

126. D. Mandler and I. Willner, *J. Am. Chem. Soc.* **109**, 7884 (1987).

127. I. Willner and D. Mandler, *J. Am. Chem. Soc.* **111**, 1330 (1989).

128. R. Maidan and I. Willner, *J. Am. Chem. Soc.* **108**, 8100 (1986).

129. I. Willner, R. Maidan, D. Mandler, H. Dürr, G. Dörr, and K. Zengerle, *J. Am. Chem. Soc.* **109**, 6080 (1987).

130. R. P. A. Sneeden, in *Comprehensive Organometallic Chemistry*, Vol. 8, G. Wilkinson, Ed., Pergamon Press, Oxford, 1982, p. 225.
131. M. E. Vol'pin and I. S. Kolomnikov, *Organometal. React.* **5**, 313 (1975).
132. D. A. Palmer and R. van Eldik, *Chem. Rev.* **83**, 651 (1983).
133. D. J. Darensbourg and R. A. Kudaroski, *Adv. Organometal. Chem.* **22**, 129 (1983).
134. F. R. Keene, C. Creutz, and N. Sutin, *Coord. Chem. Rev.* **64**, 247 (1985).
135. J.-M. Lehn and R. Ziessel, *Proc. Natl. Acad. Sci. USA* **79**, 701 (1982).
136. R. Ziessel, J. Hawecker, and J.-M. Lehn, *Helv. Chim. Acta* **69**, 1065 (1986).
137. R. Ziessel, *Nouv. J. Chim.* **7**, 613 (1983).
138. A. H. A. Tinnemans, T. P. M. Koster, D. H. M. W. Thewissen, and A. Mackor, *Recl. Trav. Chim. Pays-Bas* **103**, 288 (1984).
139. J. L. Grant, K. Goswami, L. O. Spreer, J. W. Otvos, and M. Calvin, *J. Chem. Soc. Dalton Trans.* 2105 (1987).
140. J. Hawecker, J.-M. Lehn, and R. Ziessel, *Helv. Chim. Acta* **69**, 1990 (1986).
141. C. Kutal, M. A. Weber, G. Ferraudi, and D. Geiger, *Organometallics* **4**, 2161 (1985).
142. C. Kutal, A. J. Corbin, and G. Ferraudi, *Organometallics* **6**, 553 (1987).
143. H. Ishida, K. Tanaka, and T. Tanaka, *Chem. Lett.* 1035 (1987).
144. H. Ishida, T. Terada, K. Tanaka, and T. Tanaka, *Inorg. Chem.* **29**, 905 (1990).
145. I. Willner and R. Maidan, *J. Chem. Soc. Chem. Commun.* 876 (1988).
146. P. G. Schouten, J. M. Warman, M. P. de Haas, M. A. Fox, and H.-L. Pan, *Nature* **353**, 736 (1991).
147. U. Resch and M. A. Fox, *J. Phys. Chem.* **95**, 6169 (1991).
148. Y. Degani and A. Heller, *J. Am. Chem. Soc.* **110**, 2615 (1988).
149. R. Maidan and A. Heller, *J. Am. Chem. Soc.* **113**, 9003 (1991).
150. D. J. Williams, *Angew. Chem. Int. Ed. Engl.* **23**, 690 (1984).
151. F. Effenberger, H. Schlosser, P. Bäuerle, S. Maiser, H. Port, and H. C. Wolf, *Angew. Chem. Int. Ed. Engl.* **27**, 281 (1988).
152. M. Schiavello, Ed., *Photocatalysis and Environment: Trends and Applications*, NATO ASI Ser. C, Vol. 237, Kluwer Academic Publishers, Dordrecht, The Netherlands, 1988.
153. J.-C. D'Oliveira, G. Al-Sayyed, and P. Pichat, *Environ. Sci. Technol.* **24**, 990 (1990).
154. L. B.-A. Johansson, M. Blanchard-Desce, M. Almgren and J.-M. Lehn, *J. Phys. Chem.* **93**, 6751 (1989).

INDEX

CUMULATIVE INDEX VOLUMES 1–20